AF588394

Springer Series in Materials Science

Volume 169

For further volumes:
http://www.springer.com/series/856

The Springer Series in Materials Science covers the complete spectrum of materials physics, including fundamental principles, physical properties, materials theory and design. Recognizing the increasing importance of materials science in future device technologies, the book titles in this series reflect the state-of-the-art in understanding and controlling the structure and properties of all important classes of materials.

Richard J. Bushby • Stephen M. Kelly
Mary O'Neill
Editors

Liquid Crystalline Semiconductors

Materials, Properties and Applications

Editors
Richard J. Bushby
School of Chemistry
University of Leeds
Leeds
UK

Stephen M. Kelly
Department of Chemistry
University of Hull
Hull
UK

Mary O'Neill
Departments of Physics and Mathematics
University of Hull
Hull
UK

Published by Springer,
P.O. Box 17, 3300 AA Dordrecht, The Netherlands

ISSN 0933-033X
ISBN 978-90-481-2872-3 ISBN 978-90-481-2873-0 (eBook)
DOI 10.1007/978-90-481-2873-0
Springer Dordrecht Heidelberg New York London

Library of Congress Control Number: 2012954012

Printed on acid-free paper

Springer is part of Springer Science+Business Media (www.springer.com)

Preface

This is an exciting stage in the development of organic electronics. No longer is it an area of purely 'academic interest' but increasingly real applications are being developed and some of these are beginning to come on-stream. There have been several drivers for the surge of interest shown by the electronics industry. Perhaps the greatest is the promise afforded by organic systems for low-cost processing, e.g., methods of fabrication such as roll-to-roll contact printing or ink-jet printing in which the components are 'assembled' on low-cost substrates at room temperature. The long-term vision is for the creation of integrated devices in which, for example, a photovoltaic cell, a sensor and a display, all made of organic materials, is printed onto a plastic film in a single operation. Another factor in the recent development of organic electronics is the huge advances made in organic synthesis over the last half century, which make it easy to fine-tune the properties of organics at a molecular level in a way that is simply not possible for their metallic and semi-metallic counterparts. Finally, organics are much easier to interface with biomaterials, which is perhaps the key problem in the development of new generation biosensors. Another consideration in the development of organic electronics has been 'Green Issues': not only the need to develop energy-efficient methods of manufacture but also the need to develop methods based on renewable (organic) rather than finite (inorganic and metallic) components.

Although it seems improbable that organics will ever compete with silicon-based devices in terms of high-end computing applications, where device speed is of the essence, there are many low-end areas of application, where we can look to see traditional inorganic components increasingly replaced by organics. Areas that have already been commercially developed or which are under intensive development include organic light emitting diodes (for flat panel displays and solid state lighting), organic photovoltaic cells, organic thin film transistors (for smart tags and flat panel displays) and sensors. Potentially this is a field that will affect every aspect of our lives and have an impact in every home and in every business.

Within the family of organic electronic materials, liquid crystals are relative newcomers. The first electronically conducting liquid crystals were only reported in 1988, but already a substantial literature has developed. The potential advantage of liquid crystalline semiconductors is that they have the easy processabilty of amorphous and polymeric semiconductors, but they usually have higher charge carrier mobilities. Their mobilities do not reach the levels seen in crystalline organics, but they circumvent all of the difficult issues of controlling crystal growth and morphology. Liquid crystals self-organise, they can be aligned by fields and surface forces and, because of their fluid nature, defects in liquid crystal structures readily self-heal.

Because this is a relatively young field, there are still issues which need to be understood. In particular, the theory of electronic conduction in liquid crystals is much less well developed than that of electronic conduction in other organic materials and, although the relationship between molecular structure and conductivity is mostly understood, some issues still remain to be resolved and understood.

With these matters in mind, this is an opportune moment to bring together a monograph on the subject of 'Liquid Crystalline Semiconductors'. The field is already too large to cover in a comprehensive manner, so our aim has been to bring together contributions from leading workers, which cover the main areas of the chemistry (synthesis and structure/function relationships), physics and potential applications. A general introduction to liquid crystals and the nature and kinds of their mesomorphic behaviour and structure (mesophases) is given in Chap. 1. A description of the nature and mechanisms of different kinds of charge transport in calamitic (nematic and smectic) liquid crystalline semiconductors is given in Chap. 2 followed by a similar treatment of columnar (discotic) liquid crystalline semiconductors in Chap. 3. The different approaches and methods of determining charge transport in liquid crystals are also described in detail. Chap. 4 provides a comprehensive description of the synthesis of a wide range of columnar liquid crystalline semiconductors. A series of reaction schemes are used to illustrate different synthetic strategies and approaches to the synthesis of this special type of liquid crystal. Chapter 5 gives an extensive discussion of the nature and magnitude of charge transport in reactive mesogens (monomers) and how they are determined. A similar treatment of the corresponding liquid crystal polymer networks is also given in this chapter. An insight into the nature and complexity of the optical properties of liquid crystals is provided in Chap. 6. Chapters 7, 8 and 9 describe the modes of operation, configuration and performance of a range of electrooptic devices, i.e., organic light-emitting diodes (OLEDs), especially with polarised emission, organic photovoltaics and organic field-effect transistors, incorporating the types of liquid crystalline semiconductors described in the preceding chapters. Our hope is that this book will provide a useful introduction to the field both for those in industry and for those in academia and that it will help to stimulate future developments.

Richard J. Bushby
Stephen M. Kelly
Mary O'Neill

Contents

1 **Introduction to Liquid Crystalline Phases** 1
John E. Lydon

2 **Charge Carrier Transport in Liquid Crystalline Semiconductors** 39
Jun-Ichi Hanna

3 **Columnar Liquid Crystalline Semiconductors** 65
Richard J. Bushby and Daniel J. Tate

4 **Synthesis of Columnar Liquid Crystals** 97
Sandeep Kumar

5 **Charge Transport in Reactive Mesogens and Liquid Crystal Polymer Networks** 145
T. Kreouzis and K.S. Whitehead

6 **Optical Properties of Light-Emitting Liquid Crystals** 173
Mary O'Neill and Stephen M. Kelly

7 **Organic Light-Emitting Diodes (OLEDs) with Polarised Emission** 197
E. Scheler and P. Strohriegl

8 **Liquid Crystals for Organic Photovoltaics** 219
Mary O'Neill and Stephen M. Kelly

9 **Liquid Crystals for Organic Field-Effect Transistors** 247
Mary O'Neill and Stephen M. Kelly

Index 269

Contributors

Richard J. Bushby School of Chemistry, University of Leeds, Leeds, UK

Jun-Ichi Hanna Imaging Science and Engineering Laboratory, Tokyo Institute of Technology, Yokahama, Japan

Stephen M. Kelly Department of Chemistry, University of Hull, Hull, UK

T. Kreouzis Department of Physics, Queens Mary, University of London, London, UK

Sandeep Kumar Soft Condensed Matter Group, Raman Research Institute, Bangalore, India

John E. Lydon Faculty of Biological Sciences, University of Leeds, Leeds, UK

Mary O'Neill Department of Physics and Mathematics, University of Hull, Hull, UK

E. Scheler Lehrstuhl fuer Makromolekular Chemie I, Bayreuther Institut fuer Makromolekulforschung BIMF, University of Bayreuth, Bayreuth, Germany

P. Strohriegl Lehrstuhl fuer Makromolekular Chemie I, Bayreuther Institut fuer Makromolekulforschung BIMF, University of Bayreuth, Bayreuth, Germany

Daniel J. Tate Organic Materials Innovation Centre, School of Chemistry, University of Manchester, Manchester, UK

K. S. Whitehead Department of Physics, Queens Mary, University of London, London, UK

Chapter 1
Introduction to Liquid Crystalline Phases

John E. Lydon

1.1 Introductory Comments

Introductory comments about liquid crystals usually assert that they are states of matter intermediate between solids and liquids. In some senses this is, of course, correct and, in phase diagrams, liquid crystal phases do lie between solids and liquids as far as both temperature and composition are concerned, but the statement carries many wrong implications about them. They do indeed have some degree of molecular alignment between the long-range, three-dimensional order of a crystalline solid and the total disorder of the (isotropic) liquid and they are neither as rigid as a solid nor as fluid as a liquid, but not all properties of liquid crystals can be represented by a histogram of the form shown in Fig. 1.1a. There are significant properties, which are qualitatively different to those of either classical solids or liquids. The influential general overview of liquid crystalline systems by Peter Collings had the inspired sub title – *Nature's Delicate Phase of Matter* [1] – and it is their unique sensitivity, coupled with their self-organising, self-aligning and self–healing properties, which is responsible for most of the commercial usefulness of liquid crystals. Consider, for example, the ease with which the molecules in some liquid crystalline phases can be oriented by electric or magnetic fields. A field of the order of 100 mV μm^{-1} will reorient the molecules in a typical liquid crystal display (LCD) cell. In contrast, fields of 4 or 5 orders of magnitude larger are usually unable to align the molecules in either crystalline solids or (isotropic) liquids. Similarly, the response of liquid crystalline phases to small changes of temperature or to the addition of small quantities of chiral dopants is often appreciable in liquid crystals and insignificant in solids and liquids. The interest in liquid crystals lies in the fact that their properties are not always intermediate between those of solids and those of liquids and some of their properties are unique.

J.E. Lydon (✉)
Faculty of Biological Sciences, University of Leeds, Leeds LS2 9JT, UK
e-mail: J.E.Lydon@leeds.as.uk

R.J. Bushby et al. (eds.), *Liquid Crystalline Semiconductors*, Springer Series in Materials Science 169, DOI 10.1007/978-90-481-2873-0_1,

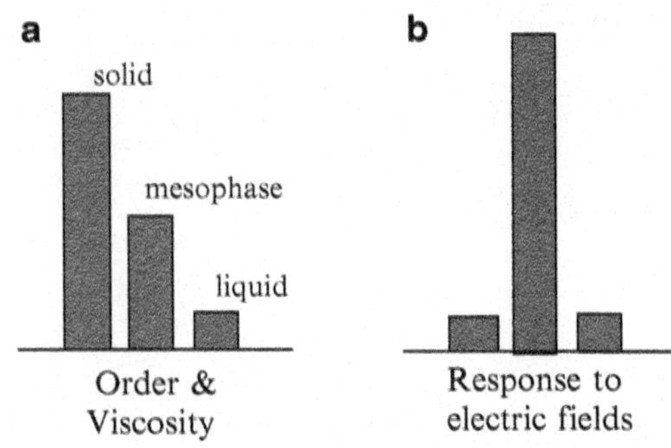

Fig. 1.1 Not all properties of liquid crystalline phases are intermediate between those of solids and liquids. (**a**) Order and viscosity, (**b**) response to electric fields

A second general point is that there are frequent references to '*the liquid crystalline phase*' as if there were only one type of mesophase, with possibly a few very minor variations in detail. This again is a misleading picture. There are dozens of thermodynamically stable liquid crystalline phases, each sufficiently different to the others, in structure and properties, to be classified as a distinct *bona fide* phase in its own right. This range of structures is perhaps, easiest to survey in terms of *dimensions of ordering*. A perfect crystalline solid has molecules lying in fixed orientations on fixed lattice sites. It could be said to have six dimensions of ordering, i.e., three in terms of the positioning along the x, y and z axes and three in terms of the orientation about the three orthogonal axes. At the other end of the scale, in the isotropic liquid there is no positional or orientational order, i.e., no dimensions of ordering at all. In the majority of cases, the melting of a crystalline solid is a synergistic process in that an individual molecule cannot move out of its lattice site or rotate out of its prescribed alignment without disrupting its neighbours and, once melting has started, the whole crystalline pattern collapses catastrophically, i.e., all six dimensions of ordering are lost simultaneously. The cause of this phenomenon lies in the fact that the potential energy wells in which the molecules sit, tend to be of more or less equal depth for translational and rotational motion. The energy barriers that prevent small-scale oscillations from becoming free translational movement and small angle oscillations from becoming completely free rotations, tend to be of comparable height.

However, this is not universally true. If, for example, the molecules are approximately spherical, and if there are no highly-directional intermolecular forces, it may be possible for the molecules to rotate freely without disturbing their neighbours enough to destroy the long-range positional ordering. Such systems are termed ***plastic crystals***. If, on the other hand, the molecules are highly asymmetric rods or discs, it may be possible for the structure to become more disordered and fluid in some dimensions than others. Such systems are termed ***liquid crystals***. Molecules which are rod-like or lath-like are said to be ***calamitic*** (from the Greek καλαμοσ for reed). They may form liquid crystal phases, which are highly fluid, but which are still able to maintain an extended pattern of alignment with the molecular long-axes all lying parallel, like a box of pencils being shaken. Conversely ***discotic*** molecules,

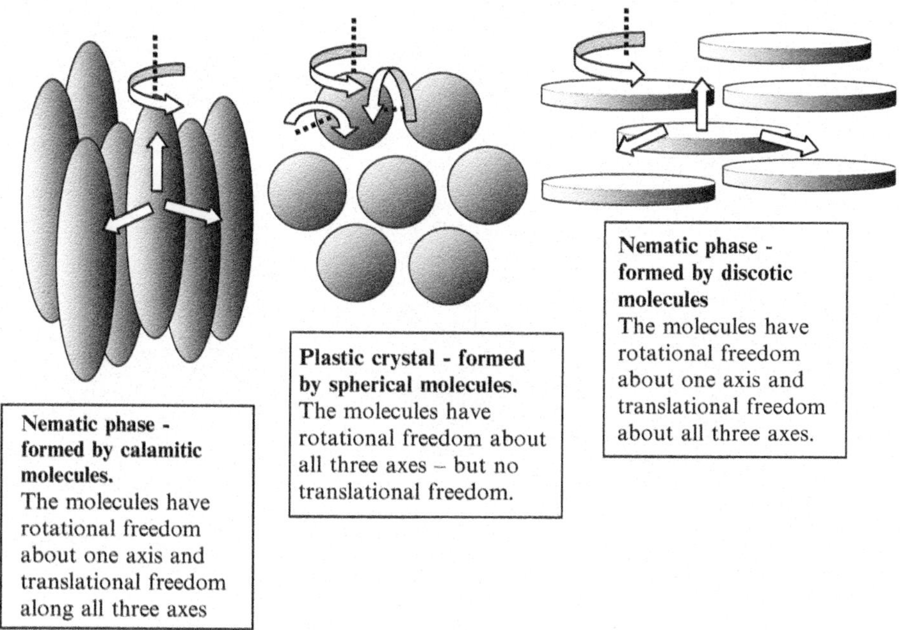

Fig. 1.2 The importance of asymmetry in the formation of liquid crystalline phases

that are flattened and disc-like, may form liquid crystal phases where the planes of the molecules tend to lie parallel, but where the molecules are free to slide over one another, e.g., like a box of coins being shaken (Fig. 1.2).

In the nematic calamitic and discotic phases, there is sufficient thermal energy to prevent the attractive forces between molecules locking them into a rigid pattern and the molecules are free to move more or less as individuals. But in the majority of types of liquid crystal phases there is a pattern of aggregation of molecules, which imposes some ordered structure on the phase, whilst at the same time, allowing considerable molecular mobility and disorder. Calamitic molecules tend to aggregate into layers, giving *smectic* phases and discotic molecules tend to form stacks, giving *columnar* phases (Fig. 1.3).

In both types of aggregation, it is possible for large amounts of thermal motion to be accommodated without destroying the structure of the phase. In particular, rotational motion about the long axes of calamitics, and about the short axes of discotics, can occur without destroying the overall pattern of molecular alignment.

In principle, there are two distinct ways of converting a solid crystalline material into a liquid crystalline phase, i.e., either by heating or by adding a solvent. Mesophases created by heating are said to be *thermotropic* and those formed by adding solvent are said to be *lyotropic* (Fig. 1.4). The usual solvent involved is water. These mesophases are therefore restricted to the temperature range over which water is liquid. Some lyotropic phases exist over narrow temperature ranges and

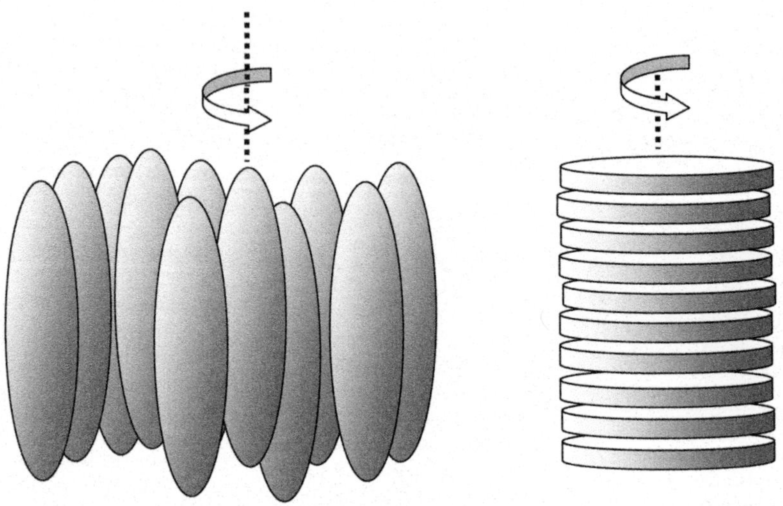

Fig. 1.3 Molecular aggregation in thermotropic liquid crystalline phases – calamitic molecules aggregating into a layer (*left*) and discotic molecules aggregating into a column (*right*)

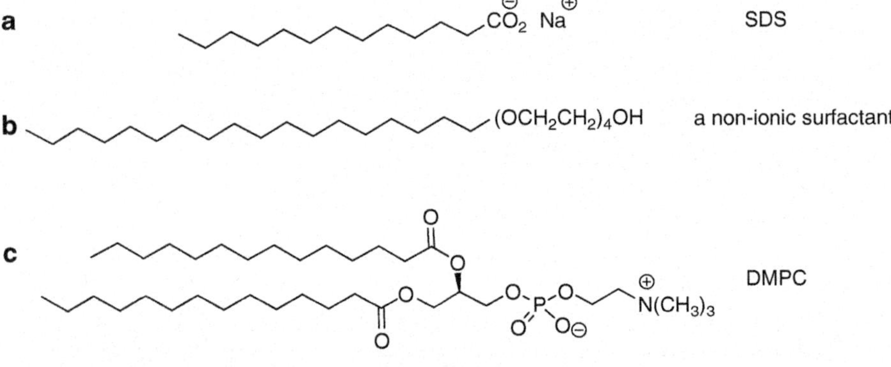

Fig. 1.4 The molecular structures of some typical lyoytropic mesogens, (**a**) sodium dodecyl-sulphonate (a sodium soap), (**b**) a non-ionic synthetic detergent, (**c**) dimyrisoylphosphatidylcholine (a biologically–occurring phospholipid)

the mesophase-forming properties of lyotropics are usually summarised in terms of a two-dimensional temperature/composition phase diagram like that shown in Fig. 1.5.

The major distinction between thermotropic and lyotropic systems lies in the patterns of aggregation of the molecules. In conventional lyotropic systems, the molecules have both hydrophobic and hydrophilic regions. This causes them to aggregate into *micelles,* allowing the hydrophobic parts to reduce their contact with the water subphase.

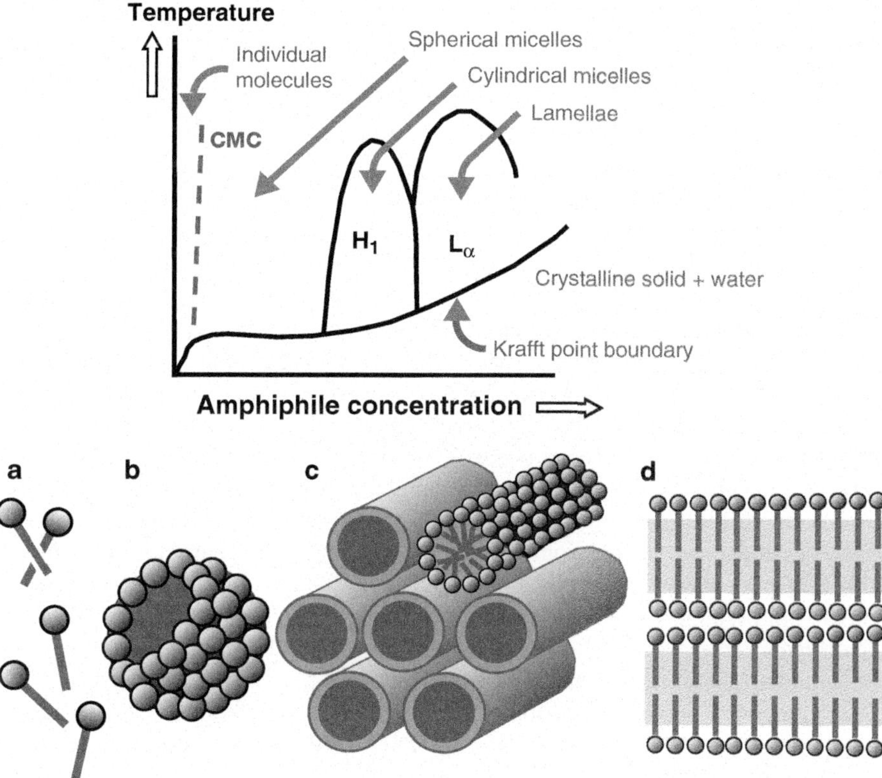

Fig. 1.5 Lyotropic phase diagram.This sketch shows the typical form of one of the simpler types of phase diagrams for lyotropic systems. As the amphiphile concentration increases, from *left to right* across the diagram, there is a progressive rise in the level of aggregation of the molecules. At extreme dilution, individual amphiphile molecules are dispersed in the solvent (**a**). At higher concentrations, small sub-micellar aggregates of three or four molecules build up. These increase in size and concentration until spherical micelles (typically composed of 60 or 70 molecules) arise (**b**). This stage (indicated by the dashed line), is termed the critical micelle concentration (CMC) and marks a drastic change in the properties of the solution The micelle is a stable arrangement because the hyrodrophobic parts of the molecules are shielded from the solvent – and because it is energetically more favourable for a single molecule to join a micelle than to exist alone, the pattern of distribution of cluster sizes changes dramatically as virtually all of the molecules present become incorporated into micelles. As these spherical assemblies increase in number they begin to fuse together into cylinders which pack in a hexagonal array in the H1 liquid crystal phase (**c**). At higher concentrations, these cylinders fuse together to produce extended sheets giving the lamellar, Lα phase (**d**). In order to fill the awkward spaces in the centres of spherical and cylindrical micelles, the alkyl chains must be fluid and disordered. As a sample is cooled, it reaches a state, where the flexibility of the chains is reduced to such an extent that micelles can no longer form and hence the mesophases, which are composed of micellar units, are no longer stable. The phase therefore separates into water and crystalline solid. The temperature at which this occurs is known as the Krafft point and the Krafft point boundary shown marks the lower limit of mesophase formation

There are some specific amphiphile systems which form exceptionally stable pancake-like micelles which, as one might expect, give anisotropic solutions analogous to discotic nematic phases. Conversely, there are other lyotropic systems containing elongated stacks of aromatic molecules, i.e., chromonic systems, which are analogous to thermotropic calamitic nematic phases.

There are also anisotropic solutions formed by hydrophilic polymers such as cellulose derivatives, collagen, α-helical polypeptides, and nucleic acids (DNA and RNA), but, as far as I am aware, there are no lyotropic liquid crystalline phases analogous to the thermotropic nematics, in which small solute molecules are free to move individually.

Lyotropic phase diagrams can be much more complex than the example given. At higher amphiphile concentrations, inverse structures can occur where the micelles are turned inside out, giving phases with a similar geometry, but of a water-in-oil type instead of an oil-in-water type. Also, there are other types of mesophase structure, which can occur between the hexagonal and lamellar and at the low concentration side of the hexagonal phase. These fall into two groups, i.e., those formed from individual micelles and those formed by extended networks of jointed cylinders or sheets. Some in the latter category are particularly complex, with very high symmetry patterns of interlocking lattices. Because both the water and the lipid regions extend continuously throughout the phase, these are known as *bicontinuous structures*. One of these, with space group Im3m, is aptly nicknamed the 'plumber's nightmare'. Many of these structures have cubic symmetry and as a consequence are optically isotropic. They are therefore difficult to miss in optical microscopy because they show up as solid black regions when samples are viewed between crossed polars.

1.2 Calamitic Mesophases

1.2.1 Nematic Phases

The word *nematic* comes from the Greek νημα (*nema*), meaning "thread". This is a reference to the linear topological defects, termed ***disclination*** lines, which can be seen with the naked eye in bulk samples of this phase. The nematic phase is one of the commonest thermotropic liquid crystalline states. It is formed by elongated rodlike molecules, such as the cyanobiophenyls shown in Fig. 1.7. In this phase, the molecules spontaneously align so that their long axes lie more of less parallel over considerable distances. When bulk samples are viewed between crossed polars in the optical microscope, the *director* (the mean direction of the molecular long axis) can be seen gradually curving over distances of the order of 0.1 mm.

This phase generally has low viscosity. The molecules are as free to flow as those in the isotropic liquid. The positions of their centres of mass are randomly distributed, but the long-range directional order is still maintained (Fig. 1.6).

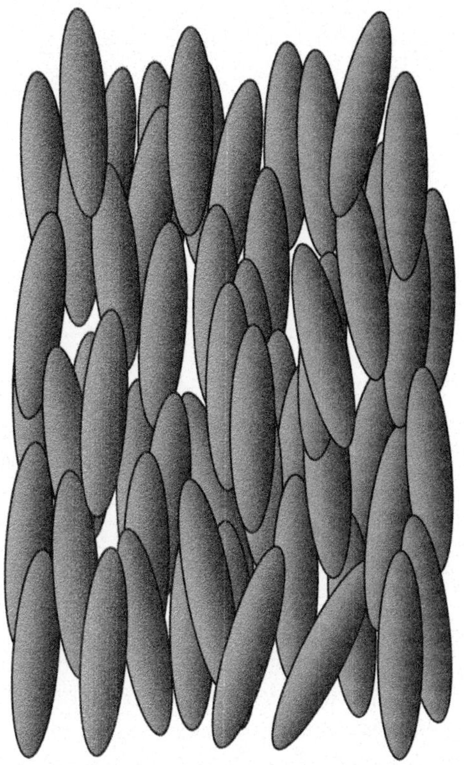

Fig. 1.6 A stylised picture of a group of molecules in a nematic array

The vital step, which turned the liquid crystal display industry from a dream into a reality, was the synthesis, in 1972, of the first stable room temperature nematic material by Gray and Harrison. This was pentylcyanobiphenyl (5CB) shown in Fig. 1.7a. Subsequently, some thousands of nematogenic compounds have been synthesised. The vast majority of these have similar structures, e.g., they are often calamitic molecules with a rigid half, consisting of two or three aromatic rings, and a flexible half of more or less the same length, consisting of a C5 or C6 alkyl chain. Some variants of this structure are shown in Fig. 1.7. These include the incorporation of a rigid aliphatic ring, replacing one of the aromatic rings for compound (b), the substitution of fluorine for hydrogen atoms, to give a molecule with negative dielectric properties (c) and the presence of chiral centres (d). In general, nematic phases of such compounds are miscible. This makes it possible to fine tune the properties for each particular type of display requirement. A typical commercial material will be a mixture of three or four components.

The majority of nematic molecules are actually more blade-like than rod-like, but the more-or-less free rotation of each molecule about its molecular long axis means that the molecules effectively see each other as circular cylinders. There has been a long search for a biaxial nematic phases, where the planes of the blades lie parallel, and, although it is thought that credible examples of such phases have been

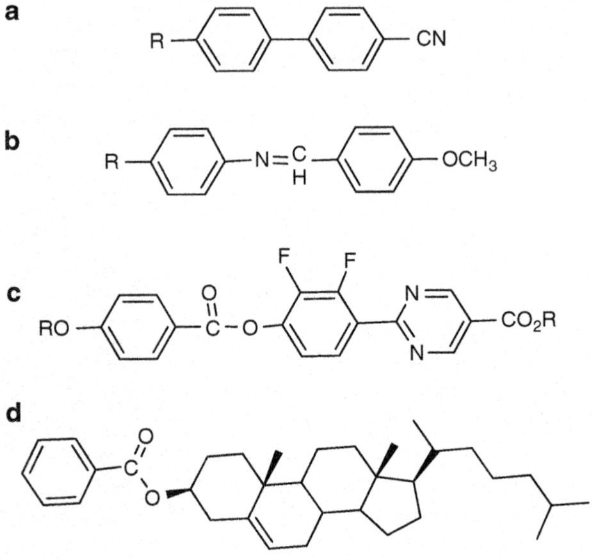

Fig. 1.7 Structures of four nematogenic molecules. (**a**) 5 CB the archetypal nematic mesogen; (**b**) MBBA; (**c**) a fluorinated mesogen with a negative dialect susceptibility; (**d**) a chiral mesogen-forming a twisted nematic (cholesteric phase) phase

found, they are very few in number. In general, nematic phases can be easily aligned by relatively weak electric fields and they usually spontaneously align on prepared substrate surfaces. It is the combination of these two properties that make them uniquely useful in liquid crystal display devices. The classical calamitic nematic phase is optically uniaxial and the prolate shape of the indicatrix usually matches the prolate shape of an individual molecule.

It is hardly surprising that the first thermotropic liquid crystal phase to be investigated, over a century ago, were cholesteryl esters. The iridescent colours, which flash out when molten material is cooled, is a beautiful phenomenon and it cried out for further investigation. These naturally-occurring sterols are chiral, optically-active molecules. Their asymmetry causes them to form a twisted nematic phase, as sketched in Fig. 1.8. When this has a pitch within the optical range, it produces iridescent interference colours. When first observed, this property was regarded as being so notable that the generic term '*cholesteric*' was used to describe all twisted mesophases of this type. The discovery that the addition of small quantities of chiral dopants to non chiral nematic phases produces similar phases, led to the feeling that they should be treated as a subdivision of nematics and they are now generally termed *twisted nematics* indicated by a symbol followed by asterisk (N*).

1.2.2 Smectic Phases

In general, as the temperature of a nematic phase is lowered, a new dimension of ordering is introduced as a phase change occurs where the molecules spontaneously

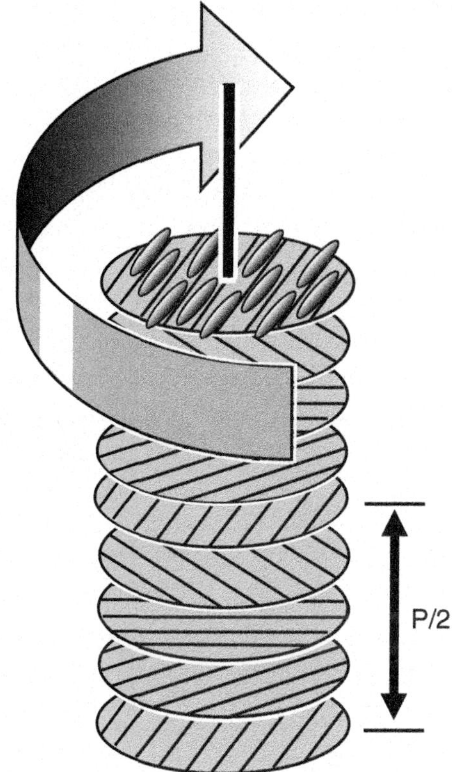

Fig. 1.8 The twisted nematic N* (cholesteric) structure, formed by chiral modifications of nematic compounds and by non-chiral nematagens doped with chiral solutes. Where the pitch of this helicoidal structure is comparable with the wavelength of light, the phase gives iridescent reflections. Larger pitch structures, where the repeat distance can be resolved in the optical microscope, appear as 'fingerprint textures' with bands separated by half a pitch. Helicoidal structures of this type are common in biological material, most obviously in the carapaces of iridescent beetles. Their electron microscope pictures show a characteristic pattern of nested arcs, known as Bouligand patterns [4]

organize themselves into layered '*smectic*' patterns. The molecular structures of commercial smectic materials are essentially those of nematics with higher transition temperatures. In the smectic A phase, abbreviated to SmA, the molecules in each layer are disordered and mobile and it can be regarded as a two-dimensional liquid. On average the molecules lie normal to the layer, but there is considerable variation of alignment, with a spread of up to 10° or 15°. There is rapid rotational motion of the molecules about their long axes on a timescale of about 10^{11} times per second. This is to be compared with the much slower motion about their short axes, which occurs on a timescale of 10^{-6}s. Although there is a tendency for the molecules to have short-range hexagonal ordering within the layers, this only extends over a range of a few Å and there is no long-range translational periodicity in the planes of the layers.

In the context of smectic phases, the word 'layer' should not be taken as implying a hard-edged, well-defined crystalline sheet. The X-ray diffraction pattern of a smectic A phase shows only a single peak relating to ordering perpendicular to the layer planes. The absence of second and higher order reflections indicates that the centres of gravity of the molecules have a more or less sinusoidal distribution, i.e., there is a *density wave* rather than a precise stacking of layers. Note the way

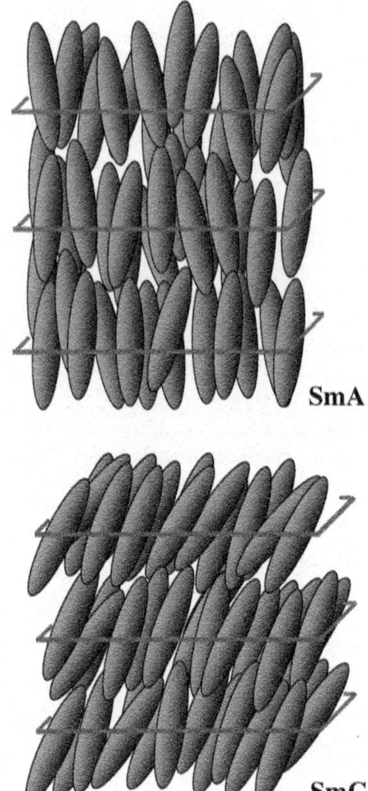

Fig. 1.9 Smectic A and C phases. These have highly mobile, disordered layers, which can be regarded as two-dimensional liquids

this pattern of long-range order coupled with appreciable short-range displacement is a recurrent feature in liquid crystal systems and can be regarded as analogous to the way in which the long-range alignment of the director in nematic phases is maintained, in spite of considerable short-range fluctuations of molecular alignment.

The smectic C phase (SmC) is the tilted analogue of the SmA phase, as sketched in Fig. 1.9. The molecules aggregate in similar fluid layers with no positional order, but the long axes of the molecules are tilted at an angle to the layer normal. Presumably, because of the degree of interpenetration of molecules of one 'layer' into the next, there is generally appreciable correlation of the tilt direction of adjacent layers. The SmC phase can be formed in a variety of ways, e.g., either by a first-order phase transition, on cooling the isotropic liquid, the nematic or smectic A phases, or by a second-order phase transition from the smectic A, as the molecular alignment gradually tilts away from the normal.

Aligned samples of SmC phases are birefringent and optical textures of un-oriented samples, viewed normal to the layers, show swirling patterns around disclinations, as the direction of the tilt vector varies from place to place. For an analogy, picture a field of corn with the eddy currents of wind causing patterns in the alignment of the stalks. This texture resembles the schlieren patterns of nematic phases, but it be distinguished by close examination of the 'strengths' of the director field patterns around the disclinations. There are over a dozen variants of smectic phases, indicated by letters of the alphabet, with a variety of combinations of short-range and long-range ordering. The elegant scheme shown in Fig. 1.10, summarises most of these. Orthogonal, i.e., non-tilted structures, are shown on the left and the tilted phases on the right. The vertical, division of the table into three levels, corresponds to different levels of ordering within the layers. The structures on the top row have mobile, disordered layers. The structures in the second row have partially-ordered layers, with long-range orientational order but only short-range positional order. The bottom two rows are the so-called '*crystal smectics*' with both extended positional and orientational order. Just as the nematic phases of chiral compounds form twisted structures, there are chiral counterparts of the smectic phases. The difficulty of reconciling a long-range layer structure with local chiral twisting leads to some remarkably complex structures, notably the twisted grain boundary phase (TGB). Note in particular, the chiral smectic C* phase, which has found extensive use in display devices.

1.3 Discotic Mesophases

It had been thought, for at least 50 years, that disc-like molecules ought to be able to form liquid crystalline phases, but their actual discovery came surprisingly late. It was in 1977, almost a century after the characterisation of nematic and smectic phases, that Chandrasekhar synthesised the first discotic mesogen [2]. This was hexaalkyl benzene, shown in Fig. 1.11. This structure, an aromatic core with a peripheral ring of flexible alkyl chains, has proved to be the pattern for the extended family of discotic mesogens. Hundreds of discoid mesogenic compounds have since been synthesised with central aromatic cores of triphenylenes, porphyrins, phthalocyanines, coronenes, etc., surrounded by alkyl chains, typically in the C5 – C8 range.

The most mobile and least-ordered mesophase formed by discotic molecules is the N_D phase, sketched in Fig. 1.12. This is the discotic analogue of the calamitic nematic phase. The planes of the molecules lie more or less parallel, but there is no positional order in any direction. In the columnar nematic phase, N_{Col} there is some degree of aggregation, with short stacks of molecules 'floating' in a nematic array. Both of these phases are relatively rare and there only are few known examples.

There is a strong tendency for aromatic molecules to aggregate face to face and the formation of columnar phases by discotic molecules is widespread. There is a cascade of discotic columnar phase structures, roughly analogous to the alphabet of smectic phases. Some of these are shown in Fig. 1.14 to give an indication of the range of possible structures. Columnar structures are characterised in terms of the pattern of stacking within the columns (ordered or disordered) and the appearance of the two-dimensional lattice, as seen in a cross-section normal to the column axes.

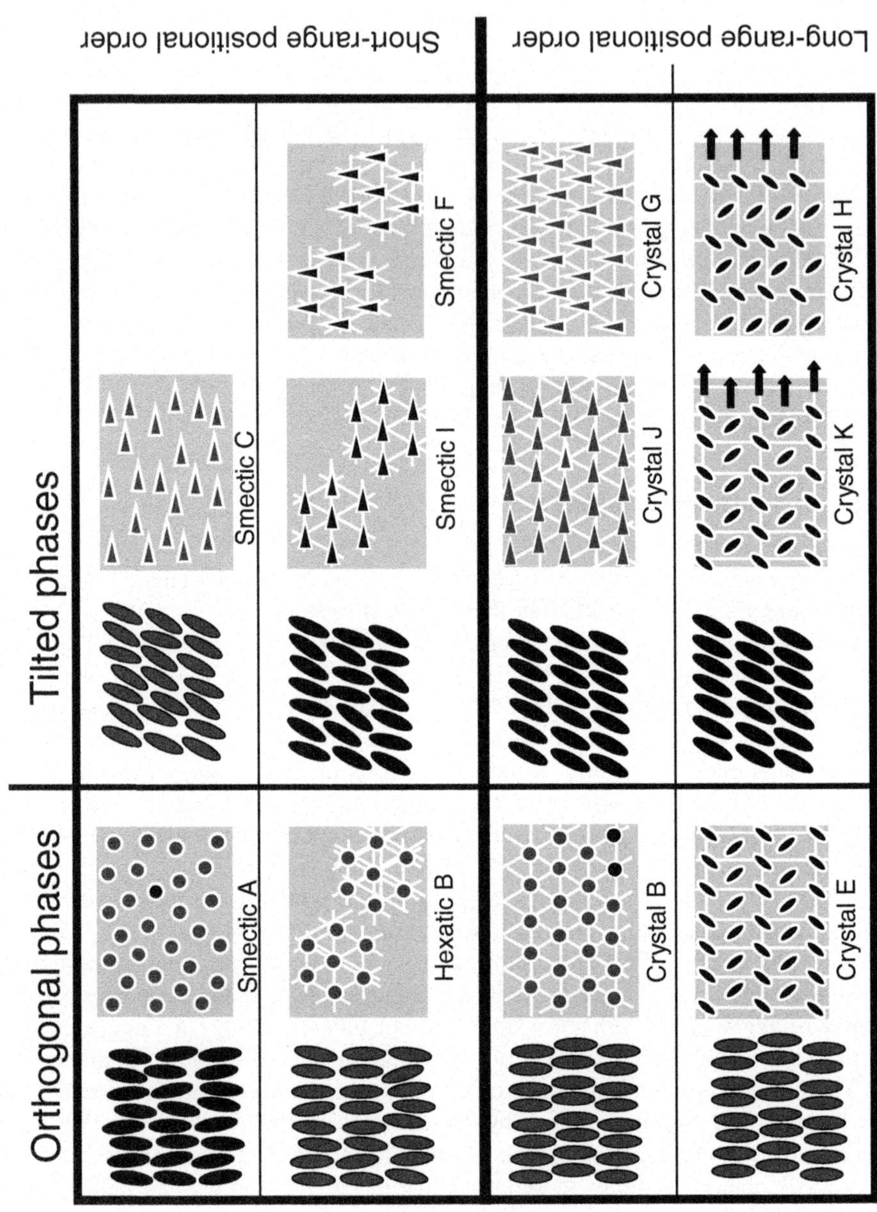
Orthogonal phases
Tilted phases
Short-range positional order
Long-range positional order
Smectic A
Smectic C
Hexatic B
Smectic I
Smectic F
Crystal B
Crystal J
Crystal G
Crystal E
Crystal K
Crystal H

Fig. 1.10 The structural hierarchy of smectic phases, see Sect. 1.2. (This scheme is redrawn (slightly) from *Smectic liquid crystals: textures and structures* by G. W. Gray and G. W. J. Goodby, and is used by permission of Professor John Goodby.) In each box, the side elevation (looking along the edges of the layers) is shown on the *left* and the plan views of an individual layer are shown on the *right*. In the *left hand column*, the **A**, **B** and **E** phases all have molecules lying more or less normal to the layers. The *right hand columns* show the tilted phase. The phases with disordered liquid-like layers are at the *top* and the ordering increases in increments *downwards*, with the lower two rows containing the three-dimensionally ordered 'crystal' smectics (which are essentially anisotropic plastic crystal phases). The second row contains phases with mixtures of short-range and long-range order. Many smectic systems pass through a succession of phases on cooling, in more or less every case, tracing a path which runs downwards – and sometimes to the *right* – through this diagram. The cross-sectional areas of the molecules are shown as circles, ellipses, isosceles triangles, (depending on which feature of the structure is being emphasised). For the **A**, hexatic **B** and crystal **B**, the circles indicate that the molecules have more or less free-rotation about their long axes. For the tilted phases, the triangles indicate the tilt direction (— i.e. they show the in-plane tilt vector). The white lines indicate hexagonal and centred rectangular lattices. In the crystal **B, E, G, J, K** and **L** phases, the molecules have both positional and rotational order. The **I**, **F** and hexatic **B** phases, are more complex – although the layers have only short-range positional order, there is extensive long-range alignment of islands of ordered structure. For the non-orthogonal hexagonal phases, the molecules can be tilted either towards the corners of the hexagons (as in **I** and **J**) or towards the edges (as in **F** and **G**). Similarly for the rectangular crystal phases, the molecules can be tilted towards the long edges of the rectangles (as in **K**) or towards the short edges (as in **H**). This scheme is by no means complete, for example, the various patterns of tilting in antiferoelectric and ferrielectric types of phase structure are not shown – and note that there are, in principle, chiral versions of every mesophase type

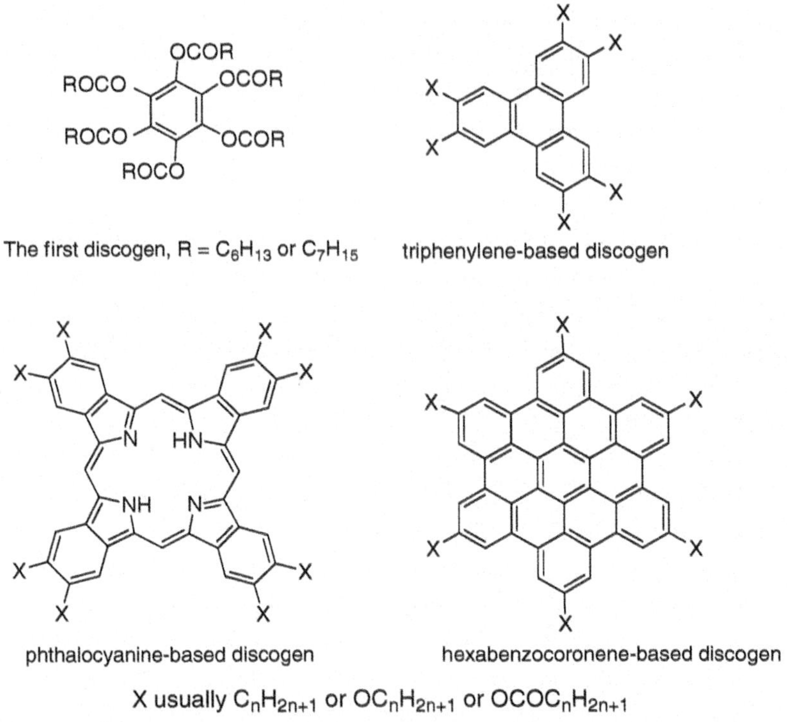

OCOR
ROCO OCOR
ROCO OCOR
ROCO

The first discogen, R = C_6H_{13} or C_7H_{15}

X X X X X X

triphenylene-based discogen

X X X X
N HN
NH N
X X X X

phthalocyanine-based discogen

X X X X X X

hexabenzocoronene-based discogen

X usually C_nH_{2n+1} or OC_nH_{2n+1} or $OCOC_nH_{2n+1}$

Fig. 1.11 Molecular structures of some discotic mesogens

Fig. 1.12 The nematic discotic, N_D phase, where individual molecules have lateral freedom of movement, and the columnar nematic phase, N_{Col} where the molecules are aggregated in short stacks of stacks of three or four molecules

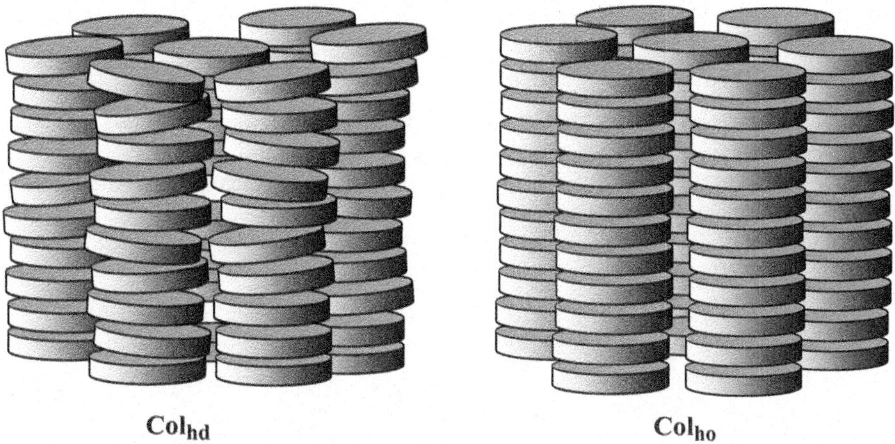

Fig. 1.13 Variants of the hexagonal discotic columnar phase with ordered and disordered column structure. This sketch is intended to suggest how, in the Colhd phase, the molecules can jostle within the cylindrical envelope of a column without disrupting the translation ordering of the hexagonal lattice. The letter h in the subscript indicates a hexagonal lattice and the letters o or d indicate ordered or disordered columns

The variety of columnar structures is every bit as complex as that for smectic systems and a number of similar themes occur. X-ray diffraction patterns indicate, that even a simple hexagonal columnar phase can have columns, which are can be ordered or disordered along the column axis see Fig. 1.13, and molecules can be normal or tilted within a column. Some of the possible patterns of lateral ordering are shown in Fig. 1.14 with the columns lying on hexagonal, rectangular, and oblique lattices. The way in which the symbols summarise the geometry of a columnar phase is illustrated in the example for Colrd shown in Fig. 1.15.

Mesophases with stacked columns of aromatic cores are attracting particular attention because of their electronic properties with the aromatic stacks transporting electrons or holes and alkyl chain regions acting as insulators. If conducting liquid-crystalline conductors can be commercially produced, they promise to have many advantageous features, such as their ease of processing and particularly their self-healing properties, i.e., the ability of mesophases to spontaneously remove any structural defects which might arise during the operation of the device.

1.4 Optical Textures

Since the earliest studies of liquid crystalline phases, the first technique that an investigator usually turns to when examining a new mesophase, has been polarising microscopy. In the vast majority of instances this will involve a more or less disordered sample of mesophase held between a glass slide and cover slip, viewed

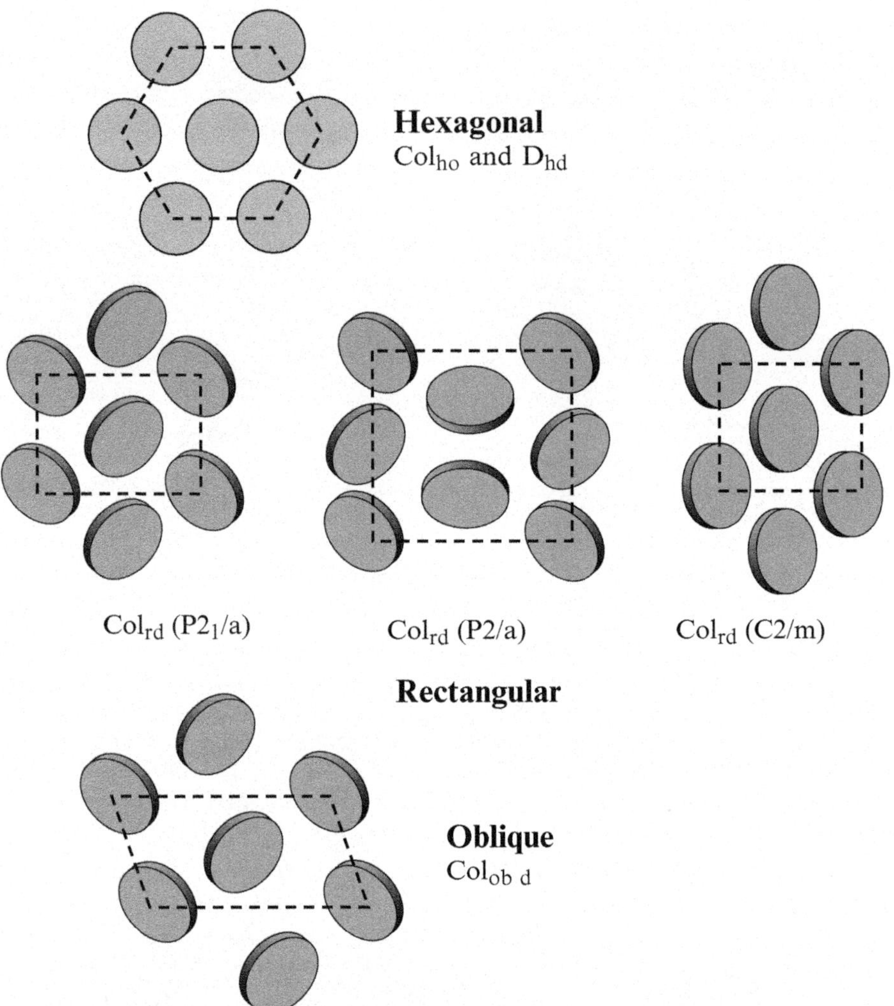

Fig. 1.14 The structural hierarchy of columnar phases. This diagram shows various columnar structures viewed down the column axis. In the high-temperature columnar structures, shown at the *top*, the thermal motion causes the molecules to lie normal to the column axis giving columns with a more-or-less circular cross-section. This in turn, gives rise to a hexagonal lattice of columns. At lower temperatures, the molecules tilt, giving the columns an elliptical cross-section. This distorts the hexagonal pattern, producing a centred rectangular lattice. At even lower temperatures, the decrease in thermal motion produces additional increments of ordering and a pattern with columns lying on an oblique lattice with parallelogram-shaped unit cells is produced

between crossed polars with a low power microscope. The appearance of a sample viewed under these conditioned is called the '*optical texture*'. Strictly speaking, one should distinguish between the structure, i.e. the three dimensional structure of the sample and the texture which is the appearance of a particular two-dimensional

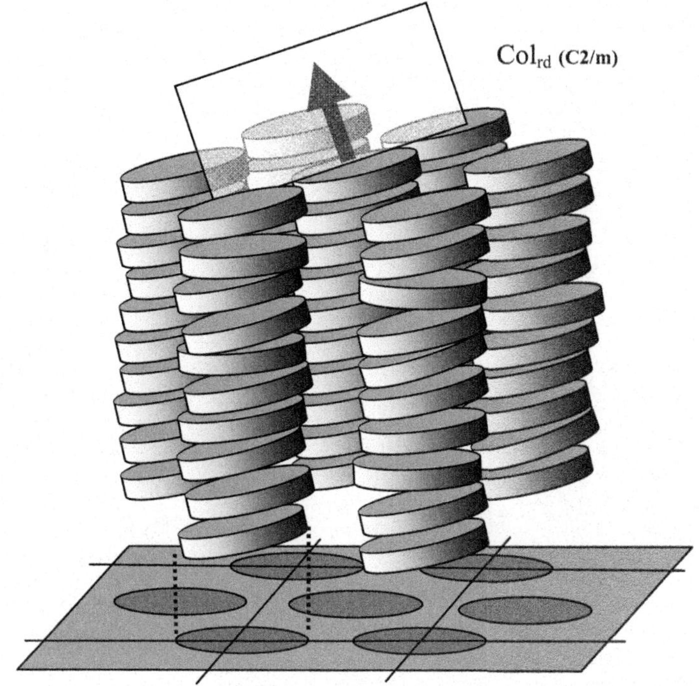

Fig. 1.15 The structure of the tilted hexagonal columnar Col_{rd} (C2/m) phase. The letter 'r' indicates that the columns lie on a two-dimensional rectangular lattice and the letter 'd' indicates that the stacking within the columns is disordered. The additional details given in brackets are the three-dimensional space group designation. It is necessary to add this, if one wishes to distinguish between this structure and that of other tilted rectangular columnar phases as shown in Fig. 1.14

slice in focus. Identifying optical textures is rather like identifying different types of timber from the grain of the wood. The sum total of the features, the colour, the pattern of annual rings, the mottling or streaking and other fine detail combine to give a characteristic pattern and the human eye (together with the information processing machinery behind it), is very good at recognising and distinguishing patterns such as these. This analogy can be pushed a little further than one might expect, when one considers the way in which the appearance of the 'figure' depends on the orientation of the cut. End-grain looks different to different to 'flat-grain' and some decorative patterns of grain can only been seen in sections cut in a particular direction. For example the 'figured' pattern of oak, valued by cabinet makers, is formed by displaying the tracheids, and is only apparent in slices cut tangentially from the log.

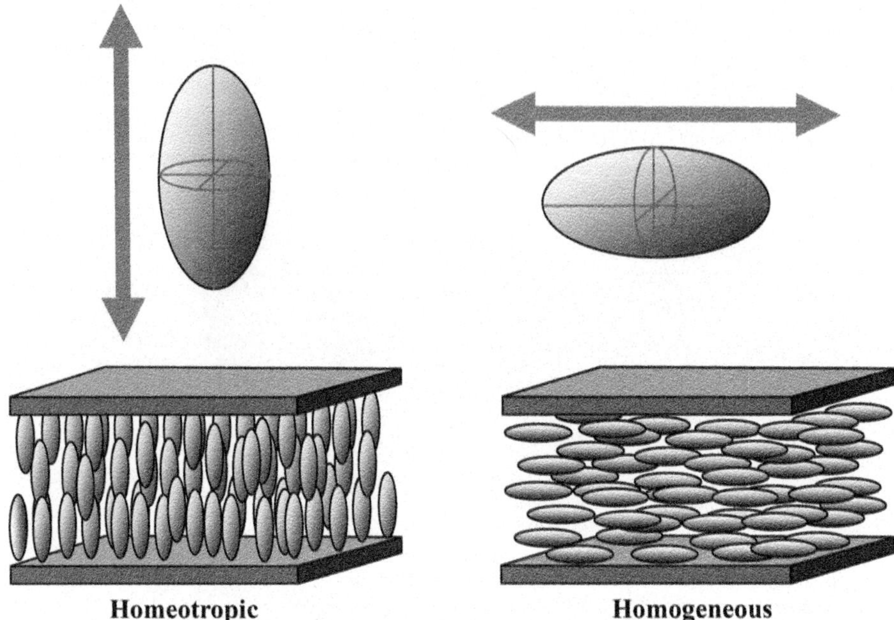

Fig. 1.16 Aligned mono-domain nematic phases

1.4.1 Nematic Textures

Mono-domain samples of a nematic phase appear to be optically isotropic when they are viewed down the director giving a '*homeotropic*' texture as shown in Fig. 1.16 which, when viewed between crossed polarisers, appears uniformly black at all orientations as the sample is rotated. In contrast, a similar sample viewed at right angles to the director is said to be '*homogeneous*' or '*planar*' and has similar optical properties to a slab of anisotropic single crystal, going from light to dark every 45° as it is rotated. Because of its importance to the display industry, there have been extensive studies of surface treatments which give homeotropic and homogenous alignments of nematic phases. In some instances it is advantageous to produce homogeneous alignments with a small tilting angle as sketched in Fig. 1.17.

Most commercial liquid crystal display systems are of the twisted nematic type. They depend on the change of optical properties of a nematic phase as it switches from homogenous to homeotropic when an electric field is applied and removed, as sketched in Fig. 1.18.

Optical textures give the most immediate and accessible information about a mesophase sample and in most cases a single texture is sufficient to identify the

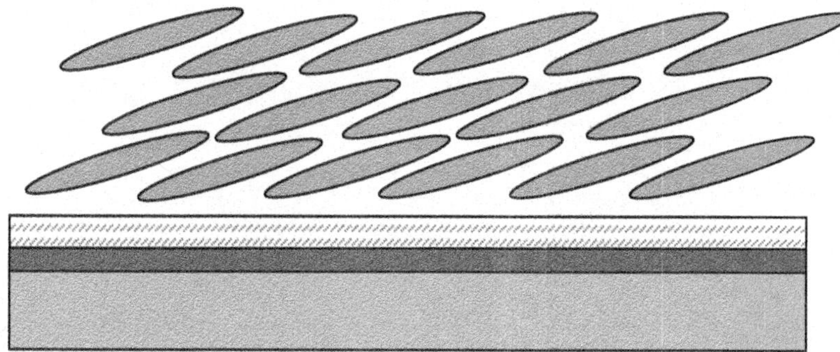

Fig. 1.17 For some display devices it is preferable to have a substrate surface treatment which gives a tilted alignment. This drawing shows a smectic A phase with a small tilt, above a glass substrate coated with a conducting layer (usually indium and tin oxide) and an alignment layer giving a small tilt angle

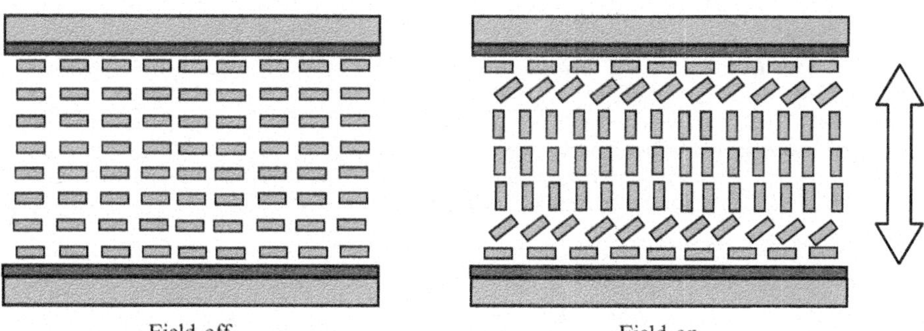

Fig. 1.18 The Fréedericksz transition. The re-alignment of a nematic material in an electric field is the basis of most commercial display devices. Note that, although the bulk of the mesophase changes alignment when the field is applied, the surface layers tend to remain anchored to the substrates

type of phase present. The reason why optical textures are so informative is a consequence of the partial fluidity of the mesophase and they are a function of both the absolute and relative values of the various elastic constants of the phase governing bend, twist and splay distortions. In systems that are fluid enough to allow considerable freedom movement to the molecules, any pattern of disorder originally present in a sample, will rapidly relax to its lowest *accessible* energy state. For example, if a single domain of aligned nematic phase is stirred, any disorder introduced will relax until it reduces to a pattern where the only abrupt discontinuities remaining are lines of *disclination*, rather than two-dimensional surfaces at domain boundaries.

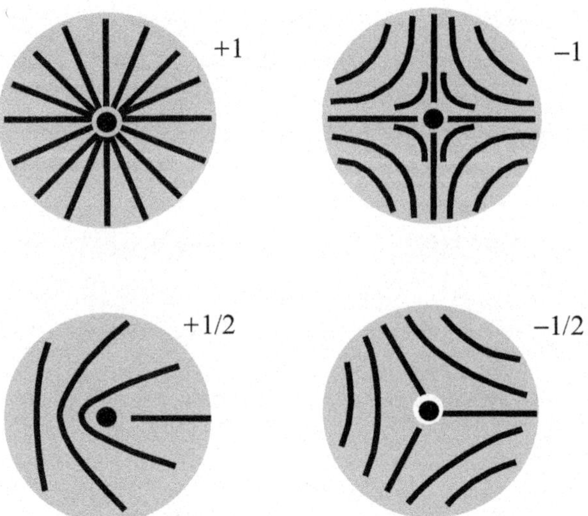

Fig. 1.19 Sign and strength of nematic disclinations. These are categorised in terms of their sign (+ or −) and strength (the number of times the local director makes a rotation of 2π in a circuit around the disclination). When nematic phases are viewed between crossed polars, disclinations of strength 1 appear as Maltese crosses with four brushes – and disclinations of strength ½ have a two-brush pattern. Although higher orders are possible, the usual director fields of nematic phases contain only the four types of disclination drawn here

The four principal types of disclination found in nematic systems are shown in Fig. 1.19. Although a director field pattern of this kind will not represent the absolute minimum energy state of a mesophase sample, the molecular patterns around the disclinations are locked-in, and there is no easy way for the system to remove them. A typical optical texture of a nematic sample held between untreated surfaces, is shown in Fig. 1.20 together with its interpretation in terms of the pattern of molecular alignment.

1.4.2 Smectic Textures

It is possible to produce extended areas of homeotropically-aligned smectic phases, but when the mesophase is formed by cooling from the melt on an unprepared surface, a disordered pattern of distinct domains is usually formed. Textures of this kind are known as 'focal conics' because of the geometry of the basic structural units. These are formed by stacking concentric, curved, cooling-tower shaped layers inside each other like Russian dolls (Fig. 1.21). The reason why such an apparently complex structure of this kind arises spontaneously is not immediately obvious, but it concerns the curvature energy of the layers. All of the layers have the same average curvature and hence the same curvature energy. The two salient geometrical features of a focal conic unit are the central disclination line and the shape of the equatorial cross-section. In the simplest case the disclination is a straight line, which passes through the centre of the circular cross-section (Fig. 1.22). In the more general case, it is a hyperbola passing through the focal point of an ellipse, hence the term 'focal conic'.

The appearance of focal conic structures depends on the direction of viewing. When the epitaxial interactions between the mesophase and the substrate surface, cause the axes of the focal conic units to lie vertical, the '*polygonal texture*' appears (Figs. 1.23 and 1.24). On the other hand, samples with focal conic units lying obliquely to the viewing direction, give the characteristic '*fan texture*' (Fig. 1.25). [A note of caution: for decades, when faced with an undecipherable optical texture with a jumbled mass of domains, workers in the liquid crystal field have often taken the easy option and simply called it a focal conic fan texture to disguise the fact that they had no idea what it was. They have usually been correct, but not always, and mesophases of very different structures have been seriously misidentified on more than one occasion.]

1.4.3 Discotic and Columnar Textures

Columnar phases tend to have low birefringence – possibly because the intrinsic birefringence of the molecule is, to some extent, opposed by the form birefringence of the column, and this often makes it difficult to interpret the appearance of the optical textures with any certainty. However, two extreme cases are readily identifiable. The homeotropic '*melting snowflake*' texture formed by columnar phases on cooling has domains with distinctive sixfold dendritic symmetry (Fig. 1.26) and corresponds to a view down the symmetry axis of the columnar assembly, as sketched in Fig. 1.27. In contrast, if the columns lie at right angles to the viewing direction, a radiating pattern of fans is often apparent (Figs. 1.28 and 1.29).

The easiest distortion of the smectic A structure is an orchestrated curvature of the layer surfaces which keeps the layer thickness constant. This leads to the Dupin cyclide structures and hence to focal conic textures. In a corresponding way, the easiest distortion of a columnar structure is generally a curvature of the director about a direction parallel to the side of a hexagon in the hexagonal lattice, as sketched in a diagrammatic way in Fig. 1.30a. This enables the inter-column spacing of the two dimensional hexagonal lattice to be conserved. Extension of this pattern in space produces a range of possible structures which have been termed '*developable domains*'. The cross-sections of some examples of these are sketched in Fig. 1.30b.

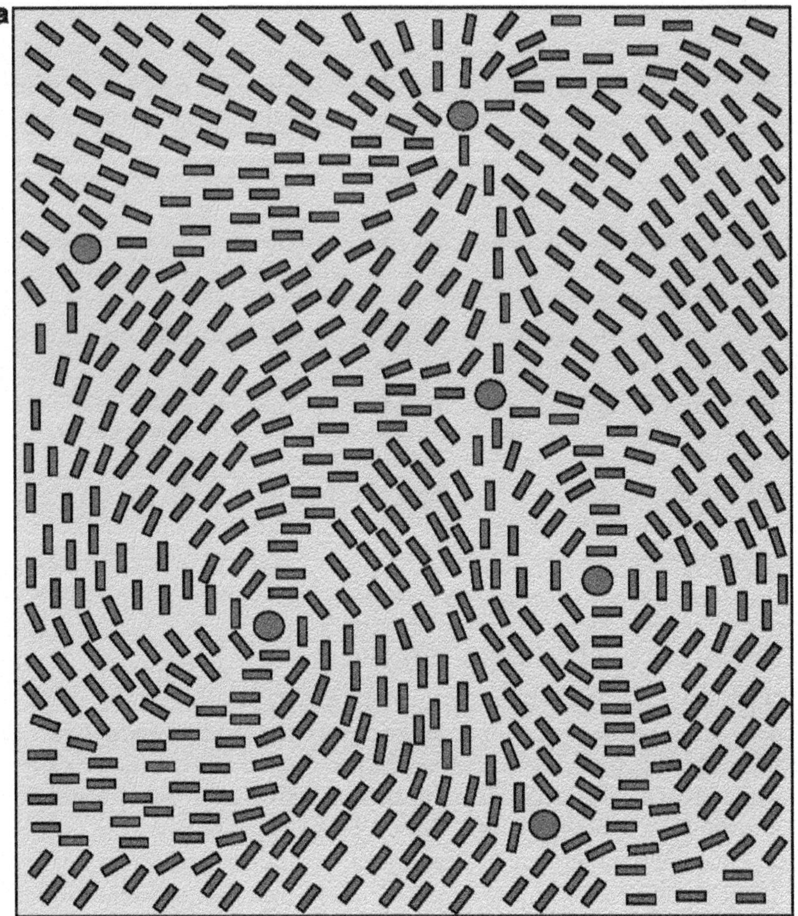

Fig. 1.20 The director field of an un-aligned sample of a nematic phase (*above*) and the corresponding optical texture as observed between crossed polars (*facing*)

Curved columnar structures frequently give rise to a pattern of radiating domains resembling those formed by smectic phases. Oswald and Piranski [5] warn

> *These textures are frequently encountered in the hexagonal phases of thermotropic and lyotropic systems. They look very much like certain focal conic textures observed in thin planar samples of smectic A. One must therefore be cautious* ——.

Call them fan textures if you wish – but don't call them focal conic fans.

The mechanical properties of a columnar phase are, in a sense, similar to those of a sheet of paper – which can easily be curved in one dimension, but which is much more difficult to then curve in another direction (because this would involve stretching as well as bending the sheet). For material with a low birefringence,

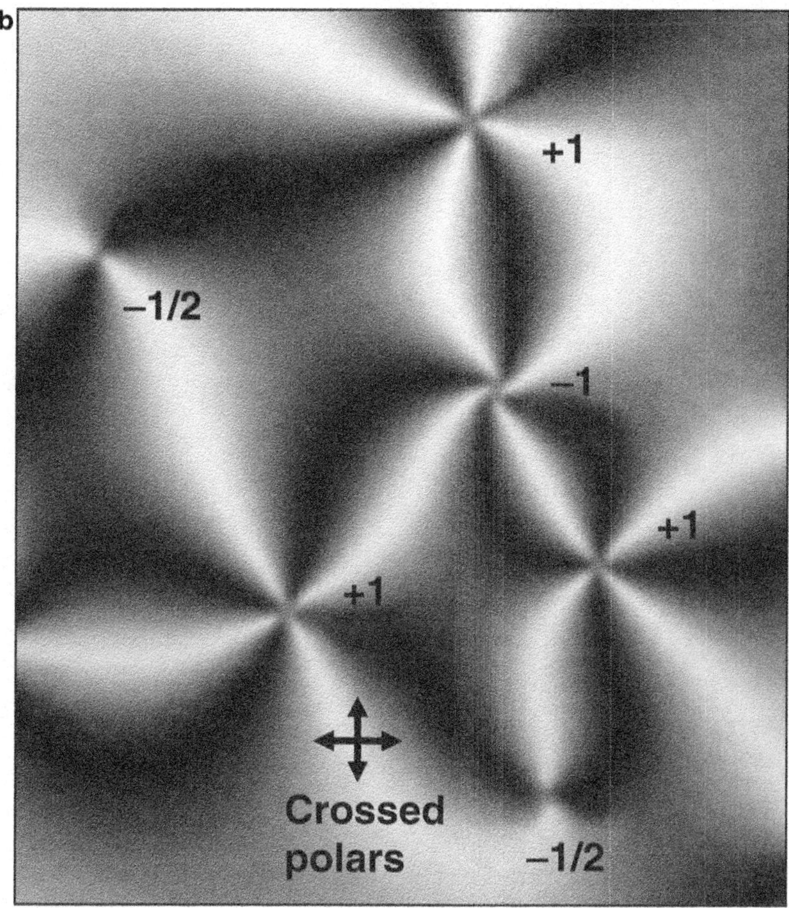

Fig. 1.20 (continued)

this curving about one axis can give rise to optical textures of domains showing first order greys, giving the appearance of pieces of medieval armour. Higher birefringence values give rise to the colourful characteristic pattern of symmetrical coloured bands reminiscent of military medal ribbons, as shown in Fig. 1.26d.

Sometimes the cores of developed columnar domains lie at right angles to the viewing direction and can be seen as extended straight lines (– clearly distinct from the hyperbolic disclinations present in the centre of focal conic units). They frequently occur along the centre of the principal dendritic spines in textures such as those shown in Fig. 1.26 – but, in contrast, in the columnar textures with petal-like domains, they sometimes occur within a 'petal' in alignments which bear no obvious geometrical relationship to the overall shape of the domain.

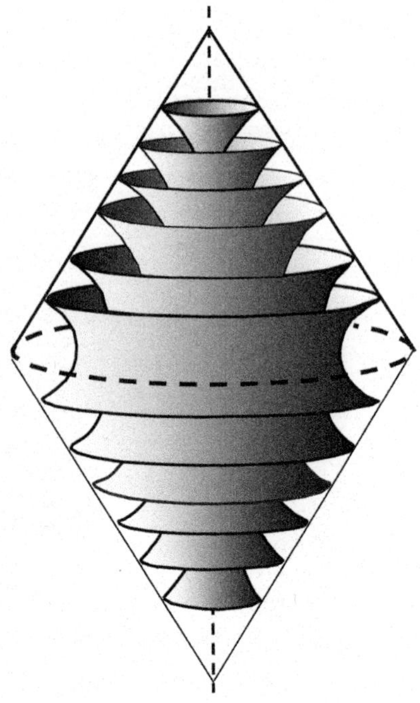

Fig. 1.21 The concentric arrangement of smectic layers within a focal conic unit

1.4.4 The Alignment of Mesophases

Thermotropic liquid crystal phases are usually easy to align and, in general, any asymmetry of their environment will cause some observable degree of alignment of the director and hence modify their optics. There are three principal techniques for alignment (a) flow, essentially by shear alignment – usually when the device is being filled; (b) interaction with a treated surface (c) electric or magnetic fields.

Samples for investigation by polarimetry are routinely aligned on treated substrates and samples for X-ray diffraction are usually aligned in a magnetic field. Most, but not all, liquid crystal display devices depend on switching between the homogeneous and homeotropic states of nematic phases. In the 'off' state, the surface interactions cause the alignment of the whole mesophase sample. In the 'on' state, the electric field dominates and realigns all but the layers immediately adjacent to the substrate surfaces. There are some devices which use smectic phases, particularly the chiral C* phase. These tend to be more expensive, but they have fast switching times and have the additional advantage that they operate between two stable states and therefore have an intrinsic long term memory.

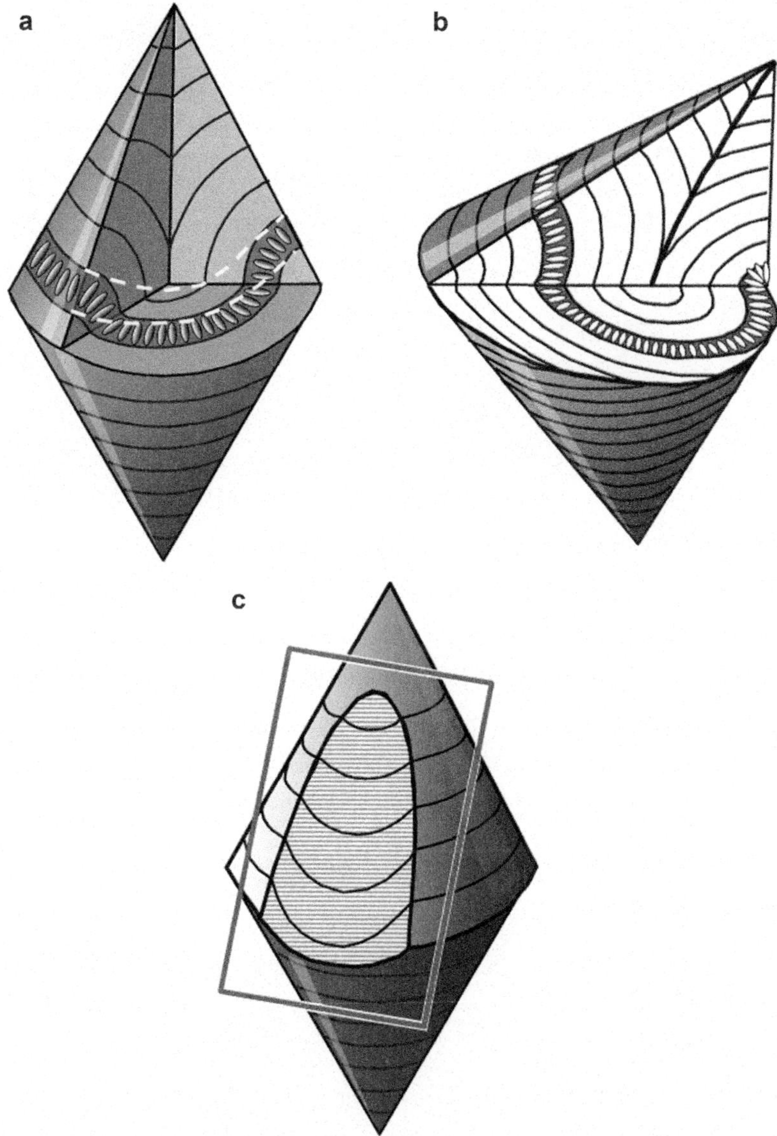

Fig. 1.22 The geometry of focal conic units. (**a**) the simplest geometry, with a circular cross-section and a straight, central discilination line, (**b**) the more general geometry, with an ellipse and hyperbola, (**c**) a slanting section of a focal conic unit, showing the origin of the characteristic focal conic fan texture

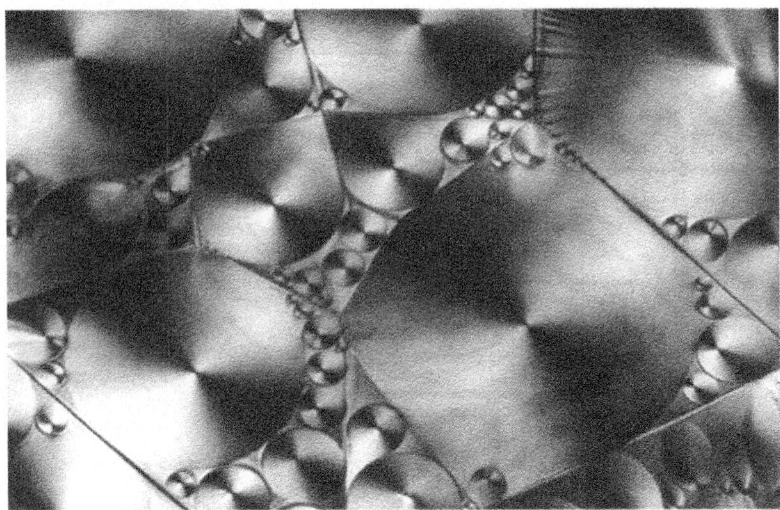

Fig. 1.23 Optical micrograph of the Sm A polygonal texture

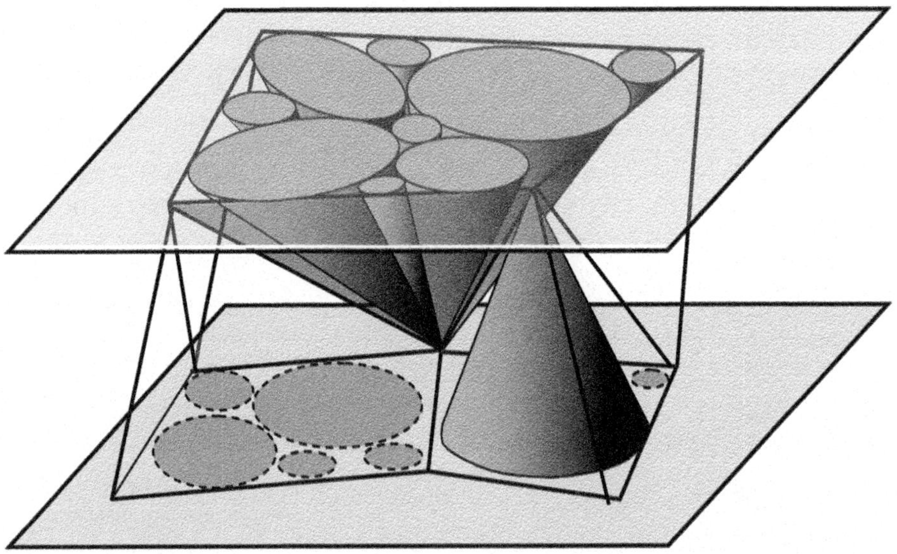

Fig. 1.24 The arrangement of focal conic units in the smectic polygonal texture

Fig. 1.25 Smectic A fan texture optical micrograph with one focal conic unit outlined

1.5 X-Ray Diffraction

It is a common misconception that an X-ray diffraction pattern is some kind of distorted picture of the real structure, but it is not a mapping or shadow of the structure, it is an analysis of the repeat distances in the structure. In mathematical terms, it is the Fourier transform of the structure and, to be more precise, since X-rays are scattered by the electrons, it is the Fourier transform of electron density pattern. The intensity of each reflection in the diffraction pattern given by a crystalline specimen is a measure of the extent to which the electron density throughout the crystal is banded into waves of electron density at a particular spacing and alignment. The process of solving crystal structures by X-rays can be visualised in terms of the reverse of the diffraction process, i.e., a Fourier summation rather than a Fourier analysis, superimposing these waves to build up the global electron density map, showing positions of the atoms. One complexity of X-ray diffraction is the reciprocal relationship between the spacing of these waves of electron density and the angles of the corresponding diffracted rays, i.e., the larger the spacing, the smaller the angles and *vice-versa*. This is a consequence of the geometry of X-ray diffraction, in fact, of any form of diffraction, and it is implicit

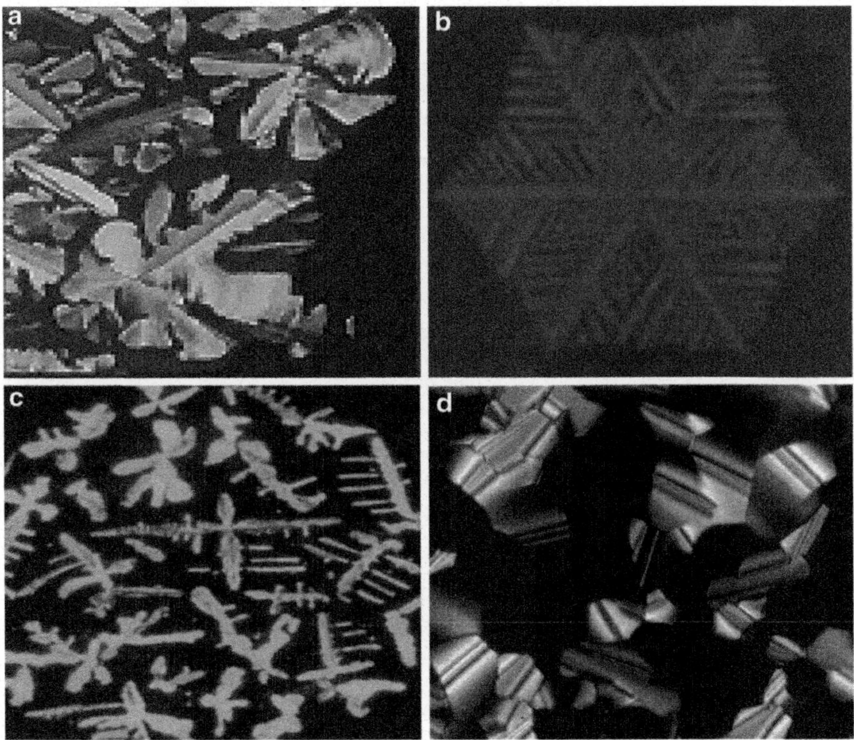

Fig. 1.26 Optical micrographs of discotic columnar systems. (**a**) A typical example of 'broken snowflake' domains, (**b**) An unusually perfect example of a single dendritic domain (of a red mesogen) with detailed hexagonal symmetry, rivaling that of an actual snowflake (From Kelber et al. [3]), (**c**) The orthogonal counterpart of snowflake domains produced by columnar phases with square or rectangular lattices – with twofold or fourfold, rather than sixfold symmetry (Copyright Wiley-VCH Verlag GmbH & Co. KGaA reproduced with permission), (**d**) the 'medal ribbon' appearance of a mosaic texture viewed between crossed polars. Each domain is gently curved about a straight line axis lying perpendicular to the viewing direction (Reproduced with permission of Professor R. J. Bushby)

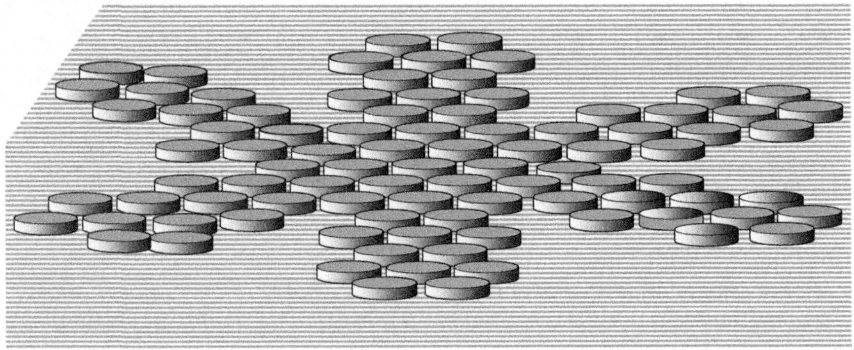

Fig. 1.27 Stylised sketch showing the arrangement of columnar stacks in the 'melting snowflake' texture of columnar mesophases

Fig. 1.28 Optical micrograph of the radiating 'fan texture' of a columnar phase

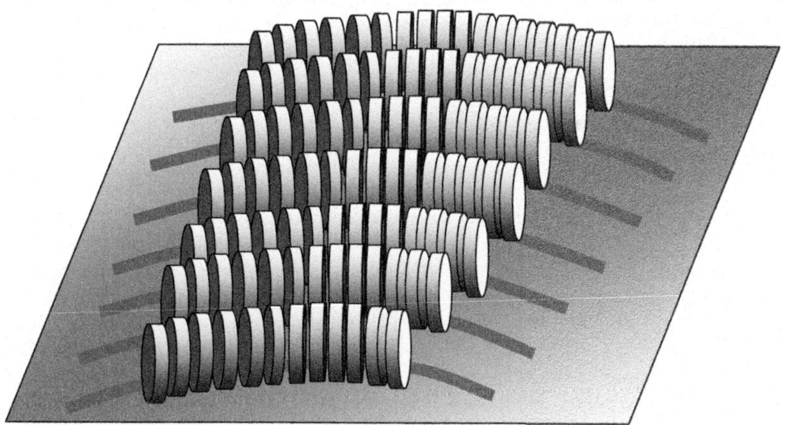

Fig. 1.29 Stylised representation of the arrangement of columns in the radiating fan texture of columnar phases

in the mathematics of the Fourier transform. Hence, X-ray diffraction patterns are said to exist in *reciprocal space.* Reflections near the middle of the diffraction pattern relate to large repeat distances in the structure and the outer reflections correspond to small repeat distances. The greater the angular spread of a diffraction pattern, the finer the detail of the information it holds. In an idealized single crystal,

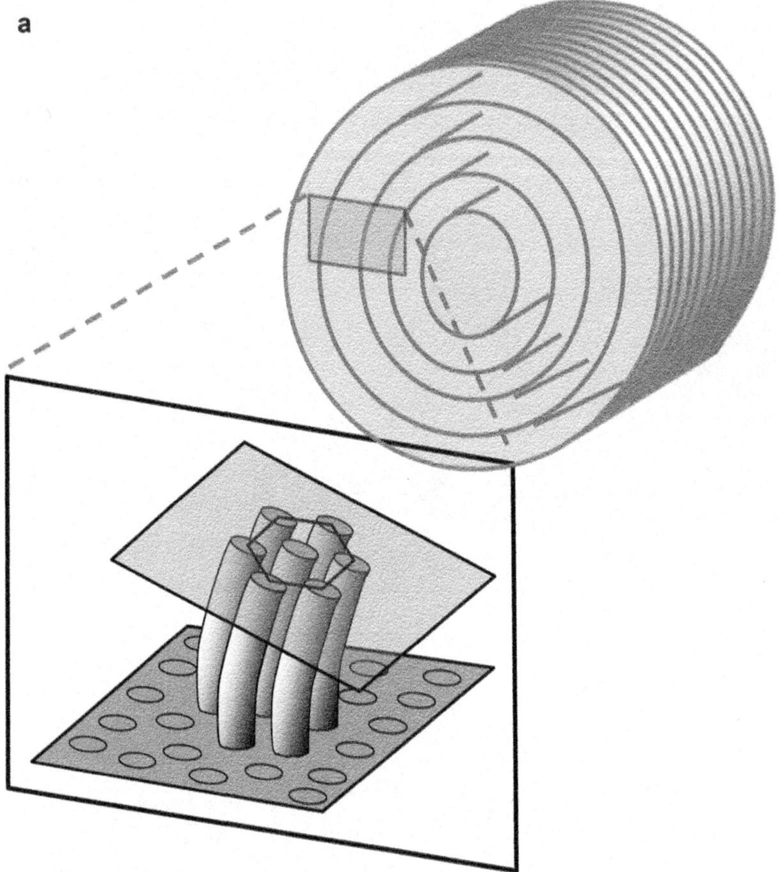

Fig. 1.30 (**a**) The easiest distortion for a hexagonal columnar structure is usually bending about a line parallel to one of the hexagon edges as shown. An extended structure which maintains the local parallelism of the columns and the lateral dimensions of the hexagonal lattice is termed a 'developable domain'. The pattern of concentric cylinders sketched in the background is the simplest structure of this kind. (**b**) Examples of developed columnar structures. These sketches show the domain cross-sections viewed down the axis of curvature. The lines indicate the orientation of the column axes, the black areas indicate regions extinguished when the sample is viewed between crossed polars, and the shaded grey areas represent the disclination cores (possibly of disordered material). Redrawn from Oswald and Piraski [5]

the molecules lie in an extended pattern, with perfect order. From one unit cell of the structure to the next, there is a perfect identity of the atomic coordinates. This long-range order, effectively infinite in terms of atomic dimensions, gives an extended Fourier transform and a corresponding diffraction pattern with sharp reflections, extending to high angles.

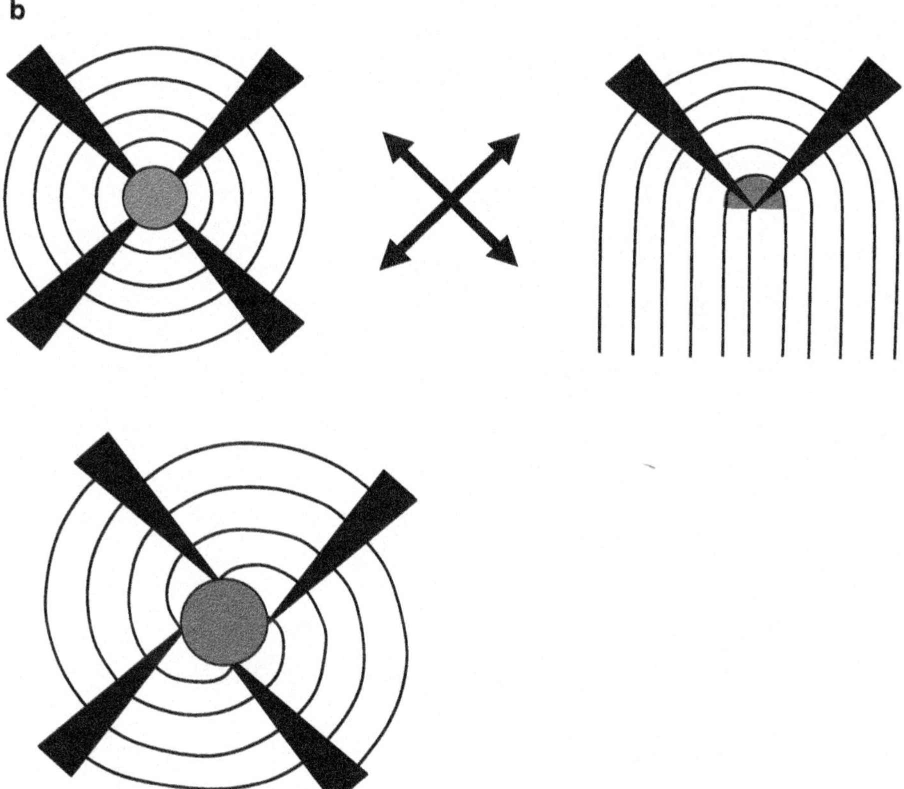

Fig. 1.30 (continued)

The kind of information about the structure of liquid crystalline phases that X-ray diffraction gives, is outlined in Fig. 1.31. This traces the development of the diffraction pattern of a calamitic crystal on heating as it passes through a succession of smectic phases, then a nematic phase and finally becomes an isotropic liquid. Reflections broaden and become diffuse as various degrees of order in the structures are lost. Note the diffraction pattern of the Smectic A phase. There is a single, fairly sharp low angle reflection corresponding to the smectic layer spacing, but there are no perceptible higher orders of reflection from this spacing. When this was first observed, it puzzled many crystallographers with lifetimes of experience of the diffraction patterns of non-mesogenic materials. How could there possibly be a structure with sufficient long-range order of one particular spacing to give reflections as strong and sharp as this, whilst at the same time giving no second or third orders? The answer lay in the particular type of partial ordering found in many mesophases (and not encountered in *bona fide* crystals) where there is

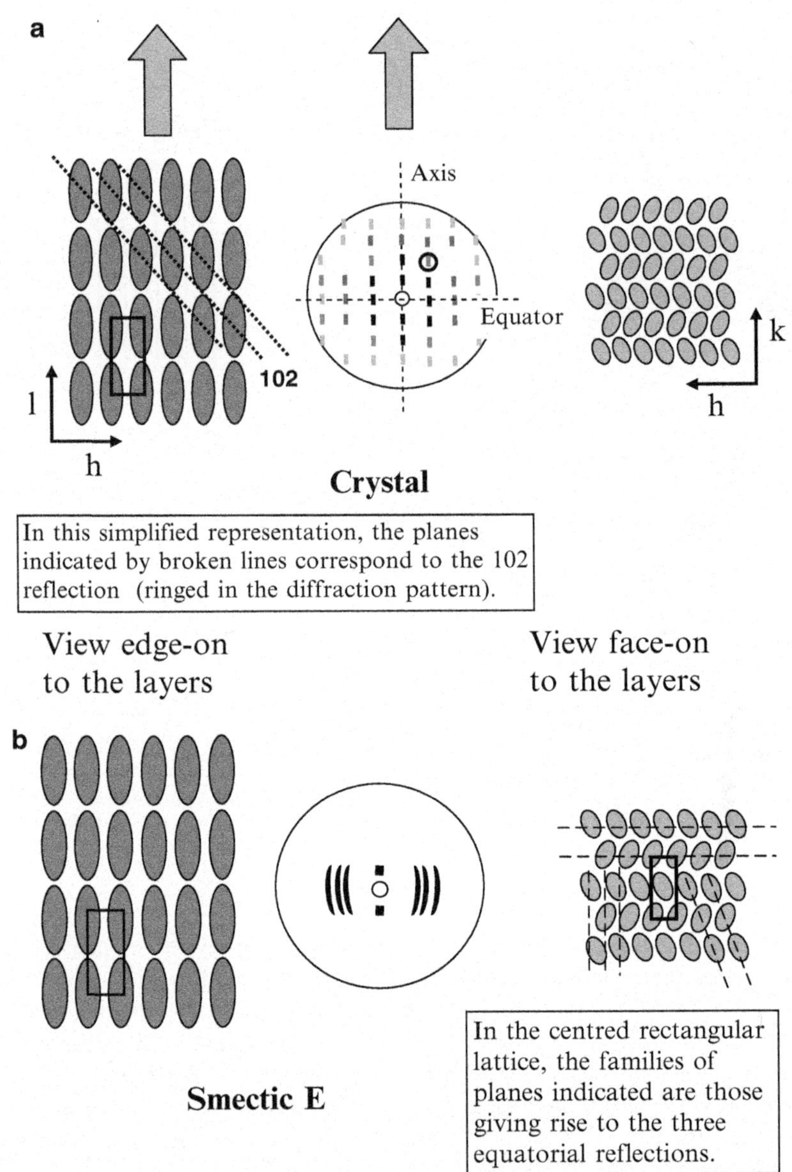

Fig. 1.31 These figures trace the development of the diffraction pattern of a sample of a calamitic mesogen as the crystalline solid (**a**) is heated and passes through a succession of mesophases (**b–d**), eventually becoming the isotropic liquid (**e**). In each figure, an edge-on view of the layers is shown on the *left* and the face-on view on the *right*. The diffraction patterns correspond to the alignment of the edge-on view on the *left*, i.e., a *vertical line* on the figure corresponds to the axial direction in the diffraction pattern. For all mesophases, the sample as a whole is taken to have a random orientation of domains around the layer normal. Note the stepwise way in which

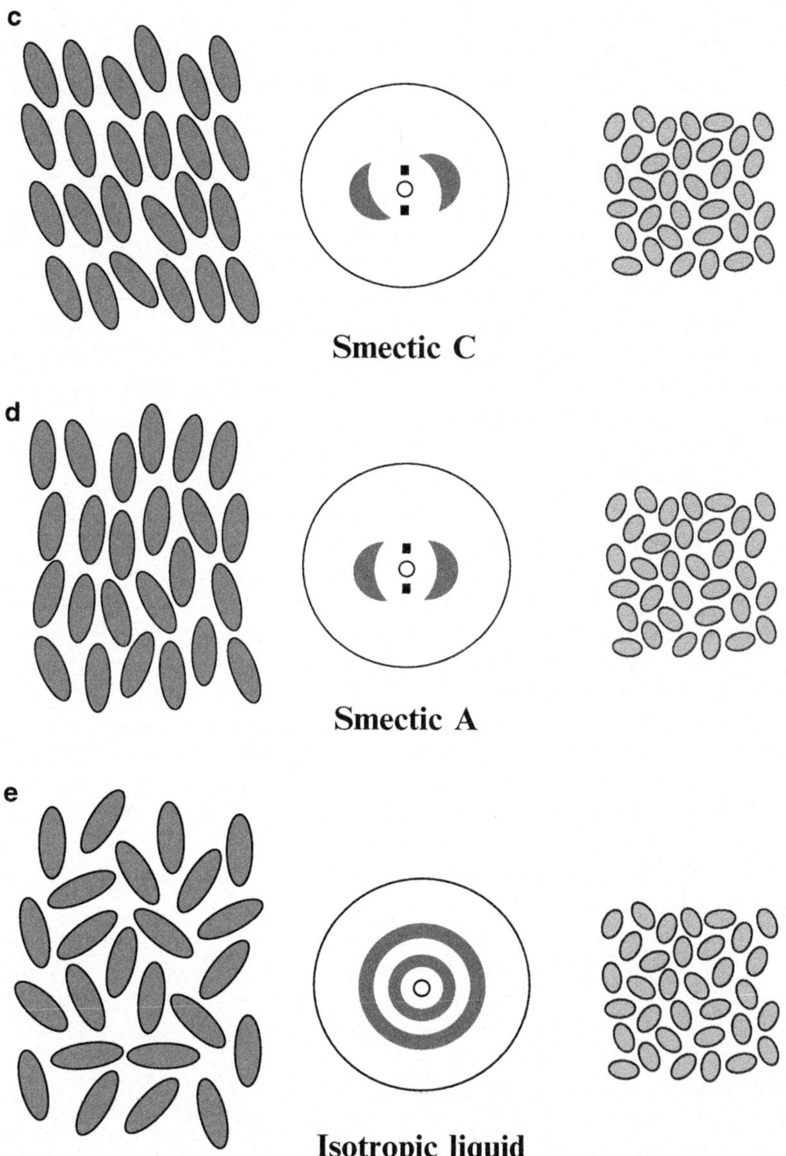

Fig. 1.31 (continued) reflections broaden, become diffuse and then disappear as various types of order in the structures are lost. In the B and E and 'higher' smectic phases there is generally some degree of long-range three-dimensional order and hence the diffraction patterns contain reflections where h and/or k >0 and l >0. In contrast, in the smectic A phase, there is no correlation between layers – and consequently, there are there only equatorial reflections, where l = 0. The local packing of molecules in the isotropic liquid is similar to that in the nematic, but there is no long-range order and different regions have random orientations

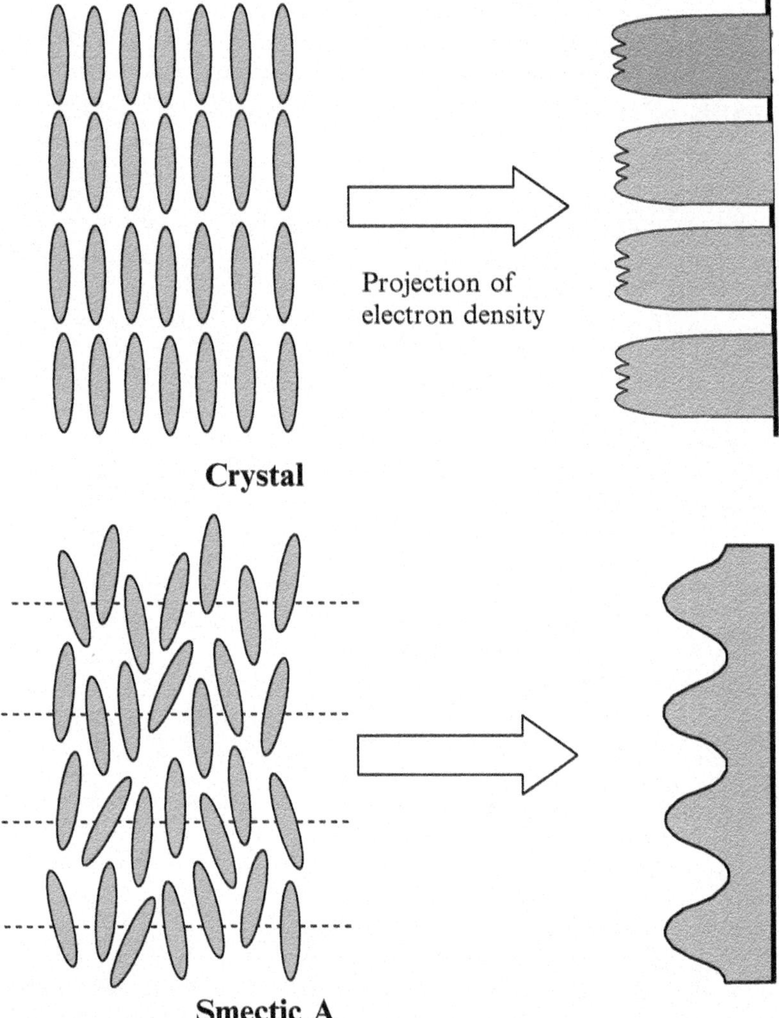

Fig. 1.32 The axial reflections of the SmA phase. In this phase the layers are disordered, thus causing the peaks to be diffuse. The projection of the electron density pattern perpendicular to the layers has the profile of a more or less sinusoidal wave, in contrast to the more detailed pattern that a perfect layer structure would give. The Fourier transform of this profile has a single dominant term and this explains why the characteristic diffraction pattern of the SmA phase has only a single strong axial reflection corresponding to the layer spacing and the higher orders, i.e., 0 0 2, 0 0 3 etc., are generally very weak

extended long-range order coupled with a degree of short-range disorder as sketched in Fig. 1.32. To emphasise this point, it is customary to refer to a *density wave* of distribution of the molecules in the higher temperature smectic phases, rather than the formation of well-defined layers.

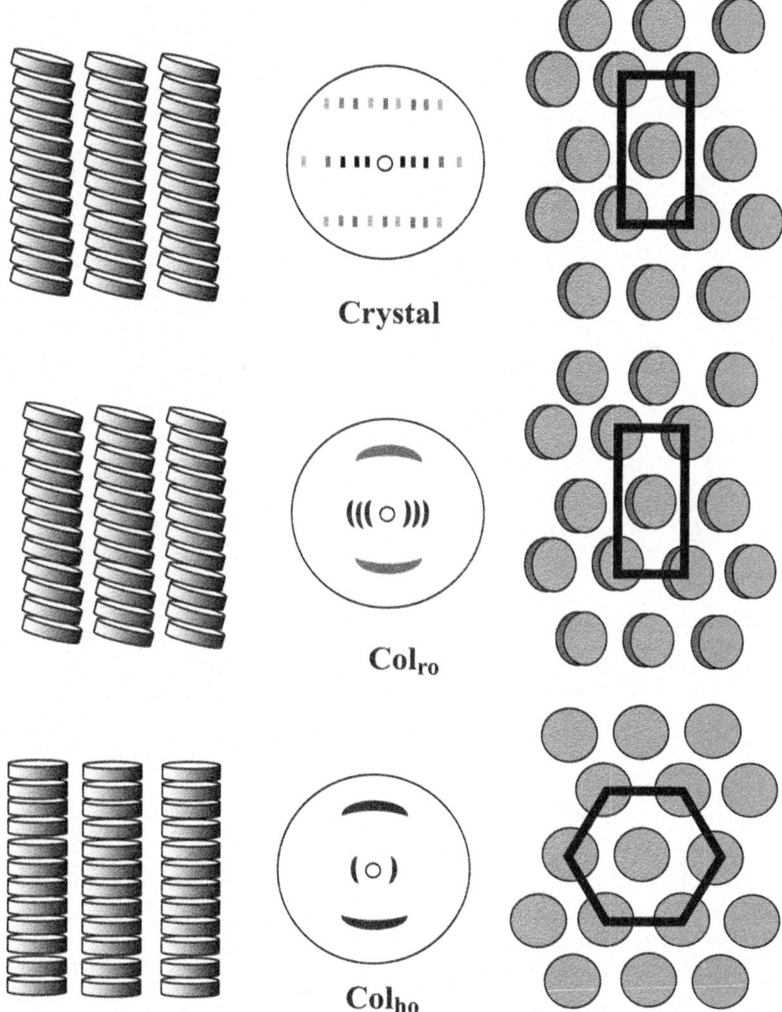

Fig. 1.33 These figures trace the development of the diffraction pattern of a sample of a discotic mesogen as the crystalline solid is heated and passes through a succession of mesophases, ending as the isotropic liquid. Note the stepwise way in which reflections broaden and become diffuse, as various types of order in the structures are lost. As in the similar scheme for smectic phases given in Fig. 1.30, the edge-on view of the molecular discs drawn on the *left* corresponds to the alignment of the diffraction pattern shown in the *centre* and again, for all mesophases, the sample as a whole is taken to have a random orientation of domains, giving rotational symmetry around the director

A similar scheme for discotic systems is shown in Fig. 1.33. The diffraction pattern of the nematic discotic phase is analogous to that of the calamitic nematic phase and contains only diffuse reflections, indicating the liquid-like distribution of face-to-face and edge-to-edge separations. For the columnar phases, the high-angle

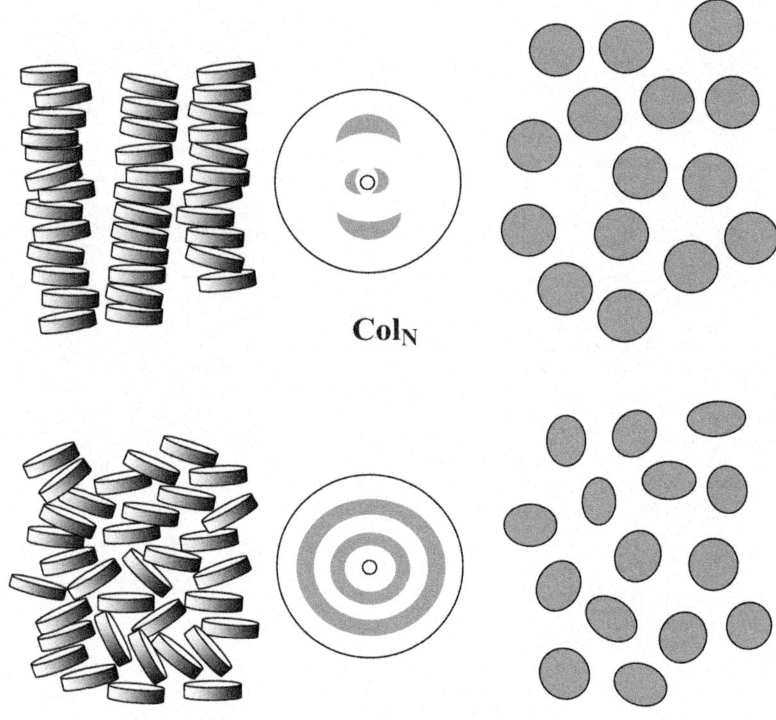

Fig. 1.33 (continued)

axial reflections indicate the degree of order within the stacked columns. The equatorial reflections indicate the progressive increase in lateral ordering of the columns for the lower-temperature phases. There are normal and tilted phases again, in loose correspondence with smectic systems.

Acknowledgements I acknowledge with gratitude, the extensive advice and help which I have received from Professor Richard Bushby whilst preparing this chapter. I am also grateful to Professor John Goodby for his advice concerning the section on smectic phases and to Professor Rob Richardson and Professor A. C. T. North for advice on X-ray diffraction patterns of mesophases.

References

1. Collings, P.J.: Liquid Crystals: Nature's Delicate Phase of Matter. Institute of Physics Publishing, Bristol (1991)
2. Chandrasekhar, S., Sadashiva, B.K., Suresh, K.A.: Liquid crystals of disc-like molecules. Pramana **9**, 471 (1977)

3. Kelber, J., Bock, H., Thiebaut, O., Grelet, E., Langhals, H.: Room-temperature columnar liquid-crystalline perylene imido-diesters by a homogeneous one-pot imidification–esterification of perylene-3,4,9,10-tetracarboxylic dianhydride. Eur. J. Org. Chem. (4), 707–712 (2011)
4. Bouligand, Y.: Twisted fibrous arrangement in biological materials and cholesteric mesophases. Tissue & Cell (4), 189–217 (1972)
5. Oswald, P., Pieranski, P.: Smectic and Columnar Liquid Crystals. Taylor & Francis, Boca Raton (2006). ISBN 10:0-8493-9840-1 & 13: 078-0-8493-9840-7

Bibliography

6. Collings, P.J., Hird, M.: Introduction to Liquid Crystals: Chemistry and Physics. Liquid Crystals Book Series. CRC Press, London (1997). ISBN ISBN 0-7484-0483-X
7. Chandrasekhar, S.: Liquid Crystals, second editionth edn. Cambridge University Press, Cambridge (1993). ISBN ISBN: 0-521-42741-X
8. Gray, G.W., Goodby, J.W.: Smectic Liquid Crystals – Textures and Structures. Leonard Hill, Glasgow/London (1984). ISBN ISBN 0-249-44168-3
9. Kumar, S.: Chemistry of Discotic Liquid Crystals: From Monomers to Polymers. Liquid Crystals Book Series. CRC Press, Hoboken (2010). ISBN 1-4398-1143-1

Chapter 2
Charge Carrier Transport in Liquid Crystalline Semiconductors

Jun-Ichi Hanna

2.1 Historical Studies of the Electrical Properties of Liquid Crystals

It is only in relatively recent years, compared with the discovery of liquid crystals, that the electrical properties of liquid crystals have attracted significant scientific interest. The first attention to them was paid in the late 1960s by two groups at almost the same time, but from different points of view. One was Kusabayashi and Labes [1], who measured the electrical conductivity in various liquid crystals including nematic, cholesteric and smectic liquid crystals. They studied the conductivity change in liquid crystals accompanying phase transitions from the point of view of extending understanding of the electrical properties of molecular crystals. Their findings, however, did not arouse further scientific attention at the time. The other early study was by Heilmeier's group [2], who invented a liquid crystal display (LCD) in the so-called *dynamic scattering mode*. The principle of this liquid crystal display was based on light scattering caused by dynamic motion of aligned liquid crystalline molecules, when ions drift in a liquid crystal cell at a given bias voltage. They studied ion transport in the nematic phase of a *p*-azoxyanisole derivative and determined its mobility to be 10^{-5} cm^2 V^{-1} s^{-1} from transient currents induced by a step-voltage technique [3]. This LCD invention attracted a lot of attention and the charge carrier transport properties in various liquid crystals, especially nematic liquid crystals, were investigated in the 1970s [4, 5]. All the results reported at that time indicated that the conduction in liquid crystals was governed by conduction of ions, whose typical mobility was 10^{-5}–10^{-6} cm^2 V^{-1} s^{-1}. It led to a general understanding that charge carrier transport in liquid crystals, taking into account

J.-I. Hanna (✉)
Imaging Science and Engineering Laboratory, Tokyo Institute of Technology, J1-2 Nagatsuta Midori-ku, Yokohama 226-8503, Japan
e-mail: hanna@isl.titech.ac.jp

R.J. Bushby et al. (eds.), *Liquid Crystalline Semiconductors*, Springer Series in Materials Science 169, DOI 10.1007/978-90-481-2873-0_2,

that discotic liquid crystals had not been discovered at that time, was governed by ionic conduction. It is quite understandable, because the liquid-like nature of liquid crystalline materials favors this idea.

In the 1980s, after the discovery of discotic liquid crystals in 1977 [6], because of the more solid-like nature of discotic liquid crystals, the focus of significant interest in the quest for electronic conduction in liquid crystals shifted from calamitic liquid crystals to the discotic ones with columnar mesophases. In 1988, Boden et al. at Leeds University studied ac-conductivity in hexakis-hexyloxytriphenylene (HAT6) chemically doped with $AlCl_3$. They demonstrated one-dimensional hole conduction along the columns through the anisotropic ac-conductivity associated with the column orientation [7]. They estimated the hole mobility to be 1×10^{-4} cm^2 V^{-1} s^{-1} in the discotic hexagonal phase of the doped HAT6 from the ac-conductivity and the carrier concentration determined by ESR [8, 9]. Furthermore, Schouten et al. estimated the charge carrier mobility to be 6×10^{-3} cm^2 V^{-1} s^{-1} in the columnar phase of an octakis-nonyloxyethyl-substituted zinc porphyrin derivative on the basis of pulsed radiolysis transient microwave conductivity measurements [10]. In 1993, Haarer et al. determined the mobility in the columnar phase of hexakis-pentyloxytriphenylene (H5T) to be 1×10^{-3} cm^2 V^{-1} s^{-1} for positive carriers and concluded that it must be the mobility attributed to hole conduction, because the value determined was too high to be ionic [11].

However, the charge carrier transport properties of calamitic mesophases was still believed to be ionic, following the historical studies in the 1970s, even after electronic conduction was established in discotic liquid crystals. In fact, Kusabayashi et al. re-investigated charge carrier transport in a calamitic liquid crystal, which was designed to have a carbazole moiety as its core for hole conduction and concluded that the conduction was ionic judging from the way its mobility depended on viscosity [12]. More recently, Closs et al. studied photoconductivity behavior in dye-sensitized thiadiazole and oxadiazole derivatives, but no conclusive result on the nature of the electronic conduction in these mesophases was reported [13]. In 1997, Funahashi and Hanna reported the discovery of electronic conduction in the smectic mesophases of two calamitic liquid crystals: a 2-phenylbenzothiazole derivative (7O-PBT-S12) [14, 15] and a 2-phenylnaphthalene derivative (8-PNP-O12) [16]. Since then, electronic conduction has been reported in various smectic liquid crystals. The materials described in this section are shown in Fig. 2.1.

Electronic conduction in the nematic phase was reported in solid-like nematic phases such as the nematic phases of polymeric liquid crystals [17, 18] and the glassy nematic phases of oligofluorene derivatives [19, 20]. However, the conduction mechanism in nematic phases of small molecules was still believed to be ionic. In fact, there were several reports of ionic conduction in the nematic phase of small molecules including 5CB [21, 22]. More recently, electronic conduction was reported in the chiral nematic (cholesteric) phase of a phenylquaterthiophene derivative, in which it was suggested that the large and extended π-conjugated core of phenylquaterthiophene favored electronic conduction thanks to a large transfer integral [23]. Indeed, there still remained a question to be answered about what the intrinsic conduction mechanism was in the nematic phase of small molecules of

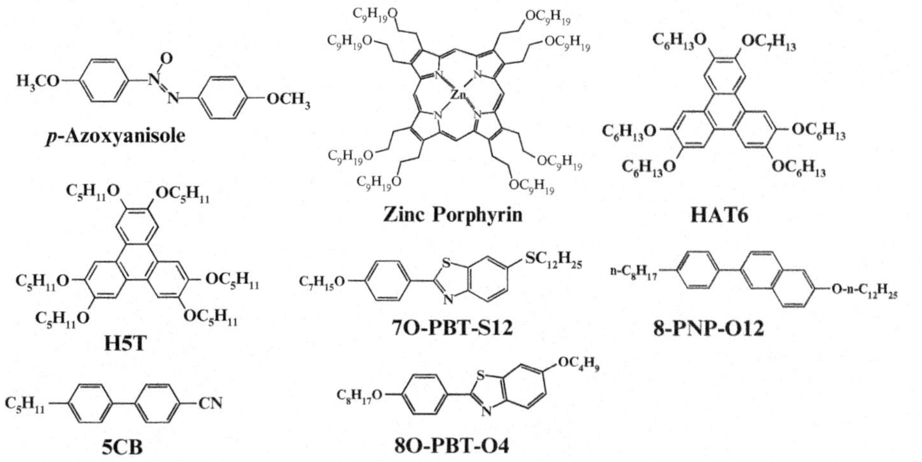

Fig. 2.1 Liquid crystalline materials described in this section

low viscosity. In 2009, Hanna et al. reported that electronic conduction took place in the nematic phase of a small molecule, a 2-phenylbenzothiazole derivative (8O-PBT-O4), following extensive purification of the material [24], indicating that the intrinsic conduction is electronic in the nematic phase of small molecules. Thus, the quest for the electronic conduction in liquid crystals over several decades, which was initiated by Kusabayashi and Labes in late 1960s, came finally to an end. Now, it is understood that the intrinsic conduction mechanism of liquid crystals having highly conjugated, aromatic cores is electronic, irrespective of the mesophase, and that ionic conduction, also often observed in liquid crystals, is extrinsic and due to chemical impurities that can be ionized by trapping a charge and/or auto-ionization under illumination.

2.2 Ionic and Electronic Conduction in Liquid Crystals

In the previous section, the historical quest for electronic conduction in liquid crystals and its discovery were briefly reviewed. The electrical properties of liquid crystals had been believed to be governed by ionic conduction for a long time. Indeed, this was supported by many experimental results, but according to present knowledge, these were due to the contamination of materials by chemical impurities. In fact, the reason why ionic conduction is so often observed in liquid crystals is very much associated with the specific nature of the mesophase. Therefore, in order to understand their charge carrier transport properties, it is very helpful to understand the nature of the mesophase, i.e., the way in which liquid crystalline molecules having an aromatic core and long flexible hydrocarbon chains aggregate in a self-organizing manner.

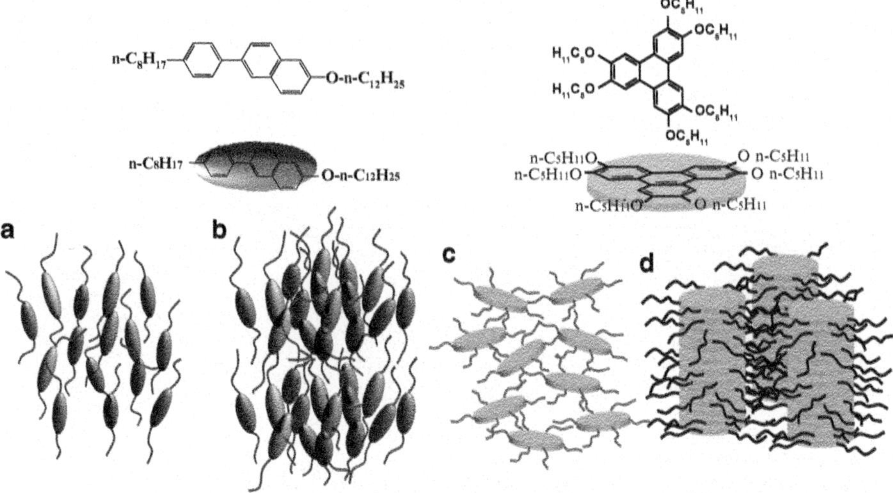

Fig. 2.2 Typical liquid crystalline molecules and their self-organized aggregates. Rod-like molecules form (**a**) the nematic phase and/or (**b**) smectic mesophases. Liquid crystal molecules orient in a given direction but have no positional order in the nematic phase, and they are arranged in layers in the smectic phases. Disc-like molecules form (**c**) the nematic phase and (**d**) are self-organized into columns in the columnar phases

A typical liquid crystalline molecule has a rigid aromatic π-conjugated core with long flexible hydrocarbon chains attached as shown in Fig. 2.2. These chains give the molecule enhanced shape anisotropy, e.g., either a calamitic (rod-like) or discotic (disc-like) shape. In addition, the long flexible hydrocarbon chains suppress the tendency for crystallization and often give the molecule an opportunity to form a mesophase. Liquid crystals are often described as a molecular aggregate, which exhibits both a liquid-like fluidity and a crystal-like molecular order at the same time. Of course, the appearance of *fluidity* in liquid crystals varies not only from material to material, but also from phase to phase. For example, *liquid-like* fluidity appears in less-ordered, calamitic mesophases such as nematic (N), smectic A (SmA) and smectic C (SmC) phases of small molecules, but the highly ordered, columnar and smectic mesophases are *solid-like*, as are polymeric and glassy liquid crystals. The fluidity, or the viscosity in a more exact physical understanding, plays a very important role in the conduction mechanism as does the purity of the materials.

Let us think of ionized molecules in a fluid media, where in principle there are two possible mechanisms of conduction, i.e., ionic and electronic. Ionic conduction, in which the ionized molecules themselves drift in a given electric field, is favored in a less-viscous medium, basically because the ionized molecules experience friction from the medium during drift. On the other hand, electronic conduction, in which charges on the ionized molecules migrate from molecule to molecule, is favored in a viscous media, because the ionic conduction is suppressed.

In general, it is not easy to distinguish electronic conduction from ionic conduction from the value of the mobility, when this is less than 10^{-4} cm^2 V^{-1} s^{-1}. In general, the mobility for electronic conduction is higher than that for ionic conduction, because no mass transport is required. A high mobility, e.g., over 10^{-3} cm^2 V^{-1} s^{-1}, is attributed to the electronic conduction without any question. However, a mobility lower than 10^{-4} cm^2 V^{-1} s^{-1} is often observed both in electronic and ionic conduction. The temperature dependence of the mobility is often a valid way to distinguish these conduction mechanisms, but it is not decisive. Ionic mobility depends on the temperature, with a typical activation energy in the range 0.3–0.5 eV. This is attributed to the activation energy of viscosity. On the other hand, the electronic mobility in mesophases hardly depends on temperature according to most experimental data accumulated so far. One of the most convenient ways to distinguish these conduction mechanisms is to examine the effect of dilution on the mobility. This gives opposite effects for ionic and electronic conduction. One expects enhanced mobility for ionic conduction in a mesophase diluted with a diluent having a low viscosity such as a hydrocarbon, because the medium where the ions drift becomes less viscous in the diluted mesophase. On the contrary, one expects reduced mobility for electronic conduction, because the average intermolecular distance between liquid crystal molecules becomes greater in the diluted mesophase [25, 26].

The reason why ionic conduction is so often observed in the mesophases of small molecules is that electronic conduction in the mesophase is quite sensitive to chemical impurities and systems easily degrade into ionic conduction [27–29]. This originates from the microphase-separated structure of mesophases as shown in Fig. 2.2. In the columnar and smectic mesophases, the flexible hydrocarbon chains are phase-separated from the rigid cores at a microscopic level and aggregate to form liquid-like regions, where the ions can move easily because of the low viscosity. That is, the inter-columns and inter-layers provide conduction channels for ions in these mesophases, respectively. It should be noted that the ions responsible for the ionic conduction are not always the ions dissociated from ionic impurities, but can also be neutral molecules of an impurity ionized by trapping a charge and/or generated by photo-ionization. The neutral impurity molecules become electrically active and cause trapping states for electrons or holes in the mesophases, if their highest occupied molecular orbital (HOMO) or the lowest unoccupied molecular orbital (LUMO) are in the midgap of those of the host liquid crystal. Therefore, the ionized molecules occasionally migrate into the liquid-like conduction channel as explained above and start to drift as ions. In fact, both electronic and ionic conduction are observed in the mesophase simultaneously when the liquid crystal is contaminated with trace amounts of such chemical impurities, e.g., a few ppm. However, the electronic conduction is completely destroyed and degrades into ionic conduction, when the host liquid crystal is contaminated more than this, e.g., tens of ppm [27–29]. On the other hand, electronic conduction can often survive in polymeric liquid crystals, even when they have this level of chemical impurities. This is because the drift of ionized impurity molecules is suppressed thanks to the high viscosity of the polymers. This is the reason why electronic conduction is relatively easier to observe in polymeric liquid crystals, although the purification of polymers is problematic in general. In the nematic phase of small molecules, no specific conduction channel for electronic charges and ions is formed, because there is no microphase-separated structure. However, the low viscosity in the nematic phases of small molecules also favors ionic conduction. Therefore, chemical impurities seriously affect the conduction mechanism in the nematic phase. Electronic conduction in the nematic phase of small molecules was established in a 2-phenylbenzothiazole derivative in 2008, after extensive purification of the material [24]. Because of the relatively small electronic mobility of 10^{-4} cm^2 V^{-1} s^{-1} or less in the less ordered mesophases, including both SmA and SmC phases in addition to the nematic phase of small molecules, we have to be very careful to judge, which kind of conduction dominates in such mesophases, i.e., electronic or ionic.

2.3 Anisotropy and Dimensionality in Carrier Transport

Liquid crystalline phases are characterized by both dynamic molecular motion and molecular order unlike crystals, molecular glasses and melts. The time scale for translational motion is quite slow compared with charge migration from molecule to molecule (10^{-10} s or less) so this hardly affects transport of the carriers. A carrier sees only a snapshot of thermally disordered molecules in terms of position and orientation and drifts among the energetically inhomogeneous molecules, that result from thermal fluctuations. On the other hand, the time scale for molecular rotation about the molecular axes is fast compared with the translational motion of the molecule (10^{11} Hz). This rate is also quite high compared with the hopping rate for carriers in less-ordered mesophases and becomes comparable to the hopping rate only when the mobility exceeds 10^{-2} cm^2 V^{-1} s^{-1}. However, such a high mobility, e.g., 10^{-2} cm^2 V^{-1} s^{-1} or more, can only be achieved in highly ordered mesophases such as SmE, SmB$_{cryst.}$ and SmG phases, where free rotation of the molecules is prohibited. Therefore, it is not possible for rotational motion to be the rate determining step for the carrier to hop to an adjacent molecule. In other words, the rotational motion hardly affects the charge transfer event in the mesophases. On the other hand, the molecular order affects the hopping rate very much. In fact, the mobility increases, when the molecular order is increased both in columnar and smectic mesophases. For smectic mesophases, the molecular order determines the packing density of molecules and the molecular orientation in the mesophase. Increases in the molecular order in smectic mesophases leads to a closer packing of molecules and a less-disordered molecular orientation in a smectic layer. Therefore, the intermolecular distances in the mesophase become shorter, e.g., from 6 to 4 Å or less, and the orientational disorder of molecules becomes smaller as the molecular order is increased. These effects enhance the charge transfer rate, because of an increase in the transfer integral between adjacent molecules, which results in enhancement of the mobility in the smectic mesophase. For columnar mesophases, the mobility also increases, when the molecular order is increased. However, the increase of mobility in the higher-ordered columnar mesophases requires a different explanation as discussed later on. The intermolecular distance in a column is almost constant (around 3.5 Å) irrespective of the nature of the columnar phase.

Another feature of liquid crystalline materials, that affects the charge carrier transport properties is the anisotropic molecular shape enhanced by the long hydrocarbon chains. When such molecules self-organize into columns or layers, the intermolecular distances are very different inter- and intra-column or inter- and intra-layer, as shown in Fig. 2.2. For example, the molecular distance intra-column or intra-smectic layer is from 3.5 to 6 Å depending on the mesophase, while the molecular distance inter-column or inter-layer corresponds to a molecular length, e.g., typically 30–40 Å. This large difference in the molecular distances results in enhanced anisotropy in the carrier transport properties. The anisotropy in mobility reaches a few orders of magnitude or even more in a typical liquid crystal. This large anisotropy in mobility results in one-dimensional carrier transport along the columns in discotic columnar phases and two dimensional carrier transport within the layers in smectic phases. For the smectic mesophases of a 2-phenylnaphthalene derivative, the two-dimensional carrier transport has been demonstrated along with the effect of chemical impurities on the carrier transport as a function of their concentration [27, 30]. For the columnar mesophase, Boden et al. reported a large anisotropy over three orders of magnitude in the ac-conductivity of hexakis-hexyloxytriphenylene (HAT6) doped with $AlCl_3$ [9]. However, the anisotropy of mobility for most discotic and smectic mesophases has not been demonstrated so far, because it is very difficult to determine the mobility inter-column or inter-layer. For example, both the small mobility due to the long hopping distances for a typical molecular length of 30–40 Å and the small photocarrier generation efficiency, due to the molecular orientation in homeotropic and homogenous molecular alignments of discotic and smectic mesophases, respectively, makes the photocurrent extremely small in time-of-flight experiments, which is the most popular method for determining the charge carrier mobility. From the device applications point of view, it is quite important to control the molecular orientation of liquid crystalline molecules in a device, because of this anisotropy. This is quite different to amorphous materials that are isotropic in terms of both molecular orientation and properties. As for the nematic phase, which is the least-ordered mesophase, characterized by orientational order of the molecular axis but no positional order of the molecules, the anisotropy in the charge carrier transport properties has been reported in polymeric liquid crystals [31]. Basically the electronic conduction in the nematic phase of small molecules is three-dimensional because there is no positional order in the nematic phase [32]. No information about the anisotropy of the mobility is available because of the recent discovery of electronic conduction in the nematic phase [24].

2.4 Carrier Transport

In amorphous aggregates, ambipolar transport, where both electrons and holes contribute to the conduction is seldom observed. Hole transport is easy to observe in various organic materials, while electron transport is very much limited to a few classes of organic materials. This is due to the chemical impurities responsible for deep-trap states for electrons, of which the last impurity can be oxygen molecules, which have a large electron affinity. Therefore, electron transport has been observed only in certain π-conjugated compounds containing electron withdrawing groups such as 2,4,7-trinitrofluorenone (TNF), 3,3′-dimethyl-5,5′-di-*tert*-butyl-4,4′-diphenoquinone and tetracyanoquinodimethane (TCNQ), which have an electron affinity higher than that of oxygen. In contrast, both electron and hole transport is often observed in the mesophases of highly purified materials, that is, the charge carrier transport is ambipolar [16, 33–36]. It is worth noting that the ambipolar charge carrier transport is reported in single crystals of various materials including not only purified polyacenes such as naphthalene and anthracene but also so-called p-type materials such as *N-iso*-propylcarbazole and copper phthalocyanine [37]. The reason why amipolar carrier transport is often observed in single crystals, is probably due to the closely packed nature of the molecules in a crystal, that does not allow oxygen molecules to diffuse into the bulk. Therefore, it is very plausible that the ambipolar charge transport often observed in the mesophases of extensively purified discotic and smectic liquid crystals [38] is due to the same reason as in the single crystals. This is one of the unique features of carrier transport in mesophases. The electron and hole mobilities of various liquid crystals reported are summarized in Tables 2.1 and 2.2.

2.5 Mesophase Structure

There are various mesophases corresponding to different orientational and positional ordering of the molecules. For discotic columnar mesophases, the geometrical order of the columns gives a variety of the columnar phases in addition to differences in the intra-columnar order. Columnar mesophases, in which the columns are aligned to form a tetragonal, rectangular, or hexagonal lattice are called the columnar tetragonal (Col_{tetr}) phase, columnar rectangular (Col_r) phase and columnar hexagonal (Col_h) phase, respectively, as shown in Fig. 2.3.

In order to describe the intra-columnar order, the terms *ordered*, *disordered* and *helical* are used. For smectic mesophases, there are two series of mesophases in terms of molecular orientation. One is a series of smectic mesophases, in which the long axis of liquid crystal molecules is perpendicular to the smectic layer. The other is a series of smectic mesophases, in which the long axis is tilted relative to the molecular layer. For these two series, the geometrical order of the molecules in the smectic layer creates a wide variety of mesophases. The names given to

Table 2.1 Amb polar mobility in various liquid crystals determined by the Time-of-Flight method

Materials	Hole Mobility (cm^2/Vs)	Electron Mobility (cm^2/Vs)
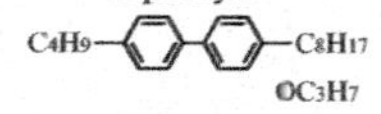 Biphenyls C_4H_9–…–C_8H_{17}	6×10^{-3} (SmB)	—
OC_3H_7	3×10^{-3} (SmB) 1×10^{-3} (SmE)	—
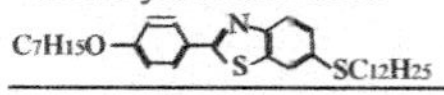 2-Phenylbenzoth azoles $C_7H_{15}O$–…–$SC_{12}H_{25}$	1×10^{-4} (SmA)	1×10^{-4} (SmA)
2-Phenylnaphthalenes 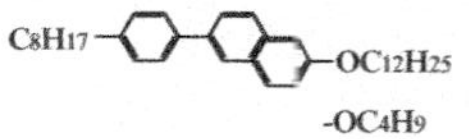 C_8H_{17}–…–$OC_{12}H_{25}$	2.5×10^{-4} (SmA) 1.7×10^{-3} (SmB)	2.5×10^{-4} (SmA) 1.6×10^{-3} (SmB)
$-OC_4H_9$	1.0×10^{-2} (SmE)	1.0×10^{-2} (SmE)
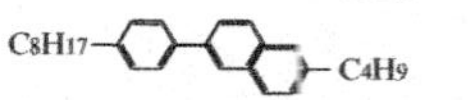 C_8H_{17}–…–C_4H_9	8×10^{-4} (SmA) 8×10^{-3} (SmB) 4.7×10^{-2} (SmE)	—
Terphenyls 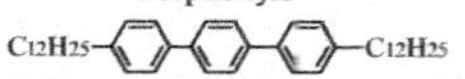 $C_{12}H_{25}$–…–$C_{12}H_{25}$	0.01 (SmX_1) 0.04 (SmX_2) 0.13 (SmX_3)	—
Terthiophenes 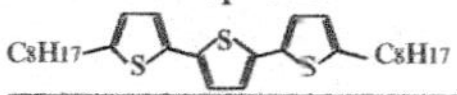 C_8H_{17}–…–C_8H_{17}	8×10^{-3} (SmC) 3.1×10^{-3} (SmF) 2.4×10^{-2} (SmG)	5×10^{-3} (SmC) 2.5×10^{-3} (SmF) 2.4×10^{-2} (SmG)
Quaterthiophenes C_6H_{13}–…–C_6H_{13}	6×10^{-2} (SmG)	—
–≡– C_4H_9	0.1 (SmG)	—
Benzothienobenzothiophenes 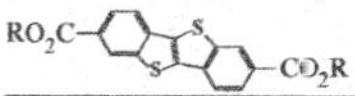 RO_2C–…–CO_2R	2.2×10^{-3} (Lamello-columnar)	2.7×10^{-3} (Lamello-columnar)
Oligofluorenes 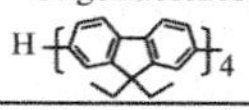 H–[…]$_4$	0.1 (Nematic Glass)	0.1 (Nematic Glass)
Polyacrylates 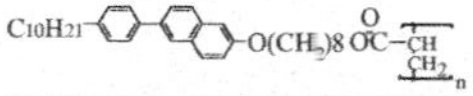 $C_{10}H_{21}$–…–$O(CH_2)_8OC$–CH–CH_2	2×10^{-4} (SmA) 1×10^{-3} (Sm Glass)	—

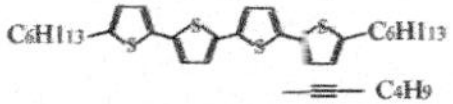

Materials	R	Hole Mobility (cm^2/Vs)	Electron Mobility (cm^2/Vs)
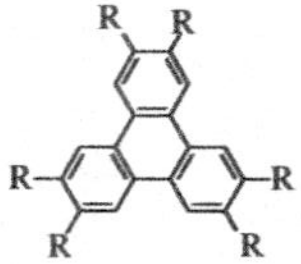 (Triphenylene, R substituents)	C_4H_9	2.5×10^{-2} (Col_{hp})	2.3×10^{-2} (Col_{hp})
	C_5H_{11}	1.9×10^{-2} (Col_{ho})	1.7×10^{-2} (Col_{ho})
	C_6H_{13}	4×10^{-4} (Col_{ho})	4×10^{-4} (Col_{ho})
	SC_6H_{13}	0.08 (Helical)	0.08 (Helical)
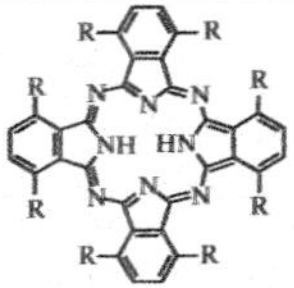 (Phthalocyanine, R substituents)	C_8H_{17}	0.1 (Col_h)	0.2 (Col_h)
		0.2 (Col_r)	0.3 (Col_r)
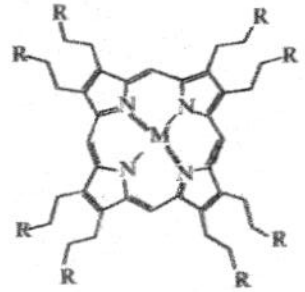 (Porphyrin, R substituents, M)	OC_9H_{19} M=Zn	1×10^{-2} (Col)	8×10^{-3} (Col)

This mobility indicates the bulk mobility for the carriers travelling over a long range

Table 2.2 Mobility in various liquid crystals determined by the Pulse-Radiolysis Time-Resolved Microwave Conductivity method

Materials	R		Phase	Σ μ Mobility (cm²/Vs)
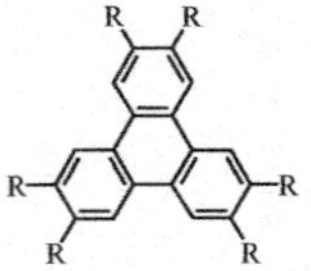 Triphenylenes	OC_4H_9		Colhp	2.5×10^{-2}
			Cryst.	6.9×10^{-3}
	OC_5H_{11}		Colh	1.0×10^{-2}
			Cryst.	3.3×10^{-1}
	OC_6H_{13}		Colhp	2.0×10^{-2}
			Cryst.	1.2×10^{-2}
	SC_6H_{13}		Colh	1.0×10^{-2}
			Helical	8.7×10^{-2}
			Cryst.	3.3×10^{-1}
	SC_8H_{16}		Colh	1.5×10^{-2}
			Cryst.	2.0×10^{-1}
	SC_8H_{16}		Colh	1.5×10^{-2}
			Cryst.	2.0×10^{-1}
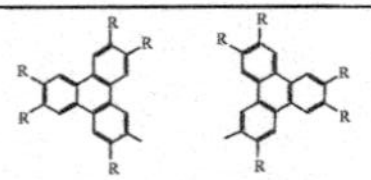 Triphenylenedimers	OC_4H_9 n=10	n=10	Colp	9.5×10^{-3}
			Cryst.	5.5×10^{-3}
	OC_4H_9	n=12	Colp	1.4×10^{-2}
			Cryst.	5.8×10^{-3}
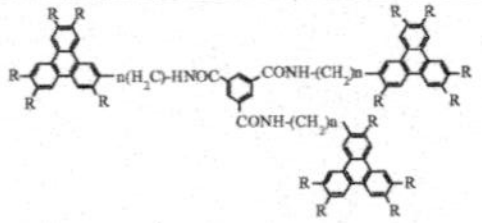 Triphenylene trimers	OC_4H_9	n=4	Colhp	6.0×10^{-2}
	OC_6H_{13}	n=3	Colh	1.7×10^{-2}
		n=6	Colobl	1.8×10^{-1}
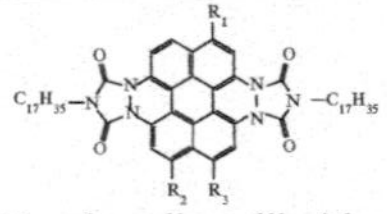 Perylenediazadiimides	H $C_{18}H_{37}$, H, $C_{18}H_{37}$ C_6H_{13}, H, C_6H_{13} C_6H_{13}, C_6H_{13}, H		Col Cryst.	$0.1\sim1.0\times10^{-1}$

Materials	R		Phase	Σμ Mobility (cm²/Vs)
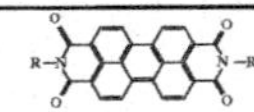 Perylene diimides	$C_{18}H3_7$		SmX	1.1×10^{-1}
			Cryst.	$1.1\sim2.1\times10^{-1}$
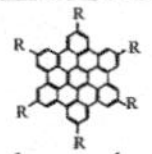 Hexabenzocolonenes	$OC_{12}H_{25}$		Colh	3.8×10^{-2}
			Cryst.	8.0×10^{-1}
	$PhOC_{12}H_{13}$		Colh	3.5×10^{-1}
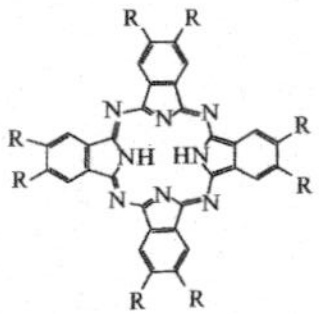 Phthalocyanines	$OC_{10}H_{21}$		Colh	3.8×10^{-1}
			Cryst.	1.3×10^{-1}
	$OC_{12}H_{25}$		Colh	3.5×10^{-1}
			Cryst.	2.1×10^{-1}
	$SC_{12}H_{25}$		Colh	2.0×10^{-1}
			Cryst.	2.5×10^{-1}
	$PhOC_{12}H_{13}$		Colh	1.8×10^{-1}
			Cryst.	3.4×10^{-1}
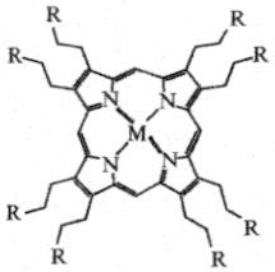 Porphyrins	OC_9H_{19}	M=Zn	Col	6.0×10^{-3}
			Cryst.	2.6×10^{-2}
	$OC_{10}H_{21}$	M=Pd	Col	5.60×10^{-3}
			Cryst.	3.9×10^{-1}
		M=Cu	Colh	6.5×10^{-2}
			Cryst.	3.6×10^{-1}
		M=Co	Colh	5.3×10^{-2}
			Cryst.	2.7×10^{-1}
		M=Zn	Colh	6.1×10^{-2}
			Cryst.	3.0×10^{-1}

This mobility indicates a sum of the mobilities for positive and negative carriers both of which contribute to the transient microwave conductivity when a pulsed electron beam of high energy in applied. It is not always the same as the mobility in the bulk material

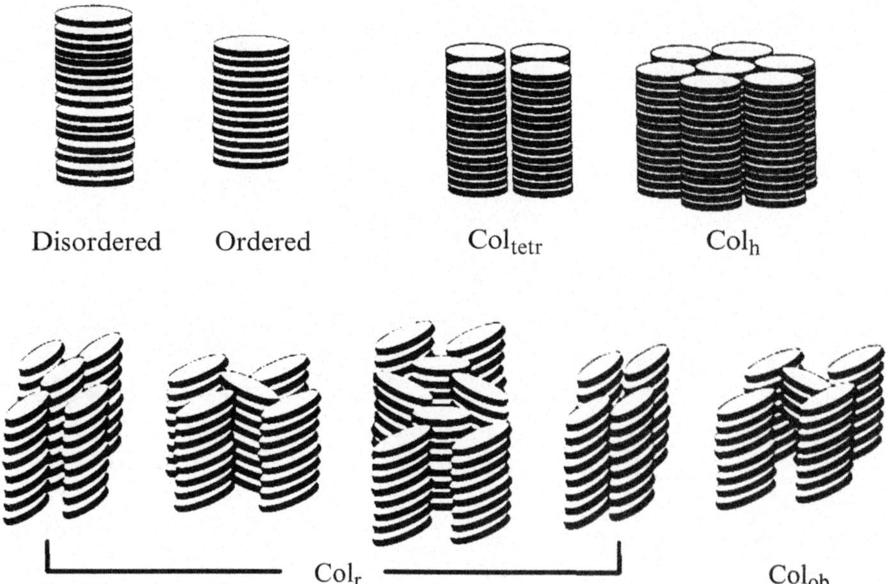

Fig. 2.3 Molecular order in the discotic columnar phase and typical examples of columnar mesophases

the smectic mesophases are A to K in the order of their historical discovery. For example, the smectic mesophase, in which the liquid crystal molecules are aligned to form rectangular and hexagonal arrays are called smectic E (SmE) and smectic B (SmB). The phase with no positional order in the smectic layer is called smectic A (SmA). Similarly, for the tilted series, the smectic mesophase, in which the liquid crystal molecules are aligned to form rectangular and hexagonal arrays, are called smectic F (SmF) and smectic G (SmG). The phase with no positional order in the smectic layer is called smectic C (SmC) as shown in Fig. 2.4.

According to the accumulated results for charge carrier transport in mesophases, this hardly depends on the mesophase, except for the mobility, see later. Hereafter, the mobility indicates the mobility within a conduction channel, i.e., in the layer for a smectic liquid crystal and in a column for a columnar phase, unless specified otherwise.

For a given class of liquid crystals having the same molecular core, the mobility is enhanced in a step-wise manner from phase to phase according to the increase in the molecular order in the mesophase. For smectic mesophases, the mobility is enhanced according to the level of the molecular order within the smectic layer as follows, SmA, SmC < SmB_{hex}, SmF < SmB_{cryst}, SmE, SmG. The mobility hardly depends on the orientation of molecular axis, i.e., whether this is perpendicular or tilted relative to the molecular layer. Figure 2.5 shows the mobility in various mesophases of 2-phenylnaphthalene and terthiophene derivatives as a function of the shortest intermolecular distance in each mesophase. It indicates clearly, that the

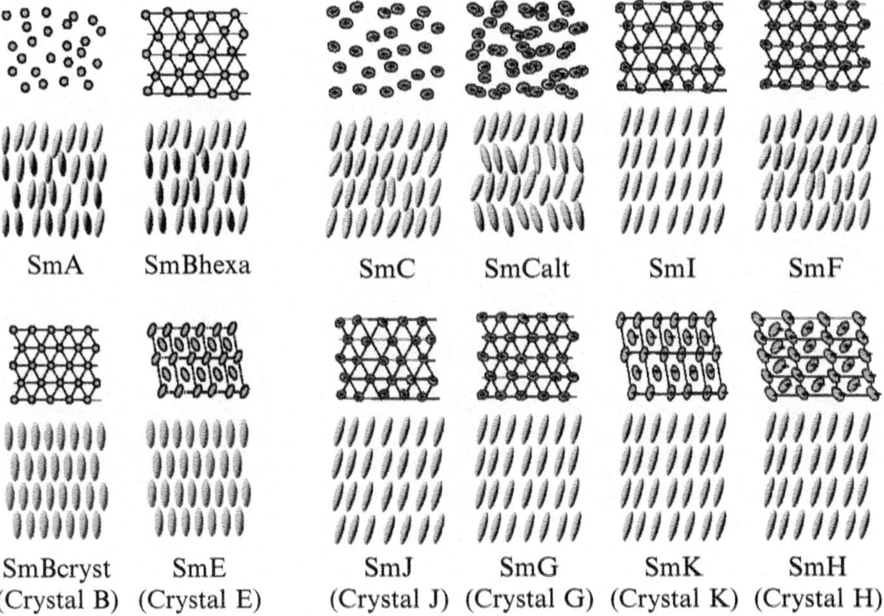

Fig. 2.4 Variety of smectic mesophases. The smectic mesophases in the *upper row* have no long range order in the layers, while the mesophases in the *bottom row* show long range order in the layers, and are sometimes called plastic crystals

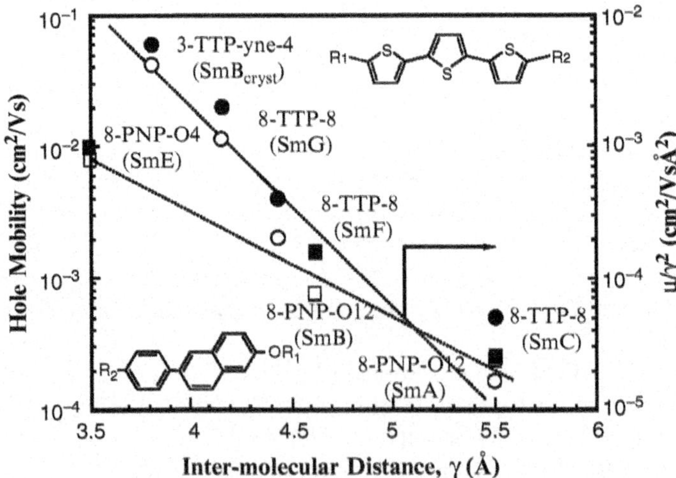

Fig. 2.5 Charge carrier mobility as a function of inter-molecular distance in 2-phenylnaphthalene (PNP) and terthiophene (TTP) derivatives. The *numbers* indicates the number of carbons in the side chain

mobility primarily depends on the intermolecular distances within the smectic layer. It gives the general idea that the typical mobility of each phases is of the order of 10^{-4} cm^2 V^{-1} s^{-1} for SmA and SmC phases, of the order of 10^{-3} cm^2 V^{-1} s^{-1} for SmB$_{hex}$ and SmF phases and of the order of 10^{-2} cm^2 V^{-1} s^{-1} for SmB$_{cryst,}$ SmE and SmG phases. According to the mobilities reported for various liquid crystalline materials, the variation for a particular mesophase seems to be within an order of magnitude from material to material, even though the core structures are different.

For columnar mesophases, the mobility is enhanced as the molecular (inter- or intra-column) order is increased as in the case of smectic mesophases. For example, the mobility is on the order of 10^{-3} cm^2 V^{-1} s^{-1} in a columnar ordered phase, of the order of 10^{-2} cm^2 V^{-1} s^{-1} in a columnar plastic phase and of the order of 10^{-2} cm^2 V^{-1} s^{-1} or higher in a columnar helical phase. However, the mobility is not decided by the intermolecular distances in the column. The intermolecular distance is almost same irrespective of columnar mesophases, i.e., 3.5 Å. In one-dimensional transport, a carrier has to move without making a detour along a given electric field. Therefore, it is very likely that the disorder of molecular alignment in a column, which results from dynamic translational motion of the liquid crystalline molecules in the column, affects the charge carrier transport properties in columnar phases, as discussed by Arikainen et al. of the University of Leeds [9]. From this point of view, there are several experimental facts that may support this idea. Boden et al. and Monobe et al. reported that the mobility decreases with increased hydrocarbon chain length in the columnar phases of triphenylene derivatives [39, 40]. Furthermore, Boden et al. reported that the hole mobility is greatly increased by the ordering imposed in complementary binary mixtures formed by addition of another large core discogen to the triphenylene molecules, which suppresses the dynamic motion of the molecules [41]. However, we have to wait to draw a conclusion about what dominates the carrier mobility in the columnar mesophases, because there is less accumulated data on the carrier mobility in discotic liquid crystals.

2.6 Temperature and Electric Field Dependence

Charge carrier transport in mesophases is quite unique compared to that in amorphous and crystalline materials. The mobility hardly depends on temperature at room temperature and above as shown in Fig. 2.6. Furthermore, it hardly depends on electric field either, as shown in Fig. 2.7. This behavior is not limited to one particular class of liquid crystals, but is probably a general characteristic of charge carrier transport in mesophases above room temperature. In fact, this behavior is also found in discotic columnar mesophases [11, 39, 40, 42].

Interestingly, it was found that at temperatures below room temperatures, the mobility does depend on both temperature and the electric field. This may be a general characteristic of charge carrier transport properties in the mesophase, although the examples known are very limited, because few liquid crystals have a wide

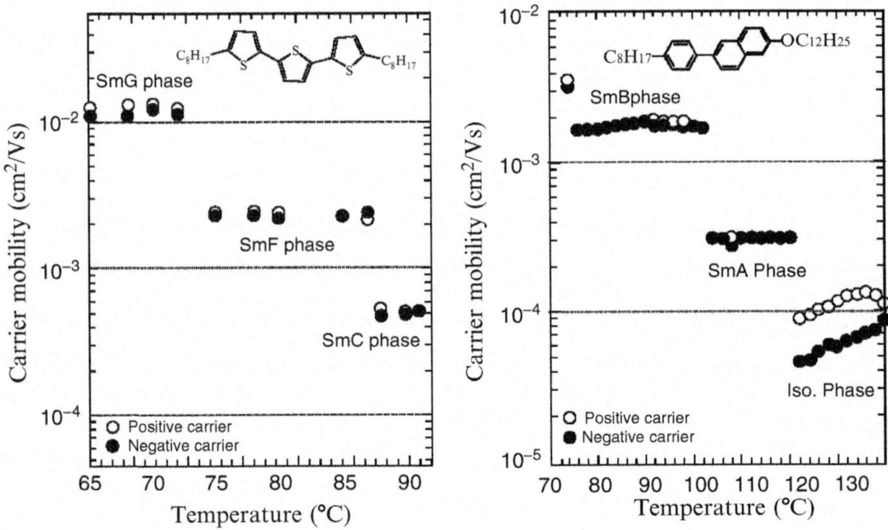

Fig. 2.6 Charge carrier mobility as a function of temperature in various smectic mesophases of typical smectic liquid crystals; 2-phenylnaphthalene and terthiophene derivatives

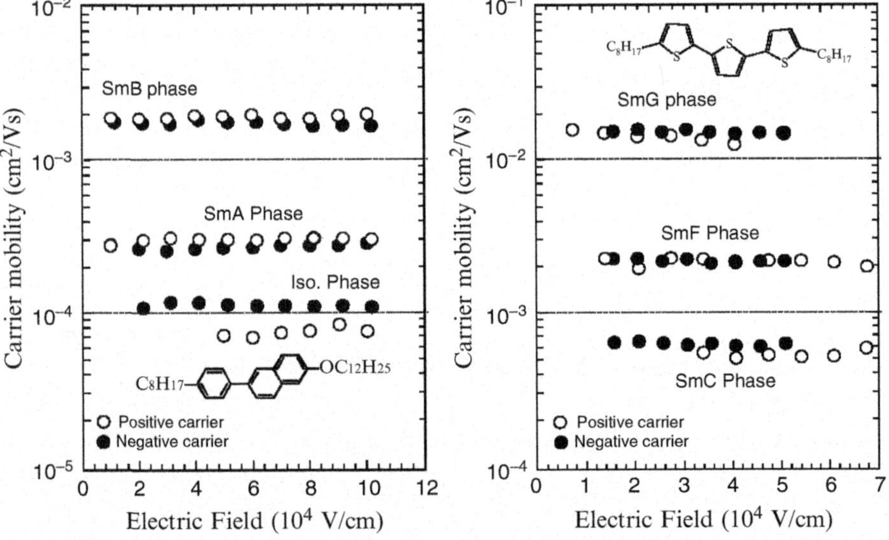

Fig. 2.7 Charge carrier mobility as a function of electric field in various smectic mesophases of typical smectic liquid crystals, 2-phenylnaphthalene and terthiophene derivatives

mesophase range below room temperature [43, 44]. Figure 2.8 shows the mobility of a quaterthiophene derivative, 5-hexyl-5′-hexenyl-2.2′:5′2′′-terthiophene, which has a wide temperature range for the smectic phase, as a function of temperature. At around room temperature, the mobility hardly depends on temperature. However,

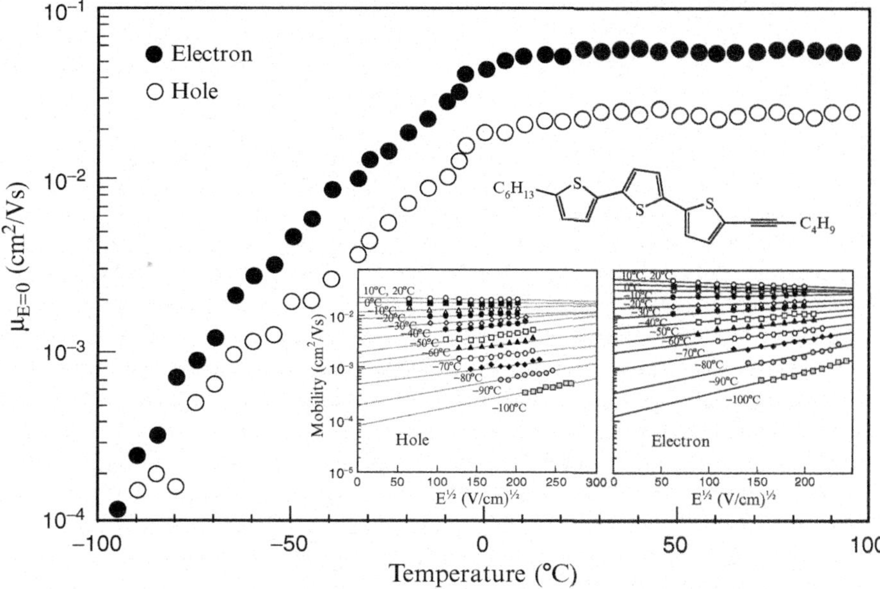

Fig. 2.8 Charge carrier mobility as a function of temperature and electric field in a smectic mesophase of the terthiophene derivative, 5-hexyl-5″-hexynyl-2,2′:5′2″-terthiophene, which has a wide temperature range for a smectic mesophase

it decreases with decreasing temperature below room temperature. In this range the mobility depends on electric filed as well. This is probably common behavior for carrier transport in mesophases at such a low temperatures. This indicates that there is a distributed density of states responsible for charge transport. These unique carrier transport properties are very important for the understanding of the conduction mechanism and the rate determining step in charge transport in mesophases. This is discussed in the following section.

As for the charge carrier transport properties of the nematic phase, there are few reports available, because of the very recent discovery of electronic conduction in the nematic phase [23, 24]. Considering the molecular alignment of the nematic phase it has no positional order, like an amorphous aggregate, so it is very possible that conduction takes place in the distributed localized states, similar to that for conduction in amorphous aggregates, and that conduction can be three-dimensional. Therefore, it is very possible that the intrinsic carrier transport properties are determined by carrier-dipole interactions as in the case of amorphous materials, i.e. there is a Poole-Frenkel type of carrier transport as is reported in the chiral nematic phase of a phenylquaterthiophene derivative [45].

2.7 Impurity Effects

In solid organic semiconductors, such as crystalline and amorphous thin films, chemical impurities with HOMO or LUMO levels in the energy gap between the HOMO and LUMO levels of the host material, work as trap states for holes and electrons, respectively. The detrapping time for trapped charges determines how each chemical impurity works as a trap state. They can either be shallow or deep trap states, depending on whether the detrapping time is shorter or longer, respectively, than the transit time of the carriers. The shallow trap states reduce mobility. This follows the Hoesterey-Letson formalism based on the multiple trapping model, as is reported for single crystal anthracene [46]. On the other hand, once carriers are trapped in deep states, the resulting trapped charges are not released from the states in the time range of the transit time and these affect the carrier transport properties of the bulk and at the interface as space charges. It is reported that the effect of impurities on the charge carrier transport of a contaminated smectic liquid crystal follows the Hoesterey-Letson formalism irrespective of the mesophases [29].

Because there are separate conduction channels for electronic charges and ions in both smectic and discotic mesophases as described in the previous section, the trapped charges contribute to the ionic conduction, when they migrate from the conduction channel, where the π-conjugated core moieties aggregate, into the inter-smectic layer, where side chains of flexible hydrocarbon carbons aggregate. This easily happens in less-ordered mesophases with a low viscosity, such as SmA and SmC phases, because the lifetime of trapped charges, i.e., the residence time of ionized molecules in the smectic layer, becomes shorter. In fact, electronic conduction often accompanies ionic conduction in the less-ordered smectic mesophases contaminated with trace amounts of chemical impurities. It is very possible that the lifetime of trapped charges in a smectic layer is not always determined by the mesophase, but that it may also be determined by the size of the ionized impurities. Ionic conduction for negatives charges, which is often observed even in highly ordered mesophases such as SmB, SmE, SmF and SmG [47], might originate from small ions such as negatively charged oxygen molecules. The time-resolved transient current measurement for mesophases proves how they are contaminated with chemical impurities. It is such a powerful technique, that trace amounts of chemical impurities as low as 0.1 ppm can be detected [28]. Figure 2.9 shows typical transient photocurrents in the SmA phase of a 2-phenylnaphthalene derivative doped with a terthiophene derivative, that is acting as the trap state. In the 1 ppm-doped sample, two shoulders are obvious: one is at around 50 μs and the other at 1,000 μs. These are attributed to the transit times for holes and positive ions, respectively [26]. The ionic conduction becomes dominant, when the concentration of impurities is increased. In fact, the fast transit in the short-time regime completely disappears in 50 ppm doped samples and the slow transit for ions becomes dominant. This indicates that electronic conduction is quite sensitive to trace amounts of chemical impurities tens of ppm are quite sufficient to degrade electronic conduction into ionic conduction.

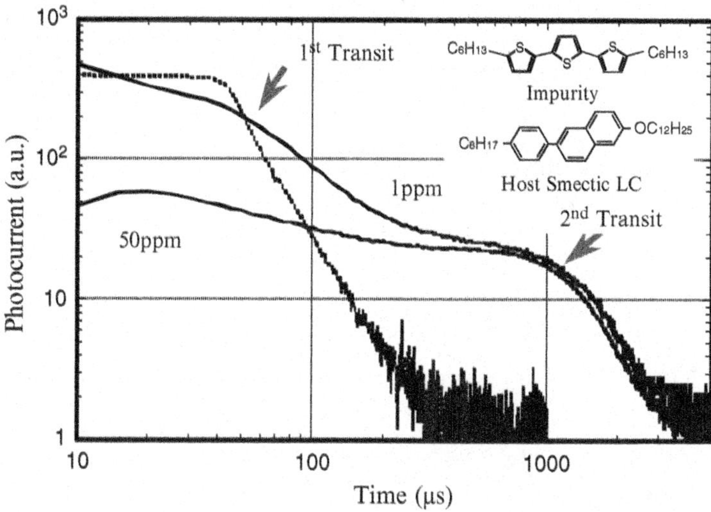

Fig. 2.9 Transient photocurrents in the SmA phase of a 2-phenylnaphthalene derivative, undoped (*dotted line*) and doped with 1 and 50 ppm of a terthiophene derivative (*solid line*), ω, ω'-dihexyl therthiophene

2.8 Structural Defects

Unlike amorphous materials, materials exhibiting molecular orientation are never free from structural defects, such as dislocations and inhomogeneous boundaries, unless they are totally uniform as in the case of single crystals in crystalline materials. For example, polycrystalline materials have grain boundaries in addition to structural defects such as grain dislocations, where chemical impurities acquired during the synthesis are often accumulated and/or there are adsorbed impurities, such as oxygen and water. These defects cause shallow or deep trap states for carriers, resulting in deteriorating carrier transport properties. Indeed, in time-of-flight experiments for polycrystalline materials, the mobility can hardly be determined because the transient photo-currents are not only quite dispersive but also they do not show transit of carriers, retaining space charges in the bulk.

In liquid crystalline mesophases, there also exist domain boundaries in polydomain samples, in addition to domain disclinations. However, judging from the fact that neither the mobility nor the $\mu\tau$-product depend on the size of the domains in a polydomain sample, these structural defects hardly affect the carrier transport properties of smectic mesophases [48–50]. Until now, the exact reason why structural defects in smectic mesophases are less harmful to carrier transport has not been explained. It is possible that the flexibility of the molecular orientation in mesophases, or the soft structure of mesophases, makes local carrier transport possible at defect sites. This is another outstanding feature of carrier transport in the mesophases, which distinguishes mesophases from crystalline materials. It provides us with a great benefit for device applications requiring large-areas. For the discotic columnar phases, there are few reports of the effect of structural defects on charge carrier transport. Unlike the smectic mesophases, where two dimensional transport takes place, it is likely that structural defects in a column may affect the carrier transport properties seriously, because a carrier has to pass along a column without detouring to adjacent columns. From this point of view, relatively low mobility in the columnar ordered phase of triphenylene derivatives, where the intermolecular distance is as small as 3.5 Å, may be explained by the structural defects or disorder of molecular alignment in the columns as described above.

2.9 A Model for Electronic Carrier Transport

As described in the previous section, both electronic and ionic conductions take place in mesophases when the material contains trace amounts of chemical impurities. These types of conduction exhibit different charge carrier transport properties, as demonstrated by the transient photocurrents the different mesophases measured under the same conditions as shown in Fig. 2.10. In these transient photocurrents, the fast transit times are shifted to shorter times as the molecular order in the mesophases is increased from SmA to SmE and from Col_h to the plastic phase, while the slow transits stay in the same time range of 1,000 μs irrespective of the mesophase in both smectics and discotics. Figure 2.11 shows Arrhenius plots of the mobilities for fast and slow transits of the 2-phenylnaphthalene derivative 8PNP-O12. The mobility for the fast transit hardly depends on the temperature, while the mobility for the slow transit does depend on temperature, with an activation energy

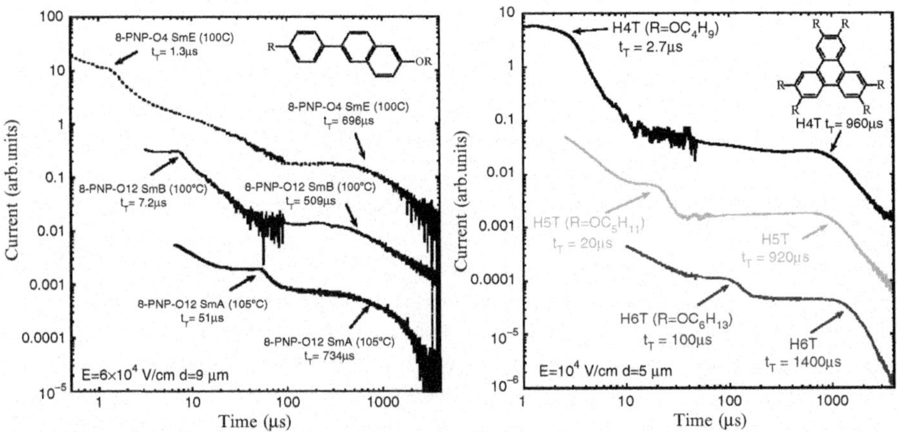

Fig. 2.10 Electronic and ionic conduction for negative carriers in smectic mesophases of 2-phenylnaphthalene derivatives and in the columnar mesophases of triphenylene derivatives. The cell thickness and applied electric field are kept constant for each material

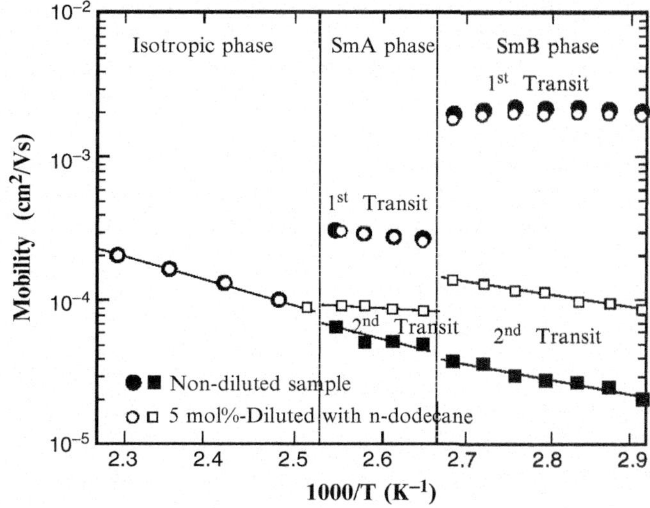

Fig. 2.11 Arrhenius plot of carrier mobility for negative carriers in various phases of a 2-phenylnaphthalene derivative, 6-(4′-octylphenyl)-2-dodecyloxynaphthalene. Two transits, i.e., fast and slow transits are observed in the SmA and SmB phases: the fast transit is attributed to electron transport and the slow one to anion transport; the ionic mobility is enhanced by 5 mol%-dilution with dodecane, while electron mobility remains unchanged or is slightly decreased

of around 0.2 eV. These facts are clear evidence that these two transits are governed by different mechanisms and that there exist two independent conduction channels in smectic and columnar mesophases. The fast and slow mobilities are attributed to the electronic charges and the ions, respectively, as discussed above.

We will focus on electronic conduction in mesophases, as the ionic conduction is not intrinsic in nature, but induced by chemical impurities. In understanding the charge transport properties in a mesophase, there are a few fundamental questions concerning the specific nature of liquid crystals. The first question is "how coherent are the energetic states through which the charge carriers are transported, i.e., extended states or molecularly localized states?" The second question is "how do the dynamic molecular motions of liquid crystal molecules, i.e., their translational and rotational motions, contribute to the charge transfer event molecule to molecule ?" This question is related to another question "can ionic conduction be an intrinsic property of the mesophases of pure materials?"According to an often-cited discussion about band conduction in organic molecular aggregates, the mobility of < 1 cm^2 V^{-1} s^{-1}, at best, in mesophases implies that the conduction is governed by a hopping mechanism through localized states, rather than the band conduction through an extended state [51]. In fact, the inter-molecular distance of 3.5 Å, at the shortest in a highly ordered mesophase, is not short enough to guarantee a band structure for energy states of electrons and holes, because of a weak interaction of molecules through the Van de Waals force. The band model, however, is discussed by Movaghar et al. as is briefly explained later on. As for the second question, we have discussed it already. We conclude that dynamic molecular motion hardly affects electronic conduction in mesophases. In the previous section, it is pointed out that the charge carrier transport properties of mesophases are unique and show neither temperature nor electric-field dependence above room temperature. These properties are quite different from the charge carrier transport properties of amorphous materials, which exhibit a Poole-Frenkel type of behavior. That is the mobility depends on both temperature and electric field as shown in the Eq. (2.1), where μ is the mobility, E is the electric field, k is the Boltzman constant, T is temperature and Δ and β are constants.

$$\log \mu(E, T) \propto -(\Delta - \beta E^{1/2})/kT \tag{2.1}$$

Several models have been proposed to explain the unique carrier transport properties of liquid crystalline mesophases in both the band and hopping conduction regimes [52, 53]. For in the band regime, for example, a thermally, non-activated process is discussed in ordered, narrow-band systems, where each charge delocalizes within a finite coherent length of a few molecules [54]. On the other hand, for the hopping regime, where charge carriers are localized by strong electron–phonon interaction to form small polarons as described by the Holstein model [25], a few models are the subject of discussion. The competition between non-thermally-activated tunneling between neighboring molecules in coherent motion coupling with intra-molecular motion T^{-n} and thermally-activated hopping process in non-coherent motion coupling with inter-molecular motion, $e^{-\delta/kT}$ results in cancellation of the temperature dependence [55] for the non-adiabatic limit in the Holstein model. A nearest neighbor transfer integral between two hopping sites smaller than a certain polaron binding energy can cancel the temperature dependence factor of the small polaron mobility in a certain temperature range [56, 57]. Taking account of the thermally fluctuating molecular alignment in the mesophase, the disorder model, which is proposed to model charge carrier transport in amorphous aggregates and which reproduces well the charge carrier transport properties in amorphous aggregates and is often cited as Poole-Frenkel behavior, successfully models the charge carrier transport in mesophases [58]. In this model, the charge carrier transport in the mesophase is described by one- or two-dimensional hopping conduction in Gaussian-distributed localized states for discotic and smectic mesophases, respectively [59, 60].

The Monte-Carlo simulation of these models reproduces well the charge carrier transport properties in the mesophases, i.e., the mobility independence of both temperature and electric field in the temperature above room temperature, if a small sigma (40 60 meV) is taken for the Gaussian width of the distribution of localized states in Eq. (2.2). In this equation μ is the mobility, σ is the Gaussian width of the distributed energy states for hopping sites, Σ is an index of the positional disorder, k is the Boltzman constant, T is the temperature, E is the electric field and C is a constant. The constants a and n depend on the type of mesophase, e.g., 0.8 and 2 for the SmB phase and 0.78 and 1.5 for the SmE phase, respectively. This value of σ (40-60 meV) is half that for typical amorphous solids [61].

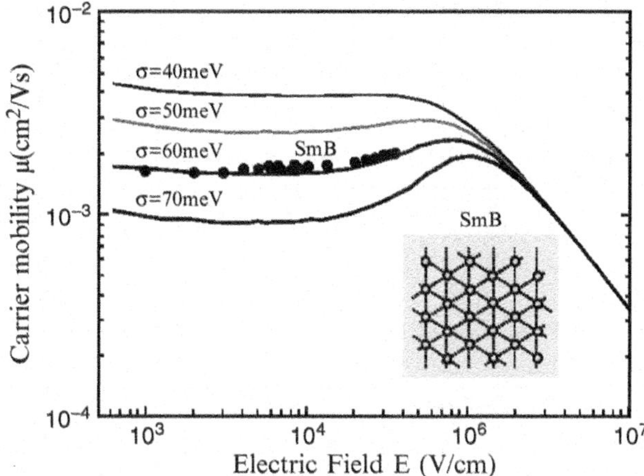

Fig. 2.12 Simulated mobility in the SmB phase of a 2-phenylnaphthalene derivative, 6-(4′-octylphenyl)-2-dodecyloxynaphthalene as a function of small sigma and molecular alignment of the SmB phase in the smectic layer. The *closed circles* indicate mobility determined by time-of-flight experiments

$$\mu_{smectic} = \mu_0 \exp\left[-\left(a\frac{\sigma}{kT}\right)^2\right] \exp\left\{C\left[\left(\frac{\sigma}{kT}\right)^n - \Sigma_0\right]\sqrt{E}\right\}$$

$$n = 2, a = 0.8 \text{ for SmB}, n = 1.5, a = 0.78 \text{ for SmE} \tag{2.2}$$

Fig. 2.12 illustrates the density of states for hole and electron conduction in a smectic mesophase.

On the other hand, in the temperature range below room temperature, the mobility does depend on both the temperature and the electric field, i.e., there is a Poole-Frenkel type of behavior, as demonstrated for terthiophene derivatives [43, 44], which show a wide temperature range for the mesophase from 0 to 100°C as shown in Fig. 2.9. This is good evidence that the localized states responsible for the hopping conduction have energetic disorder. Application of the disorder model for this experimental data gives us a small sigma of 40 meV, which shows a good agreement with the small value of σ derived from Monte-Carlo simulation for the temperature range, where the mobility depends neither on temperature nor on electric field [60]. The disorder model for discotic columnar phases gives a similar value of a small sigma of several tens of meV as is found in smectic mesophases [59]. Therefore, the charge carrier transport properties in liquid crystals can be understood in the framework of the disorder model. However, the energetic disorder of 40–60 meV in the mesophase is comparable to the reorganization energy for charge transfer to adjacent molecules, e.g., tens of meV. Therefore, we have to pay attention to the small polaron model for exact modeling of the charge carrier transport in mesophases.

2.10 Chemical Structure

The hopping rate in a molecular aggregate, from one adjacent molecule to another, is determined by a transfer integral, a reorganization energy and an energetic distribution of each molecule. The transfer integral is governed by the spatial overlap of molecular orbitals for the molecules concerned, which is determined by their relative configurations in space and intermolecular distance. The reorganization energy and the energetic distribution of each molecule depend on the molecular structure. Therefore, the charge carrier transport properties in the mesophases should depend on the molecular structure. However, there are few experimental results on the carrier transport properties of the mesophases linked to the chemical structure of the liquid crystals so far. In general, it is well accepted that the larger the π-conjugated system in the molecular core in a liquid crystal, the higher the mobility in the mesophase becomes, because of an increase in the transfer integral. For discotic liquid crystals, an empirical formulation for the highest mobility in the Col_h phase is proposed in terms of a number carbon atoms consisting of π-conjugate system in the core [62] as shown in Eq. (2.3), where μ_{max} is the highest mobility for a particular core moiety, n is the number of carbon atoms in the core moiety and the constant b is found to be 83.

$$\sum \mu_{max} = 3 \exp[-b/n]\,(cm^2V^{-1}s^{-1}) \qquad (2.3)$$

In this formulation, the highest mobility of 3 $cm^2\ V^{-1}\ s^{-1}$ corresponds to that across molecular planes of large two-dimensionally extended π-conjugated system in graphite, which is thought to be an extreme case of the columnar phase in discotic liquid crystals, as shown in Fig. 2.13.

The idea that the size of the π-conjugate system in the molecular core of a liquid crystalline compound affects the mobility in a mesophase is also valid in calamitic liquid crystals. For example, terphenyls exhibit higher mobility in the same mesophase, compared to biphenyls, as do quaterthiophenes compared with terthiophenes. The highest mobility achieved so far in mesophases is 0.1 0.3 $cm^2\ V^{-1}\ s^{-1}$ for quaterthiophenes and phthalocyanine derivatives determined by time-of-flight experiments [42, 63, 64] and 1 $cm^2\ V^{-1}\ s^{-1}$ for hexabenzocoronens determined by time-resolved transient microwave conductivity measurements [65], which indicate that the bulk mobility in a mesophase and a microscopic mobility in a column, respectively (Fig. 2.14).

As for the chemical structure of molecules, it has been shown by experiment and by theory, that the dipole moment of a molecule dominates the extent of distribution in the density of localized states associated with hopping conduction for amorphous aggregates and so the mobility seriously depends on the dipole moment of the molecule [66]. For meosphases, however, any clear relation between the chemical structure of the molecular core and the mobility in a mesophase has not yet been clarified in either discotic or calamitic liquid crystals. This omission is because the data for the mobility in various liquid crystals is still too

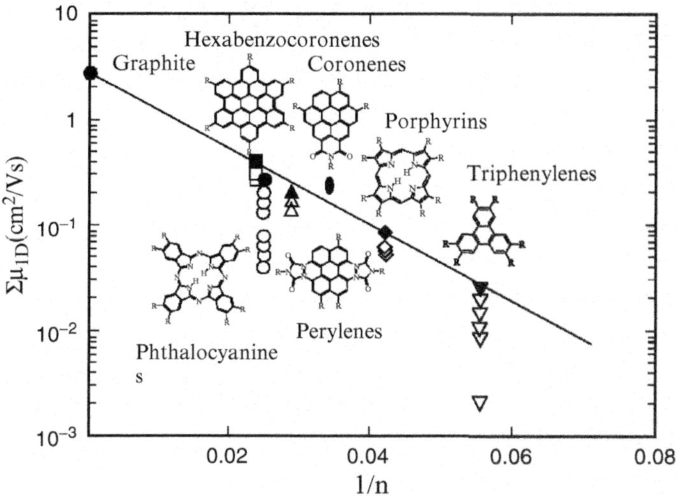

Fig. 2.13 Intra-columnar mobility in the hexagonal columnar phases (D_h) of various discotic materials. The n indicates the number of carbon atoms in a discotic core of the materials. The mobility was determined by pulsed radiolysis transient microwave conductivity (PR-TRMC) measurements

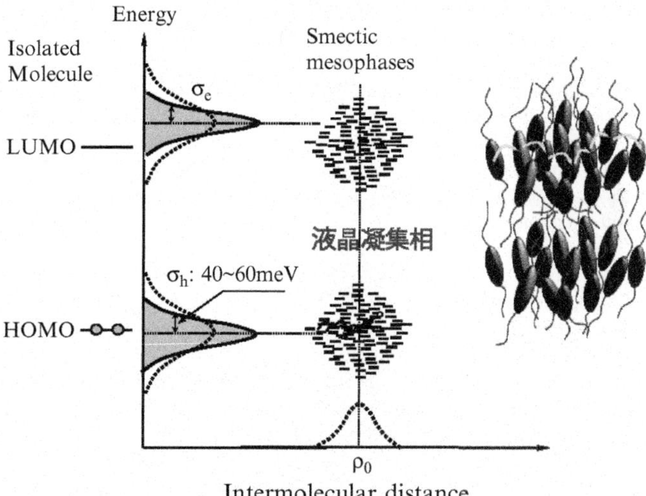

Fig. 2.14 Schematic diagram of the density of states in the smectic mesophase. The Gaussian width of the density of states for electrons and holes is about 40-60 meV, which is a half of that found in amorphous materials which is indicated by the *dotted line*

limited. Taking account of molecular alignment in mesophases, where the molecular alignment is not as perfect as in molecular crystals and is disordered to some extent, this carrier-dipole interaction could be valid in mesophases, because carriers experience variable carrier-dipole interactions during drift at a given electric field.

Therefore, accumulation of experimental data on mobilities in various classes of liquid crystals are needed before any structural relationship with charge carrier transport properties is established, leading to a rational design of self-organizing molecular semiconductors.

References

1. Kusabayashi, S., Labes, M.M.: Conductivity in liquid crystals. Mol. Cryst. Liq. Cryst. **7**, 395–405 (1969)
2. Heimeier, G.H., Zanoni, L.A., Burton, L.A.: Dynamic scattering: a new electrooptic effect in certain classes of nematic liquid crystals. Proc. IEEE **56**, 1162–1171 (1968)
3. Heilmeier, G.H., Heyman, P.M.: Note on transient current measurement in liquid crystals and related systems. Phys. Rev. Lett. **18**, 583–585 (1967)
4. Yoshino, K., Tanaka, N., Inuishi, Y.: Anomolous carrier mobility in smectic liquid crystal. Jpn. J. Appl. Phys. **15**, 735–736 (1976)
5. Derfel, G., Lipinski, A.: Charge carrier mobility measurements in nematic liquid crystals. Mol. Cryst. Liq. Cryst. **55**, 88–99 (1979)
6. Chandrasekhar, S., Sedaschiva, B.K., Suresh, K.A.: Pramana **9**, 471–480 (1977)
7. Boden, N., Bushby, R.J., Clements, J., Jesudason, M.V., Knowles, P.F., Williams, G.: One-dimensional electronic conductivity in discotic liquid crystals. Chem. Phys. Lett. **152**, 94–99 (1988)
8. Boden, N., Bushby, R.J., Clements, J., Movaghar, B., Donovan, K.J., Kreuozis, T.: Mechanism of charge transport in discotic liquid crystals. Phys. Rev. B **52**, 13274–13280 (1995)
9. Arikainen, E.O., Boden, N., Bushby, R.J., Clements, J., Movaghar, B., Wood, A.: Effects of side-chain length on the charge transport properties of discotic liquid crystals and their implications for the transport mechanism. J. Mater. Chem. **5**, 2161–2165 (1995)
10. Shouten, P.G., Warman, J.M., de Haas, M.P., Fox, M.A., Pan, H.-L.: Charge migration in aggregates of peripherally substituted porphyrins. Nature **353**, 736–737 (1991)
11. Adam, D., Closs, F., Frey, T., Funhoff, D., Haarer, D., Schuhmacher, P., Siemensmeyer, K.: Transient photoconductivity in a discotic liquid crystal. Phys. Rev. Lett. **70**, 457–460 (1993)
12. Okamoto, K., Nakajima, S., Ueda, M., Itaya, A., Kusabayashi, S.: Electrical dark conductivity and photo-conductivity of 2-(paradecyloxybenzylideneamino)-9-fluorenone in the nematic state. Bull. Chem. Soc. Jpn. **56**, 3830–3832 (1983)
13. Closs, F., Siemensmeyer, K., Frey, T., Funhof, D.: Liquid crystalline photoconductors. Liq. Cryst. **14**, 629–634 (1993)
14. Funahashi, M., Hanna, J.: Fast hole transport in a New calamitic liquid crystal of 2-(4′-heptyloxyphenyl)-6-dodecylthiobenzothiazole. Phys. Rev. Lett. **78**, 2184–2187 (1997)
15. Tokunaga, K., Iino, H., Hanna, J.: Reinvestigation of carrier transport properties in liquid crystalline 2-phenylbenzothiazole derivatives. J. Phys. Chem. B **111**, 12041–12044 (2007)
16. Funahashi, M., Hanna, J.: Fast ambipolar carrier transport in smectic phases of phenylnaphthalene liquid crystal. Appl. Phys. Lett. **71**, 602–604 (1997)
17. Redecker, M., Bradley, D.D.C., Inbasekaran, M., Woo, E.P.: Nondispersive hole transport in an electroluminescent polyfluorene. Appl. Phys. Lett. **73**, 1565–1567 (1998)
18. Farrar, S.R., Contoret, A.E.A., O'Neill, M., Nicholls, J.E., Richards, G.J., Kelly, S.M.: Nondispersive hole transport of liquid crystalline glasses and a cross-linked network for organic electroluminescence. Phys. Rev. B. **66**, 125107–125111 (2002)
19. Chen, L.-Y., Hung, W.-Y., Lin, Y.-T., Wu, C.-C., Chao, T.-C., Hung, T.-H., Wong, K.-T.: Enhancement of bipolar carrier transport in oligofluorene films through alignment in the liquid-crystalline phase. Appl. Phys. Lett. **87**, 112103–112103-3 (2005)

20. Yasuda, T., Fujita, K., Tsutsui, T., Geng, Y., Culligan, S.W., Chen, S.H.: Carrier transport properties of monodisperse glassy-nematic oligofluorenes in organic field-effect transistors. Chem. Mater. **17**, 264–268 (2005)
21. Murakami, S., Naito, H., Okuda, M., Sugimura, A.: Transient photocurrent in amorphous selenium and nematic liquid crystal double layers. J. Appl. Phys. **78**, 4533–4537 (1995)
22. Sawada, A., Manabe, A., Naemura, S.: Comparative studies on the attributes of ions in nematic and ionic phases. Jpn. J. Appl. Phys. Part 1 **40**, 220–224 (2001)
23. Funahashi, M., Tamaoki, N.: Electronic conduction in the chiral nematic phase of an oligothiophene derivative. Chem.Phys.Chem. **7**, 1193–1197 (2006)
24. Tokunaga, K., Takayashiki, Y., Iino, H., Hanna, J.: Electronic conduction in nematic phase of small molecules. Phys. Rev. B **79**, 033201–033205 (2009)
25. Iino, H., Hanna, J., Haarer, D.: Electronic and ionic carrier transports in discotic liquid crystalline photoconductor. Phys. Rev. B **72**, 193203–193206 (2005)
26. Iino, H., Hanna, J.: Electronic and ionic transports for negative charge carriers in smectic liquid crystalline photoconductor. J. Phys. Chem. B **109**, 22120–22125 (2005)
27. Funahashi, M., Hanna, J.: Impurity effect on charge carrier transport in smectic liquid crystals. Chem. Phys. Lett. **397**, 319–323 (2004)
28. Ahn, H., Ohno, A., Hanna, J.: Detection of trace amount of impurity in smectic liquid crystals. Jpn. J. Appl. Phys. **44**, 3764–3768 (2005)
29. Ahn, H., Ohno, A., Hanna, J.: Impurity effects on charge carrier transport in various mesophases of smectic liquid crystals. J. Appl. Phys. **102**, 093718 (2007)
30. Kurotaki, K., Hanna, J.: Carrier transport in molecularly diluted liquid crystalline photoconductor. J. Imaging Sci. Technol. **43**, 237–241 (1999)
31. Redecker, M., Bradley, D.D.C., Inbasekaran, M., Woo, E.P.: Mobility enhancement through homogeneous nematic alignment of a liquid-crystalline polyfluorene. Appl. Phys. Lett. **74**, 1400–1402 (1999)
32. Tokunaga, K., Takayashiki, Y., Iino, H., Hanna, J.: One-dimensional to three-dimensional electronic conduction in liquid crystalline mesophases. Mol. Cryst. Liq. Cryst. **510**, 250/[1384]–258/[1392] (2009)
33. Funahashi, M., Hanna, J.: High ambipolar carrier mobility in self-organizing terthiophene derivative. Appl. Phys. Lett. **76**, 2574–2576 (2000)
34. Mery, S., Haristoy, D., Nicoud, J.-F., Guillon, D., Diele, S., Monobe, H., Shimizu, Y.: Bipolar carrier transport in a lamello-columnar mesophase of a sanidic liquid crystal. J. Mater. Chem. **12**, 37–41 (2002)
35. Iino, H., Hanna, J., Haarer, D., Bushby, R.J.: Fast electron transport in discotic columnar phases of triphenylene derivatives. Jpn. J. Appl. Phys. **45**, 430–433 (2006)
36. Iino, H., Takayashiki, Y., Hanna, J., Bushby, R.J.: Fast ambipolar carrier transport and easy homeotropic alignment in a metal-free phthalocyanine derivative. Jpn. J. Appl. Phys. **44**, L1310–L1312 (2005)
37. Schein, L.B.: Temperature independent drift mobility along the molecular direction of As_2S_3. Phys. Rev. B **15**, 1024–1034 (1977)
38. Iino, H., Hanna, J.: Ambipolar charge carrier transport in liquid crystals. Optoelectron. Rev. **14**, 295–302 (2005)
39. Boden, N., Bushby, R.J., Lozman, O.R., Lu, Z.B., McNeill, A., Movaghar, B., Donovan, K., Kreouzis, T.: Enhanced conductivity in the discotic mesophase. Mol. Cryst. Liq. Cryst. **410**, 541–549 (2004)
40. Monobe, H., Shimizu, Y., Okamoto, S., Enomoto, H.: Ambipolar charge carrier transport properties in the homologous series of 2,3,6,7,10,11-hexaalkoxytriphenylene. Mol. Cryst. Liq. Cryst. **476**, 277–287 (2007)
41. Wegewijs, B.R., Siebbeles, L.D.A., Boden, N., Bushby, R.J., Movaghar, B., Lozman, O.R., Liu, Q., Pecchia, A., Mason, L.A.: Charge-carrier mobilities in binary mixtures of discotic triphenylene derivatives as a function of temperature. Phys. Rev. B **65**, 245112-1–245112-8 (2002)

42. Iino, H., Takayashiki, Y., Hanna, J., Bushby, R.: Fast ambipolar carrier transport and easy homeotropic alignment in a metal-free phthalocyanine derivative. Jpn. J. App. Phys. **44**, L1310–1312 (2005)
43. Funahashi, M., Hanna, J.: Mesomorphic behaviors and charge carrier transport in terthiophene derivatives. Mol. Cryst. Liq. Cryst. **410**, 529–540 (2004)
44. Funahashi, M., Zhang, F., Tamaoki, N., Hanna, J.: Ambipolar transport in the smectic E phase of 2-propyl-5′′-hexynylterthiophene derivative over a wide temperature range. ChemPhysChem **9**, 1465–1473 (2008)
45. Funahashi, M., Tamaoki, N.: Electronic conduction in the chiral nematic phase of an oligothiophene derivative. ChemPhysChem **7**, 1193–1197 (2006)
46. Hoesterey, D.C., Letson, G.M.: The trapping of photocarriers in anthracene by anthraquinone, anthrone and naphthacene. J. Phys. Chem. Solid. **24**, 1609–1615 (1963)
47. Iino, H., Hanna, J.: Optoelectron. Rev. **13**, 295–302 (2005)
48. Maeda, H., Funahashi, M., Hanna, J.: Effect of grain boundary on carrier transport of calamitic liquid crystalline photoconductive materials. Mol. Cryst. Liq. Cryst. **346**, 183–192 (2000)
49. Maeda, H., Funahashi, M., Hanna, J.: Electrical properties of domain boundaries in photoconductive smectic mesophases and their crystal phases. Mol. Cryst. Liq. Cryst. **366**, 369–376 (2001)
50. Zhang, H., Hanna, J.: High mu tau product in a smectic liquid crystalline photoconductor of a 2-phenylnaphthalene derivative. Appl. Phys. Lett. **85**(22), 5251–5253 (2004)
51. Friedman, L.: Transport properties of organic semiconductors. Phys. Rev. **133**, 1668 (1964)
52. Kreouzis, T., Donovan, K.J., Boden, N., Bushby, R.J., Lotzman, O.R., Liu, Q.: Temperature-independent hole mobility in discotic liquid crystals. J. Chem. Phys. **114**, 1797–1802 (2001)
53. Shiyanovskaya, I., Singer, K.D., Twieg, R.J., Sukhomlinova, L., Gettwert, V.: Electronic transport in smectic liquid crystals. Phys. Rev. E **65**, 041715 (2002)
54. Pecchia, A., Siebbeles, L., Movaghar, B.: Charge carrier transport in higly ordered smectic and discotic mesophases. Proc. SPIE **4991**, 253–273 (2003)
55. Silinsh, E.A., Shlihta, G.A., Jurgis, A.J.: A model description of charge carrier transport phenomena in organic molecular crystals. 1. Polyacene crystals. Chem. Phys. **138**, 347–363 (1989)
56. Holstein, T.: Studies of polaron motion. 1. The molecular-crystal model. Ann. Phys. (N.Y.) **8**, 325–342 (1959)
57. Holstein, T.: Studies of polaron motion. 1. The small polaron. Ann. Phys. (N.Y.) **8**, 343–389 (1959)
58. Bässler, H.: Charge transport in disordered organic photoconductors – a Monte-Carlo simulation study. Phys. Status Solid. B **175**, 15–56 (1993)
59. Bleyl, I., Erdelen, C., Schmidt, H.W., Haarer, D.: One-dimensional hopping transport in a columnar discotic liquid-crystalline glass. Philos. Mag. B **79**, 463–475 (1999)
60. Ohno, A., Hanna, J.: Simulated carrier transport in smectic mesophase and its comparison with experimental result. Appl. Phys. Lett. **82**, 751–753 (2003)
61. Bosenberger, P.M., Weiss, D.S.: Organic Photoreceptors for Xerography. Marcel Deccer, Inc., New York (1998)
62. Van de Craats, A.M., Warman, J.M.: The core-size effect on the mobility of charge in discotic liquid crystalline materials. Adv. Mater. **13**, 130 (2001)
63. Funahashi, M., Hanna, J.: High carrier mobility up to 0.1 $cm^2\ V^{-1}\ s^{-1}$ at ambient temperatures in thiophene-based smectic liquid crystals. Adv. Mater. **17**, 594–598 (2005)
64. Takayashiki, Y., Iino, H., Shimakawa, T., Hanna, J.: Ambipolar carrier transport in terphenyl derivative. Mol. Cryst. Liq. Cryst. **480**, 295–301 (2008)
65. van de Craats, A.M., Warman, J.M., Fechtenkotter, A., Brand, J.D., Harbison, M.A., Mullen, K.: Record charge carrier mobility in a room-temperature discotic liquid-crystalline derivative of hexabenzocoronene. Adv. Mater. **11**, 1469–1472 (1999)
66. Borsenberger, P.M., Bässler, H.: Concerning the role of dipolar disorder on charge transport in molecularly doped polymers. J. Chem. Phys. **95**, 5327–5331 (1991)

Chapter 3
Columnar Liquid Crystalline Semiconductors

Richard J. Bushby and Daniel J. Tate

3.1 Introduction

In a typical discogen the aromatic core of the molecule is surrounded by an annulus of alkyl chains. The most common liquid crystal phases that they form are columnar in which the molecules are stacked one on top of another like a pile of coins (Fig. 3.1). The columns are arranged on a regular two-dimensional lattice. Whereas there may be short-range positional order of the discs along the column, there is no correlation of the positions of the discs between the columns and thus, no three-dimensional order. The alkyl chains are fluid and disordered and a static image like that shown in Fig. 3.1, fails to convey the highly dynamic nature of the phase. For the Col_h phase of the ethers of triphenylene shown in Fig. 3.1 the interconversion of the *trans-gauche* conformations of the alkyl chains occurs with a frequency of the order of 10^{10} s^{-1}, the rotation of the aromatic cores has a time constant of the order of 10^8 s^{-1} [1] and the exchange of discs between neighboring columns with a frequency of the order of 10^5 s^{-1} [2]. Like most other molecular materials, pure discotic liquid crystals are insulators because the intrinsic concentration of charge carriers is very low. To make them conducting, holes or electrons have to be injected. This can be achieved by chemical doping, radiolysis, photolysis or (at sufficiently high potentials) from the surface of an electrode. The holes or electrons that are injected are essentially localized on the aromatic cores so that each column shown in Fig. 3.1 acts like an insulated molecular wire. The charge carriers can hop from the aromatic core of one molecule to the aromatic core of another along the stack, but

R.J. Bushby (✉)
School of Chemistry, University of Leeds, Leeds, LS2 9JT UK
e-mail: R.J.Bushby@Leeds.ac.uk

D.J. Tate
Organic Materials Innovation Centre, School of Chemistry, University of Manchester, Manchester, M13 9PL UK
e-mail: Daniel.Tate@Manchester.ac.uk

R.J. Bushby et al. (eds.), *Liquid Crystalline Semiconductors*, Springer Series in Materials Science 169, DOI 10.1007/978-90-481-2873-0_3,

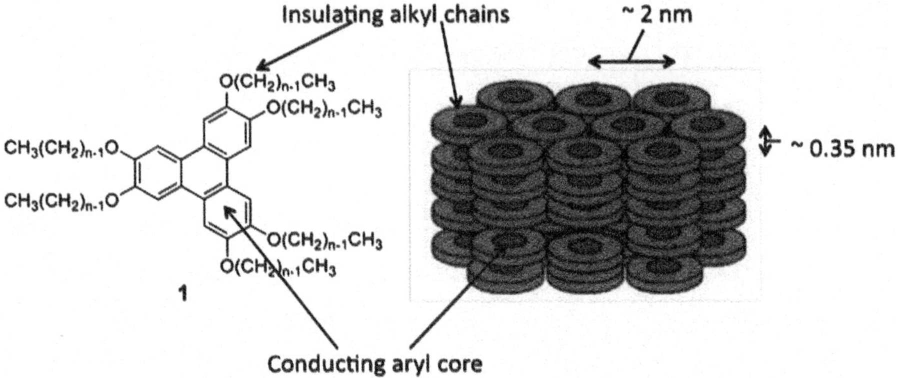

Fig. 3.1 Molecular structure and schematic of the conducting columnar Col_h phase of a HATn triphenylene-based discogen **1** (n is the number of carbons in each side-chain. Hence in HAT5 there are six pentyloxy side-chains)

the columns are insulated from each other by the sheath of alkyl chains. For the Col_h phase of HATn, the hopping frequency is $>10^9\ s^{-1}$. An important consequence of the highly dynamic, fluid nature of columnar phases is that defects that appear within a given 'molecular wire' rapidly disappear: the column self-heals. However, there are other advantages in using columnar liquid crystalline semiconductors. Liquid crystals are partially ordered molecular materials whose properties, including their conduction properties, lie between those of (fully ordered) crystalline and (fully disordered) amorphous molecular materials. They combine the best properties of both. Like amorphous molecular materials they are easily processed and they lack the grain boundaries that, in their crystalline counterparts, act as deep traps for the charge carriers. On the other hand, most are sufficiently well ordered that they share some of the advantages of crystalline molecular solids. The charge carrier mobilities can be almost as high as those in the crystal. Furthermore, at least in the most ordered liquid crystalline phases, impurities that could act as charge traps have limited solubility and so are relatively 'forced out of the system'.

3.2 Methods of Study of Conductivity and of Measurement of the Charge Carrier Mobilities

3.2.1 Chemically Doped Systems

Historically, the first attempts to create conducting discotic liquid crystals introduced the charge carriers by chemical doping: oxidation (*p*-doping) or reduction (*n*-doping). These experiments mimicked the chemical doping procedures that had been used for some years for other organic molecular and polymeric materials. Hence, Drenth [3] showed that when HATn compounds were oxidized (*p*-doped)

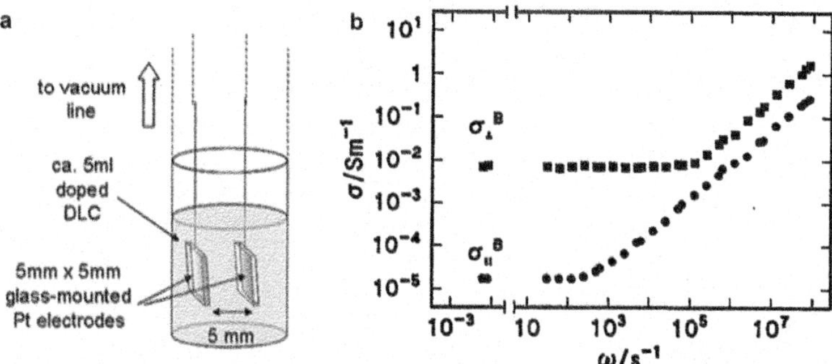

Fig. 3.2 (**a**) Schematic of the cells used for the chemical doping studies of conducting liquid crystals (**b**) the frequency dependence of the conductivity of HAT6 doped with 0.5 mol% of aluminium trichloride in the Col_h phase (parallel and perpendicular to the axis of the column)

with iodine vapor a solid (approximate composition 1.0 HATn : 0.7 I_2) was obtained which was conducting but, the phase transitions were strongly depressed and, since conduction was apparently independent of orientation, they concluded that 'electron transport through the columns was not a major route of charge transport'. At about the same time we showed that by using only 1 mol% of oxidant, conducting materials could be obtained that showed slightly lowered and broadened phase transitions and in which the conductivity was strongly dependent on orientation [4, 5]. In our first experiments HATn discotics were doped on a vacuum line using antimony pentachloride vapor [6] but heating the discogen under vacuum with a weighed amount of aluminium trichloride or aluminium tribromide or doping in solution using nitrosonium salts ($NO^+ X^-$) proved experimentally simpler and more reproducible. The AC conductivities were determined in a commercial conductivity cell with a capacity of about 5 ml and with 5 × 5 mm platinum electrodes 5 mm apart. The number of (spin-bearing) charge carriers per unit volume was measured by EPR spectroscopy.

Given the conductivity (σ) and number of charge carriers (n) the mobilities of the charge carriers (μ) can be determined directly.

$$\sigma = ne\mu$$

The mobilities we obtained were clearly too high to be accounted for by ion migration and (critically) the conductivity did not show dramatic breaks at the crystal/Col_h or Col_h/isotropic phase boundaries (if the conduction had been due to ions, because of the viscosity differences between the phases, there would have been dramatic changes at these transitions). The conductivity shows the same qualitative frequency dependence (Fig. 3.2) as that seen in other hopping conductors. In the low frequency range it is more or less independent of frequency but above a critical value, related to the slowest hopping event, the conductivity shows a power law increase with frequency.

$$\sigma(\omega) \sim \omega^s, \; s \sim 0.7 - 0.8$$

In our first paper [4] we presented a mixture of conductivity and EPR studies. We interpreted the temperature/phase dependence of the EPR line shapes of aluminium trichloride doped HAT6 in terms of Dysonian theory, but this later proved to be incorrect. We later showed that both the g value and the EPR line width are dependent on the orientation of the director relative to the field and this leads to a complex, line shape in powder samples [7]. Just as p-doped, hole conductors can be produced by oxidation of discogens with pi excessive cores using aluminium trichloride, aluminium tribromide or NO^+X^-, n-doped, electron conductors can be produced by reduction of discogens with pi deficient cores using alkali metals [8].

The importance of these chemical doping studies and the advantages of chemical doping are that: (1) They established the basic hopping nature of conduction in liquid crystal systems [5, 9] and the way that the order of the liquid crystal determines the mobility [5]. (2) Because these are thick samples, their behavior tends to be dominated by the bulk of the liquid crystal rather than by the liquid crystal/electrode interface. (3) It is possible to investigate the nature of the charge carrier by spectroscopic methods. Studies of HAT6 doped with aluminium trichloride suggested the formation of monomeric molecule radical cations $HAT6^+$ paired with $AlCl_4^-$ counterions [10]. (4) It provides the only easy method by which the conductivity can be measured both parallel to and perpendicular to the director. In small cells the alignment of columnar phases of discotic liquid crystals is often totally dominated by surface interactions and it is not usually possible to manipulate the alignment by using external magnetic fields. In the large cells used in the doping studies, however, provided fields > 4T are employed, the alignment within the cell can be controlled using an external magnetic field.

Despite these advantages this method of studying conducting liquid crystals has been almost entirely replaced by PR-TRMC (Sect. 3.2.2) and ToF (Sect. 3.2.3) methods. The main reasons are that: (1) because of the air instability of the organic radical ions produced samples need to be prepared and studied using glove box and/or vacuum line methods; (2) because under DC fields slow ion (e.g. $AlCl_4^-$ or BF_4^-) migration results in polarization of the cell the method is limited to AC studies; (3) because of the dimensions of commercial conductivity cells large samples (5–10 g) of pure discogen are required for each experiment. In many cases such large samples are not available. Even in the case of the HAT discotics, which are fairly easy to make, it proved a real problem. This problem was exasperated by the fact that it proved impossible to recycle HAT materials from old cells. After doping with a few mol% of aluminium trichloride or tribromide the recovered HAT was found to contain trace (~1 %) amounts of dehydrogenation (**1,** one chain converted to $-OCH{=}CH(CH_2)_{n-3}CH_3$) and dealkylation (**1**, one chain replaced by –OH) products [11]. These co-chromatographed and co-crystallised with the parent HAT and could not be removed meaning that fresh material had to be made and purified for each experiment. (4) Side-reactions leading to trace impurities are another drawback of this method.

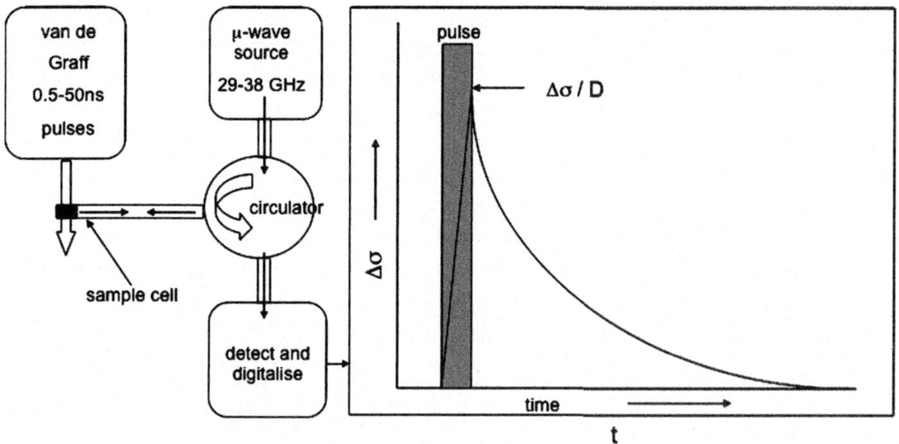

Fig. 3.3 Schematic of a PR-TRMC experiment

3.2.2 Pulse-Radiolysis Time-Resolved Microwave Conductivity (PR-TRMC)

Historically, the second method used to study conducting discotics was pulse-radiolysis, time-resolved, microwave conductivity (PR-TRMC) [12]. This is another method, which had already proved very effective in studying the conductivity of other classes of organic molecular materials. A schematic of the experimental setup is shown in Fig. 3.3 [13]. Charges are injected from a van der Graff generator, radiolysis producing hole-electron pairs. A dosimeter is used to directly measure the energy dissipated by the electron beam per unit volume allowing the initial concentration of charge carriers to be estimated (typically in the micro-molar regime). The conductivity, induced is measured from the increase in microwave power consumption. The charge carrier mobility (μ_{TRMC}) can then be determined directly from this estimation and the measured radiation induced conductivity ($\Delta\sigma/D$).

$$\sum \mu_{TRMC} = \frac{E_p}{P} \cdot \frac{\Delta\sigma}{D}$$

Where, E_p is the average energy required to produce an electron-hole pair and P is the fraction of charge carriers that survive until the end of the pulse. The values of P are estimated according to a model that describes scavenging of the charge carriers by the aromatic cores of the columnar structures formed by the discotic molecules, see for example ref [14]. Since $\sum\mu_{TRMC}$ is measured in three dimensions, the intracolumnar mobility parallel to the columns $\sum\mu_{1D}$ is three times as big.

The advantages of PR-TRMC are: (1) that the high penetrating power of the ionising radiation means that it is insensitive to the color or morphology of the sample; (2) that no electrodes are required by this technique so interface effects are avoided; (3) that any sample can be used regardless of is alignment; (4) that the samples required are small.

There are also some problems with this method. Because of the very high frequency employed the charge carriers only migrate over microdomains avoiding the deeper traps. As can be seen from Fig. 3.2 the mobility measured at such high frequencies, over such small domains is always going to be higher, sometimes much higher, than the low frequency or DC mobility. It can be argued that this is the 'fundamental value' of the material free from grain boundary etc. effects but it does not relate in any simple way to the bulk DC or low frequency AC value which is the value that is relevant to device production. From that standpoint, PR-TRMC mobilities are seen as 'very optimistic values'. The other disadvantages of the method are (1) that, at present it is only carried out at one centre in the world and (2) that the sum of electron and hole mobilities is obtained.

A laser pulse can be used instead of the van der Graff generator to produce the electron-hole pair, which is a technique known as flash photolysis, time-resolved, microwave conductivity or FP-TRMC. The main disadvantage is that the concentration of electron-hole pairs formed in this way is more difficult to estimate. Usually, it can only be used to measure the product of the quantum yield and the sum of the mobilities ($\phi\Sigma\mu$). However, the technique has been used to good effect to study the anisotropy of the conductivity in oriented films [15, 16]. More recently a variation on this method has been introduced in which the transient absorption spectrum is measured simultaneously. This provides structural information on the charge carriers and (provided assumptions are made) allows their concentration to be determined yielding $\Sigma\mu$ [17, 18].

3.2.3 *Time of Flight (ToF) Conductivity*

Although historically not the first, perhaps the most important method for studying the conductivity of discotic liquid crystals is the time of flight (ToF) method first applied to these systems by Haarer and coworkers [19, 20]. A schematic of the experimental setup is shown in Fig. 3.4 [21, 22]. A thin film (typically 10–20 μ) of a homeotropically-aligned sample (one in which the columns run electrode to electrode) is sandwiched between transparent ITO electrodes. A laser pulse with the minimum intensity required to produce the photocurrent, is use to produce a thin layer ($\ll 1\ \mu$) of electron-hole pairs close to the electrode. According to the polarity of the cell one charge carrier is rapidly captured by the adjacent electrode and a thin sheet of the counter charges is driven through the sample. Although there are many trapping/detrapping events, on average, each charge experiences almost the same transit. As a result, even though the charge package broadens (its migration is a biased random walk) the current remains more or less constant until the counter-electrode is reached. Transport of this kind is known as non-dispersive or Gaussian transport and is usually limited to liquids, liquid crystals and amorphous

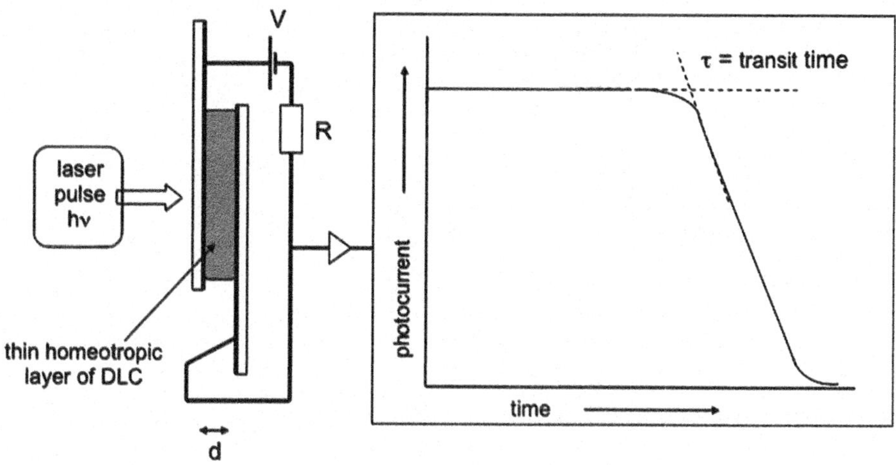

Fig. 3.4 Schematic of a ToF experiment and (*right*) of a Gaussian transit

materials. Gaussian transport produces a characteristic TOF photocurrent with a clear flat region (in the short time domain) that decays rapidly to zero (at long times) (Fig. 3.4). It gives the most direct possible method for measuring the mobility. The TOF transit time τ is used to determine the velocity of the charge carriers (v_D):

$$v_D = \frac{d}{\tau}$$

The mobility of the carriers (μ_{TOF}) in the applied field E can be calculated from the transit time.

$$\mu_{TOF} = \frac{d^2}{V.\tau} = \frac{v_D}{E}$$

Where d is the sample thickness and V is the applied voltage.

In polycrystalline or heterogeneous samples there is deep trapping and the transport is dispersive. No inflection is seen in the photocurrent-time plot, which decays in a monotonic fashion. The advantages of the ToF method are that: (1) The samples are small. (2) In principle, both the hole and electron mobilities can be obtained independently. The disadvantages of the method are that: (1) For columnar phases the sample must be homeotropically aligned and for some of the most interesting discogens, particularly the benzocoronene discogens, this is all but impossible to achieve. (2) Unless the sample is very pure and very well aligned, the transit time can be rather indistinct and difficult to determine accurately. As samples are repeatedly recrystallised and chromatographed or the samples are annealed the sharpness of the inflection in the transit often (but not always) improves. Hence, this is a method that places a premium on the skill of the experimentalist and

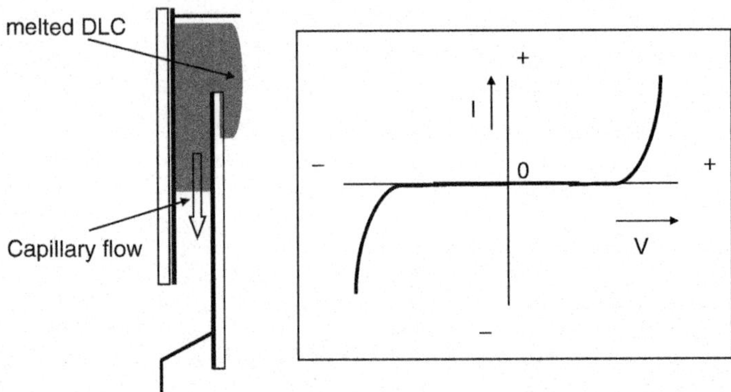

Fig. 3.5 Schematic a the filling procedure of a 'sandwich cell' and of the idealised I/V response of the filled cell

particularly on their ability to purify the material. As a result, values even for the same compound reported in the literature often differ quite widely (Table 3.1). In general errors in reported values are probably underestimated. (3) If the melting point of the sample is too high, filling the cell can be impossible and/or the prolonged heating needed to anneal the sample can lead to chemical decomposition. (4) The use of too high a laser pulse energy leads to space charge effects, distorted transits and apparently shortened transit times (and hence overestimated mobilities). (5) Too high a laser pulse energy can also lead to photodegradation of the sample.

3.2.4 'Sandwich Cell' Studies

The cells used for ToF studies ('sandwich cells') are usually made from a staggered pair of ITO-coated glass slides separated by PTFE spacers. They are filled by capillary action from a melted blob of the discogen placed over one edge as shown in Fig. 3.5. When a low DC field is applied to the filled cell no conduction is (or should be) observed: the material is an insulator. Once, however, the applied field is increased above a critical value V_c charge carriers are injected from the electrode surface and conduction is seen [23]. Hence conduction in sandwich cells is dominated by the behavior of the electrode/liquid crystal interface and this critical field for charge carrier injection. The DC I/V response has the typical sigmoid form shown schematically in Fig. 3.5. Such plots should be symmetric and free from hysteresis. The non-conducting region below the critical potential for charge injection arises because of the mismatch between the work functions of the electrode and the liquid crystal. One of the most useful pieces of information obtained from studies of this kind is an indication of the presence or otherwise ofppm inorganic salt impurities; usually a 'relic' of the organic synthesis. As in the pharmaceutical industry and the liquid crystal display industry, so in the study of conducting discotics, the presence of ppm ionic salt impurities such as sodium chloride, and the difficulty of removing these by recrystallisation and/or chromatography presents a problem. If such salts are present, then the 'sandwich cell' I/V response shows significant conduction below the critical field for charge injection, particularly in the isotropic phase. Furthermore, because slow ion, e.g., Na^+, Cl^- etc., migration results in polarisation of the cell, the I/V response of the cell can become time-dependent leading to hysteresis as the applied field is slowly increased and then slowly decreased again. Above the critical voltage for charge injection one should see a space-charge limited current and a quadratic dependence of the current on the voltage. If the contacts are ohmic and the material is trap-free the Mott-Gurney equation [24] should apply:

$$I = \frac{9}{8}\varepsilon_0\varepsilon_r\mu\frac{(V - V_c)^2}{d^3}$$

In principle, all of the parameters in this equation are measurable so it should be possible to determine the mobility from the voltage dependence of the current. However, the actual behaviour of cells of this type is critically dependent on the nature of the interface. The currents obtained can also depend on the 'history' of the cell. One example of this and of another way that ppm levels of ionic impurities can cause problems is provided by the 'field anneal' phenomenon [25, 26]. If the sample is heated into the isotropic phase and a DC current applied to the cell for a period of time ionic impurities become adsorbed on the electrode surface where they help to facilitate charge injection. Thereafter the I/V characteristics of the Col phase are found to be totally different and, in extreme cases, the measured conductivity can (apparently) be orders of magnitude higher! Because of the extreme sensitivity of 'sandwich cell' measurements to small difference in the electrode surface and, because experimentally the voltage dependence of the current often deviates quite markedly from the expected $(V–V_c)^2$ behaviour, reliable determination of mobilities from the I/V curves is difficult and this approach is rarely used [27].

3.2.5 *OFET Studies*

Unlike the ToF experiment, which requires a homeotropic alignment of the discotic liquid crystal, OFET experiments require a planar alignment (columns running parallel to the surface) with control of the azimuthal angle of the director, so that the preferred conduction pathway is from source to drain electrode (Fig. 3.6). This can be more difficult to achieve than homeotropic alignment, but in recent years a number of methods have been described including, Langmuir-Bloggett dipping [13, 28], zone crystallisation [29], shearing (zone casting) [30], the use of rubbed

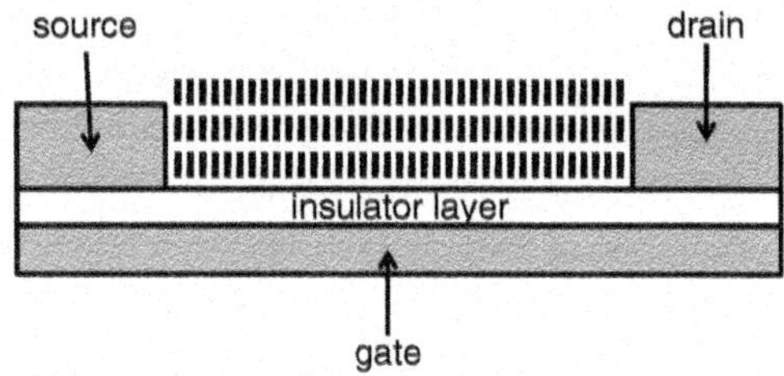

Fig. 3.6 Simple schematic of a DLC-based bottom-gate FET showing the need for the columns of discogen to be aligned parallel to the substrate from source to drain electrode

Teflon surfaces [31], magnetic fields [32] and isotropic phase dewetting of patterned surfaces [33]. The best method for extracting the FET mobility from the data depends on the characteristics of the system.

3.3 *n*-Type and *p*-Type Conductors

Whereas there is a superfluity of organic *p*-type conductors there are far fewer useful n-type conductors and particularly few n-type molecular conductors, which have high electron mobilities. Part of the reason for this is that there are many more molecular impurities, that can act as electron traps compared to the number that can act as hole traps so that the study of *n*-type conductors places a very high premium on material purity. In particular, most organic radical anions react rapidly, often at a diffusion controlled rate, with both oxygen and water. Since the standard redox potentials for the reduction of water is −0.658 V vs SCE and that of oxygen is −0.920 V vs SCE it has been argued that (neglecting overpotentials) stable *n*-type conductors will only be obtained if the molecular reduction potential is less negative than −0.658 V [34]. It is difficult to design organic molecules such as this and it may not be necessary. In some recent experiments it has been shown that, even with simple triphenylene and phthalocyanine derivatives (systems for which the radical anions should certainly be oxygen and water unstable) good electron ToF transits can be observed, even in an ambient atmosphere, provided care is taken to remove all other impurities [35–37]! There are two possible explanations of this surprising observation. Firstly, in the thin cells used for ToF experiments diffusion of oxygen and water is very slow and once it is depleted from a region it is very slow to return. More importantly the 'redox potential' argument was developed for amorphous materials in which the oxygen solubility and diffusion rates are relatively high. In the more ordered liquid crystal phases, as in the crystalline phase, the solubility and diffusion rate of oxygen are expected to be low and so they are effectively excluded from the bulk.

As can be seen from the Tables (below), when electron and hole mobilities have been independently measured for the same compound under the same conditions, it is usually found that the electron mobility is a little (<50%) bigger than the hole mobility.

3.4 Charge Carrier Mobilities

3.4.1 Triphenylene Derivatives

3.4.1.1 Symmetrical Triphenylene Derivatives

In most symmetrical triphenylenes the mobility is essentially temperature- and field-independent even when a particular temperature is quoted in the table the temperature dependence is generally very weak: almost all of these could be classified as having essentially temperature independent mobilities (Table 3.1).

3.4.1.2 Unsymmetrical Triphenylene Derivatives

In contrast to the symmetrical triphenylene derivatives, for many of the unsymmetrical derivatives, the mobility is noticeably field and/or temperature dependent. In such systems it is possible to fit the data to a model in which there is a Gaussian distribution of trap energies characterized by σ (Table 3.2).

3.4.1.3 'Improved' Triphenylene Derivatives

The conduction in triphenylene-based columnar phases can be increased by factors, which increase the coherence length of the molecules within each column. This can be done by: (1) designing hydrogen-bonding disc-to-disc as in **9** and **10**; (2) by adding a gelling agent such as **11** to 'stiffen up the intercolumnar space; (3) or by making CPI compounds in which the alternating triphenylene and hexaphenyl triphenylene (**12** or **13**) lock together a bit like 'Lego' bricks (Table 3.3).

3.4.2 Phthalocyanine Derivatives

Like the symmetrical triphenylenes, the mobility in octa-substituted phthalocyanines is usually temperature- and field-independent. In part because of the high clearing temperatures, they can be difficult to align in a homeotropic manner in a ToF cell, so most data has been obtained by the PR-TRMC method (Table 3.4).

3.4.3 Hexabenzocoronenes Derivatives

Although the hexabenzocoronenes have some of the highest PR-TRMC mobilities, they generally align in a planar manner and have proved frustratingly difficult to align in a homeotropic manner. As a result there is very little ToF data to compare with that of other discogens. When FET mobilities have been measured these have usually been at room temperature and in the crystalline phase (Table 3.5).

Table 3.1 Symmetrical triphenylene derivatives

Compound	Phase	Temperature (°C)	Mobility ($cm^2\ V^{-1}\ s^{-1} \times 10^3$)	Method	Ref.
1a, HAT3	Col_p	130	$\mu^+ = 15$	ToF	[38]
1b, HAT4	Col_p	88	$\Sigma\mu_{1D} = 25$	PR-TRMC	[13]
1b, HAT4	Col_p	'Independent'	$\mu^+ \sim 18$	ToF	[39]
1b, HAT4	Col_p	'Independent'	$\mu^- = 20$	ToF	[37]
1c, HAT5	Col_p	77	$\Sigma\mu_{1D} = 10$	PR-TRMC	[13]
1c, HAT5	Col_h	'Independent'	$\mu^+ = 1.6$	ToF	[40]
1c, HAT5	Col_h	'Independent'	$\mu^+ \sim 1.0$	ToF	[19]
1c, HAT5	Col_h	'Independent'	$\mu^- = 2.0$	ToF	[37]
1d, HAT6	Col_h	87	$\mu^+ = 0.12$	Doping (5% $AlCl_3$)	[5, 8, 9]
1d, HAT6	Col_h	68	$\Sigma\mu_{1D} = 2.0$	PR-TRMC	[13]
1d, HAT6	Col_h	71,68	$\Sigma\mu_{1D} = 2.0$	PR-TRMC	[13]
1d, HAT6	Col_h	71	$\mu^+ = 0.71$	ToF	[41]
1d, HAT6	Col_h	86	$\mu^+ = 0.44$	ToF	[5]
1d, HAT6	Col_h	'Independent'	$\mu^+ = 0.26$	ToF	[42]
1d, HAT6	Col_h	'Independent'	$\mu^+ = 0.44$	ToF	[40]
1d, HAT6	Col_h	'Independent'	$\mu^+ = 0.45$	ToF	[43]
1d, HAT6	Col_h	'Independent'	$\mu^- = 0.40$	ToF	[37]
1d, HAT6	Col_h	'Independent'	$\mu^- = 0.28$	ToF	[42]
1e, HAT7	Col_h	'Independent'	$\mu^+ = 0.15$	ToF	[40]
1f, HAT8	Col_h	'Independent'	$\mu^+ = 0.087$	ToF	[40]
1g, HAT11	Col_h	54	$\Sigma\mu_{1D} = 2.0$	PR-TRMC	[41]
1g, HAT11	Col_h	54	$\mu^+ = 0.10$	ToF	[41]
1h	Col_h	180	$\mu^+ = 0.37$	ToF	[44]
1h	Col_h	180	$\mu^- = 0.37$	ToF	[44]
2a, HHTT	Col_h	70	$\Sigma\mu_{1D} = 5.0$	PR-TRMC	[45]
2a, HHTT	Col_h	93	$\mu^+ = 1.0$	ToF	[45]
2a, HHTT	Col_h	80	$\mu^- \sim 1.0$	ToF	[36]
2a, HHTT	H	40	$\Sigma\mu_{1D} = 80$	PR-TRMC	[45]
2a, HHTT	H	65	$\Sigma\mu_{1D} = 87$	PR-TRMC	[13]
2a, HHTT	H	45	$\mu^+ = 80$	ToF	[20, 36]
2a, HHTT	H	40	$\mu^+ = 100$	ToF	[45]
2b, HOTT	Col_h	60	$\Sigma\mu_{1D} = 15$	PR-TRMC	[13]
2c, HDTT	Col_h	70	$\Sigma\mu_{1D} = 20$	PR-TRMC	[13]

HAT6, $R = C_6H_{13}$

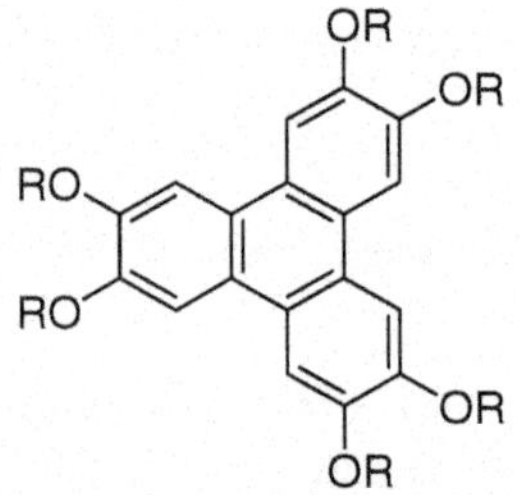

1a, HAT3, $R = C_3H_7$
1b, HAT4, $R = C_4H_9$
1c, HAT5, $R = C_5H_{11}$
1d, HAT6, $R = C_6H_{13}$
1e, HAT7, $R = C_7H_{15}$
1f, HAT8, $R = C_8H_{17}$
1g, HAT11, $R = C_{11}H_{23}$
1h, $R = (CH_2)_3(CF_2)_5CF_3$

(continued)

Table 3.1 (continued)

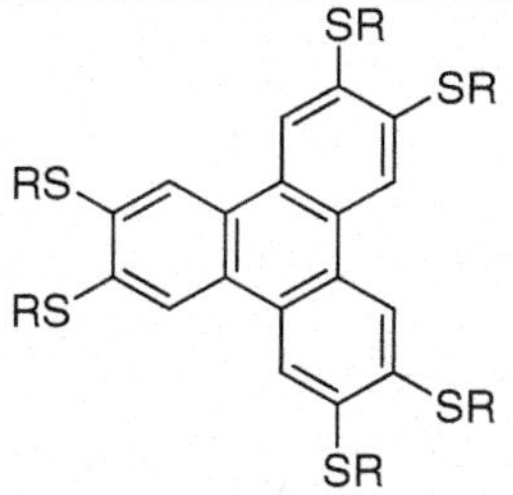

SR, SR, RS, RS, SR, SR

2a, HHTT, R = C_6H_{13}
2b, HOTT, R = C_8H_{17}
2c, HDTT, R = $C_{10}H_{21}$

Col_h – Hexagonal columnar liquid crystal phase
Col_p – Plastic columnar liquid crystal phase
H – Helical phase

There have also been studies of systems, that can be thought of as hexabenzo-coronenes in which the outer rings have been reduced **17** or hexabenzocoronenes with extended polyaryl cores **18** and **19**. In these cases our knowledge of the mobilities is limited to PR-TRMC studies (Table 3.6).

3.4.4 Other Discogens

Discogens based on other nuclei that have been studied include those given in Table 3.7.

3.5 Interpretation of Charge Carrier Mobility Values

3.5.1 Comparison of Mobilities Obtained by the Different Methods

The confusing nature of this field arises partly from the variety of methods that have been used for measuring charge carrier mobilities and the fact that these are not directly comparable. Because of the nature of the frequency dependence of the conductivity in liquid crystals (Fig. 3.2) it is obvious that high-frequency AC (PR-TRMC) measurements are going to give a higher 'mobility' than a low-frequency AC or a DC study. This is fundamentally the result of the fact that, at PR-TRMC frequencies, the charge carriers only move a small distance and the probability of them encountering a 'deeper' trap is correspondingly small Furthermore PR-TRMC yields a sum of positive and negative charge-carrier mobilities which adds a further factor of $\sim$2 to PR-TRMC values. As can be seen from the Tables, for any given compound the PR-TRMC mobility values are generally larger than those

Table 3.2 Unsymmetrical triphenylene derivatives

Compound	Phase	Temperature (°C)	Mobility ($cm^2 V^{-1} s^{-1} \times 10^3$)	σ (meV)	Method	Ref.
3, HAT6-NO_2	Col_h	20–135	$\mu^+ = 0.020–0.20$ (field 1×10^5 Vcm^{-1})	100	ToF	[46]
4, 2FHAT6	Col_h	25	$\mu^+ = 1.6$		ToF	[47]
Quinone **5**	Col_h	20–126	$\mu^+ = 0.011$		ToF	[48]
6a, HAT5-pivaloate	Col_h		Field and temperature dependent	84	ToF	[49]
6b, HAT5-pentenoate	Col_h		Field and temperature dependent	104	ToF	[49]
6c, HAT5-cyclohexanoate	Col_h		Field and temperature dependent	108	ToF	[49]
6d, HAT5-cyanobenzoate	Col_h		Field and temperature dependent	124	ToF	[49]
6e, HAT5-nitrobenzoate	Col_h		Field and temperature dependent	127	ToF	[49]
7a, HAT4-acryate	Col_h	93	$\mu^+ = 1.5$ (field 2×10^4 Vcm^{-1})		ToF	[50]
7b, HAT4-acryate	Col_h	80	$\mu^+ = 0.6$ (field 2×10^4 Vcm^{-1})		ToF	[50]
6f, HAT5-acrylate	Col_h	80	$\mu^+ = 0.3$ (field 2×10^4 Vcm^{-1})		ToF	[50]
6g, HAT5-acrylate	Col_h	60	$\mu^+ = 0.2$ (field 2×10^4 Vcm^{-1})		ToF	[50]
6h, HAT5-acrylate	Col_h	40	$\mu^+ = 0.05$ (field 2×10^4 Vcm^{-1})		ToF	[50]
8a,H4TD10	Col_p	−103 to 127	$\Sigma\mu_{1D} = 15 \pm 5$		PR-TRMC	[13, 14]
8a, H4TD10	Col_p	−145 to 127	$\mu^+ = 0.002–10$ ('zero field')	48	ToF	[14, 51]
8b, H4TD12	Col_p	75	$\Sigma\mu_{1D} = 14$		PR-TRMC	[13]

(continued)

Table 3.2 (continued)

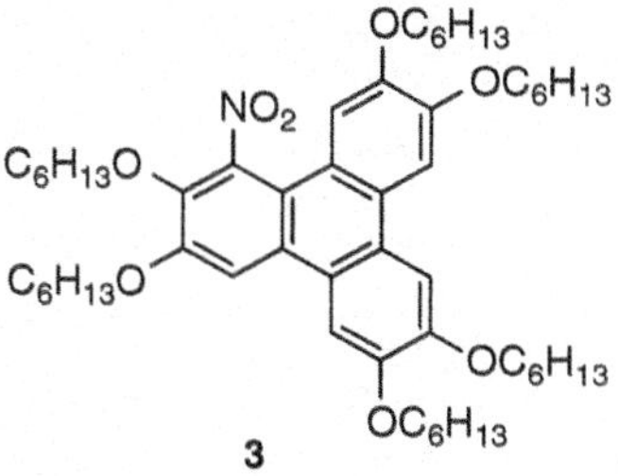

OC_6H_{13}, OC_6H_{13}, NO_2, $C_6H_{13}O$, $C_6H_{13}O$, OC_6H_{13}, OC_6H_{13}

3

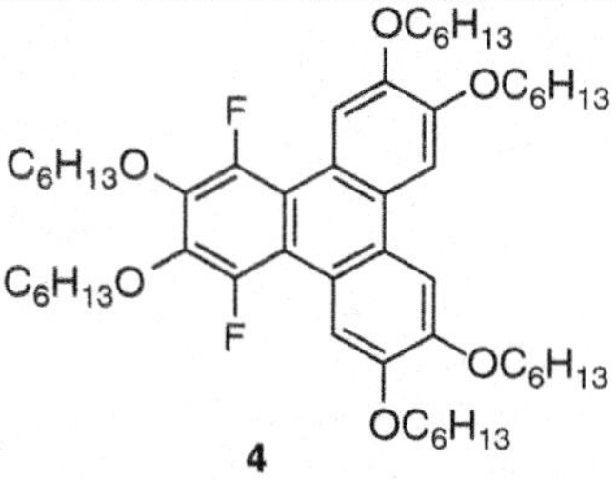

OC_6H_{13}, OC_6H_{13}, F, $C_6H_{13}O$, $C_6H_{13}O$, F, OC_6H_{13}, OC_6H_{13}

4

OC_5H_{11}, X, $C_5H_{11}O$, $C_5H_{11}O$, OC_5H_{11}, OC_5H_{11}

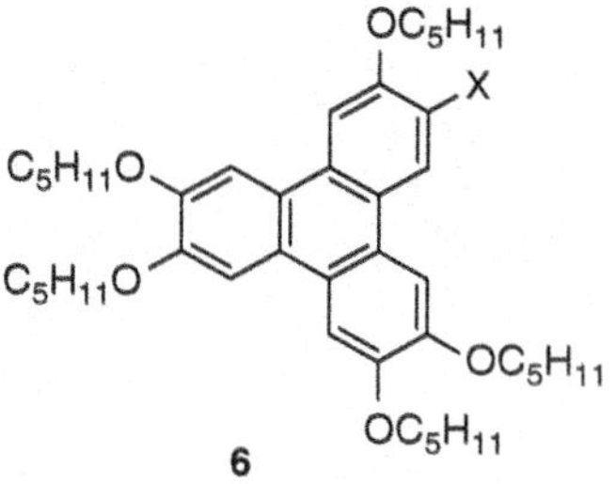

6

6a, X = $OCOC(CH_3)_3$
6b, X = $OCO(CH_2)_2CH{=}CH_2$
6c, X = $OCOC_6H_{11}$
6d, X = $OCOC_6H_4CN$
6e, X = $OCOC_6H_4NO_2$
6f, X = $O(CH_2)_2OCOCH{=}CH_2$
6g, X = $O(CH_2)_4OCOCH{=}CH_2$
6h, X = $O(CH_2)_6OCOCH{=}CH_2$

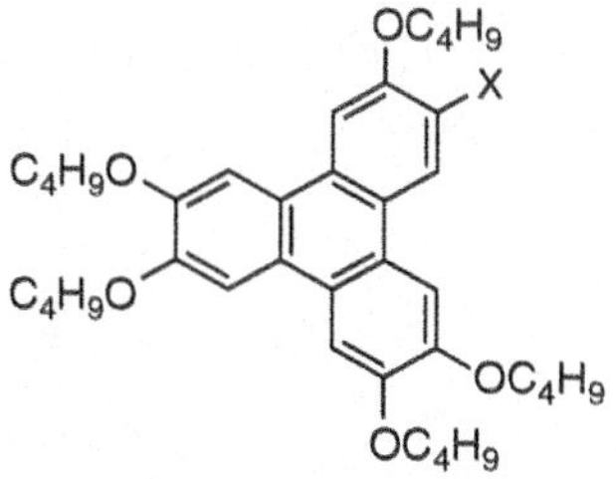

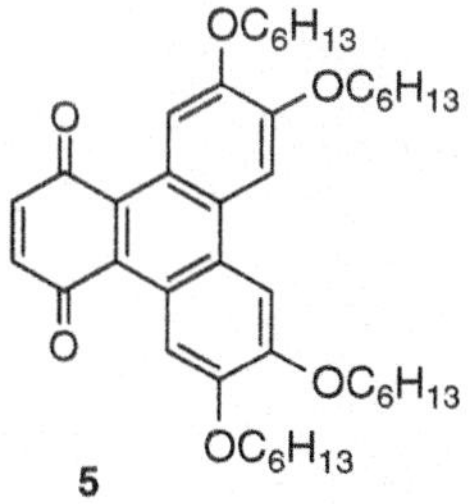

OC_4H_9, X, C_4H_9O, C_4H_9O, OC_4H_9, OC_4H_9

7a, X = $O(CH_2)_2OCOCH{=}CH_2$
7b, X = $O(CH_2)_3OCOCH{=}CH_2$

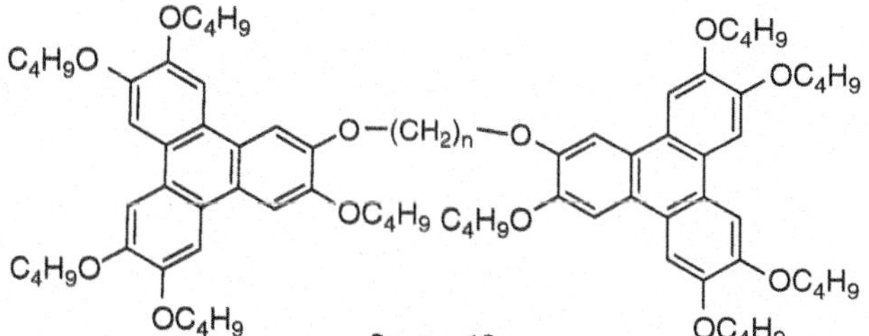

OC_6H_{13}, OC_6H_{13}, O, O, OC_6H_{13}, OC_6H_{13}

5

OC_4H_9, C_4H_9O, $O{-}(CH_2)_n{-}O$, OC_4H_9, C_4H_9O, C_4H_9O, OC_4H_9, OC_4H_9, OC_4H_9, OC_4H_9, OC_4H_9

8a, n = 10
8b, n = 12

Table 3.3 'Improved' triphenylene derivatives

Compound	Phase	Temperature (°C)	Mobility ($cm^2\ V^{-1}\ s^{-1} \times 10^3$)	Method	Ref.
H-bonded amide T-3-hex, **9a**	Col_h	150	$\Sigma\mu_{1D} = 17$	PR-TRMC	[52, 53]
H-bonded amide T-4-Bu, **9b**	Col_p	150	$\Sigma\mu_{1D} = 60$	PR-TRMC	[52, 53]
H-bonded amide T-4-MeBu, **9c**	Col_p	150	$\Sigma\mu_{1D} = 90$	PR-TRMC	[52, 53]
H-bonded amide T-4-hex, **9d**	Col_r	150	$\Sigma\mu_{1D} = 230$	PR-TRMC	[52, 53]
H-bonded amide T-5-hex, **9e**	Col_{obl}	150	$\Sigma\mu_{1D} = 16$	PR-TRMC	[52, 53]
H-bonded amide T-6-hex, **9f**	Col_{obl}	150	$\Sigma\mu_{1D} = 18$	PR-TRMC	[52, 53]
H-bonded azatriphenylene, **10a**	Col_h	−80 to 200	$\Sigma\mu_{1D} = 40–80$	PR-TRMC	[54]
Gel-stabilised HAT5 + 3% **11**	Col_h		$\mu^+ = 3.1$	ToF	[40]
Gel-stabilised HAT6 + 3% **11**	Col_h		$\mu^+ = 1.2$	ToF	[40, 43]
Gel-stabilised HAT7 + 3% **11**	Col_h		$\mu^+ = 2.3$	ToF	[40]
Gel-stabilised HAT8 + 3% **11**	Col_h		$\mu^+ = 2.1$	ToF	[40]
CPI compound 1:1 HAT6:PTP9, **1d:12a**	Col_h	67	$\Sigma\mu_{1D} = 23$	PR-TRMC	[41]
CPI compound 1:1 HAT6:PTP9, **1d:12a**	Col_h	67	$\mu^+ = 23$	ToF	[41, 55]
CPI compound 1:1 HAT6:PDQ9, **1d:13**	Col_h		$\mu^+ = 16$	ToF	[56]
CPI compound 1:1 HAT6:PTPO11, 1d:12b	Col_h		$\mu^+ = 20$	ToF	[56]
CPI compound 1:1 HAT11:PTP9, **1g:12a**	Col_h	61	$\Sigma\mu_{1D} = 17$	PR-TRMC	[41]
CPI compound 1:1 HAT11:PTP9, **1g:12a**	Col_h	61	$\mu^+ = 20$	ToF	[41, 55]
CPI compound 1:1 HAT6:PTPO6, **1d:12c**	Col_h	161	$\Sigma\mu_{1D} = 32$	PR-TRMC	[41]
CPI compound 1:1 2F-HAT6:PTP9, **4:12a**	Col_h		$\mu^+ = 15$	ToF	[56]

(continued)

Table 3.3 (continued)

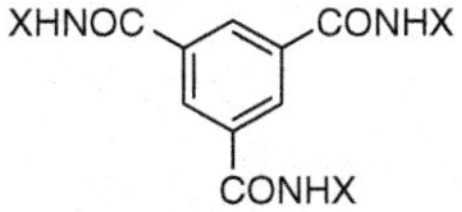

XHNOC CONHX CONHX

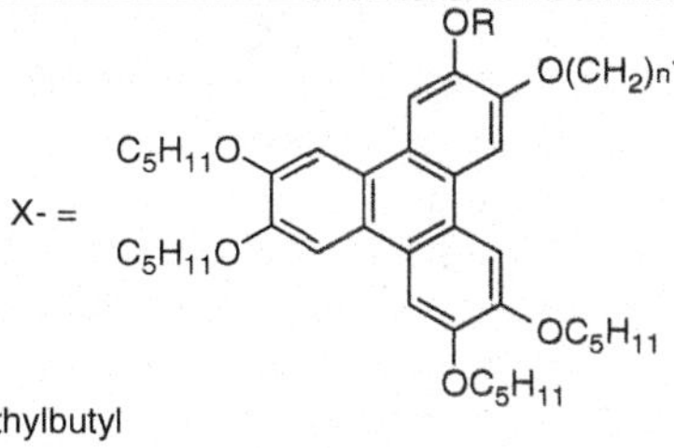

X- =

OR $O(CH_2)_n$- $C_5H_{11}O$ $C_5H_{11}O$ OC_5H_{11} OC_5H_{11}

9a, T-3-Hex, n = 3, R = hexyl
9b, T-4-Bu, n = 4, R = butyl
9c, T-4-MeBu, n = 4, R = (S)2-methylbutyl
9d, T-4-Hex, n = 4, R = hexyl
9e, T-5-Hex, n = 5, R = hexyl
9f, T-6-Hex, n = 6, R = hexyl

$C_6H_5CH_2OCONHCH(Pr^i)CONH(CH_2)_{10}NHCOCH(Pr^i)NHCO_2CH_2C_6H_5$

11

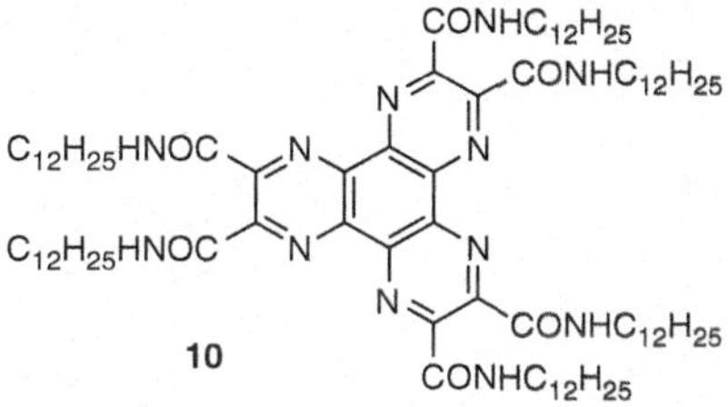

$CONHC_{12}H_{25}$ $CONHC_{12}H_{25}$ $C_{12}H_{25}HNOC$ $C_{12}H_{25}HNOC$ $CONHC_{12}H_{25}$ $CONHC_{12}H_{25}$

10

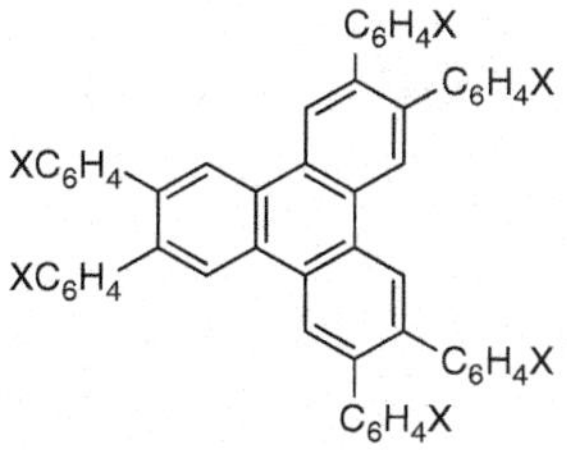

C_6H_4X C_6H_4X XC_6H_4 XC_6H_4 C_6H_4X C_6H_4X

12a, X = C_9H_{19}
12b, X = $OC_{11}H_{23}$
12c, X = OC_6H_{13}

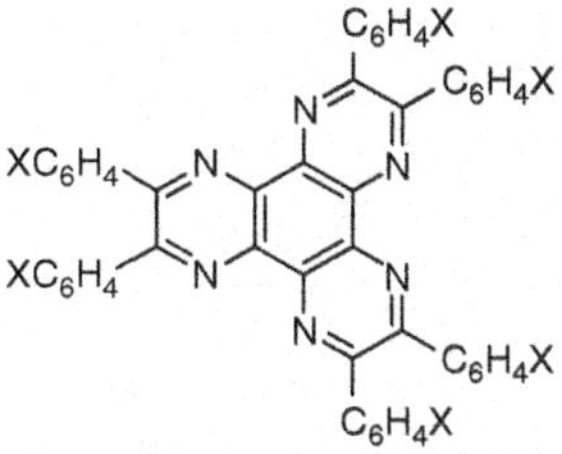

C_6H_4X C_6H_4X XC_6H_4 XC_6H_4 C_6H_4X C_6H_4X

13, X = C_9H_{19}

Col_r – Rectagonal columnar liquid crystal phase
Col_{ob} – Oblique columnar liquid crystal phase

obtained by other methods. However, the difference between PR-TRMC and ToF mobilities only becomes really large for the most disordered materials. This is also easy to understand. For a relatively ordered material there are few ‘deeper’ traps so the length scale associated with the measurement (and probably the method used) become relatively unimportant. Although the amount of data is limited, it seems that the mobilities obtained from doping studies and from FET measurements are a little smaller than those obtained by ToF.

Table 3.4 Phthalocyanine derivatives

Substituents	Phase	Temperature (°C)	Mobility ($cm^2\,V^{-1}\,s^{-1} \times 10^3$)	Method	Ref.
$\alpha-C_8H_{17}$	Col_r	85	$\mu^+ = 200$	ToF	[35]
$\alpha-C_8H_{17}$	Col_r	85	$\mu^- = 300$	ToF	[35]
$\alpha-(CH_2)_4CHMe_2$	Col_h	170	$\mu^+ = 140$	ToF	[48, 57]
$\beta-C_{10}H_{21}$	Col_h	~200	$\Sigma\mu_{1D} = 120$	PR-TRMC	[13, 58, 59]
$\beta-OC_6H_{13}$	Col	~200	$\Sigma\mu_{1D} = 60$	PR-TRMC	[13, 58]
$\beta-OC_8H_{17}$	Col	87	$\Sigma\mu_{1D} = 80$	PR-TRMC	[13, 58]
$\beta-OC_9H_{19}$	Col	~200	$\Sigma\mu_{1D} = 50$	PR-TRMC	[13, 58]
$\beta-OC_{11}H_{23}$	Col	76	$\Sigma\mu_{1D} = 70$	PR-TRMC	[13, 58]
$\beta-OC_{12}H_{25}$	Col_h	~200	$\Sigma\mu_{1D} = 60$	PR-TRMC	[13, 58]
$\beta-OC_{18}H_{37}$	Col	~200	$\Sigma\mu_{1D} = 50$	PR-TRMC	[13, 58]
$\beta-O[(CH_2)_2CHMeCH_2]_2H$	Col	~200	$\Sigma\mu_{1D} = 70$	PR-TRMC	[13, 58]
$\beta-O[(CH_2)_2C^*HMeCH_2]_2H$	Col	~200	$\Sigma\mu_{1D} = 60$	PR-TRMC	[13, 58]
$\beta-O[(CH_2)_2CHMeCH_2]_3H$	Col	~200	$\Sigma\mu_{1D} = 40$	PR-TRMC	[13, 58]
$\beta-SC_8H_{17}$	Col	~200	$\Sigma\mu_{1D} = 240$	PR-TRMC	[13, 60]
$\beta-SC_8H_{17}$, Cu complex	Col_h	~200	$\Sigma\mu_{1D} = 390$	PR-TRMC	[61]
$\beta-SC_{12}H_{25}$	Col_h	~200	$\Sigma\mu_{1D} = 280$	PR-TRMC	[13, 59, 60]
$\beta-SC_{12}H_{25}$, 0.5 Lu complex	Col_h	38	$\Sigma\mu_{1D} = 350$	PR-TRMC	[61]
$\beta-SC_{16}H_{33}$	Col	~200	$\Sigma\mu_{1D} = 150$	PR-TRMC	[13, 60]
$\beta-SC_{18}H_{37}$, 0.5 Lu complex	Col_h	100	$\Sigma\mu_{1D} = 130$	PR-TRMC	[61]
$\beta-C_6H_4OC_{12}H_{25}$	Col_h	~200	$\Sigma\mu_{1D} = 190$	PR-TRMC	[13, 59, 60]
$\beta-C_6H_4OC_{18}H_{37}$	Col	~200	$\Sigma\mu_{1D} = 240$	PR-TRMC	[13, 60]
$\beta-C_6H_3(OC_{12}H_{25})_2$	Col	~200	$\Sigma\mu_{1D} = 240$	PR-TRMC	[13, 60]
$\beta-COC_{12}H_{25}$	Col	~200	$\Sigma\mu_{1D} = 80$	PR-TRMC	[13, 58]
Phthalocyanine **15**	Col_r	110	$\mu^+ = 2.2$	ToF	[62]
Phthalocyanine **15**	Col_r	110	$\mu^- = 2.4$	ToF	[62]

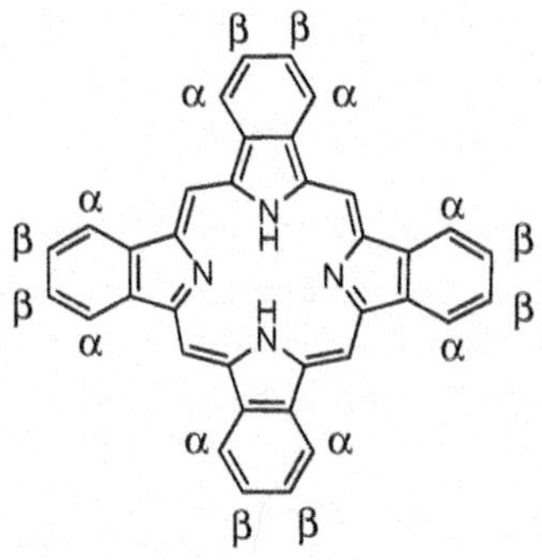

14, position of the eight α and eight β sites

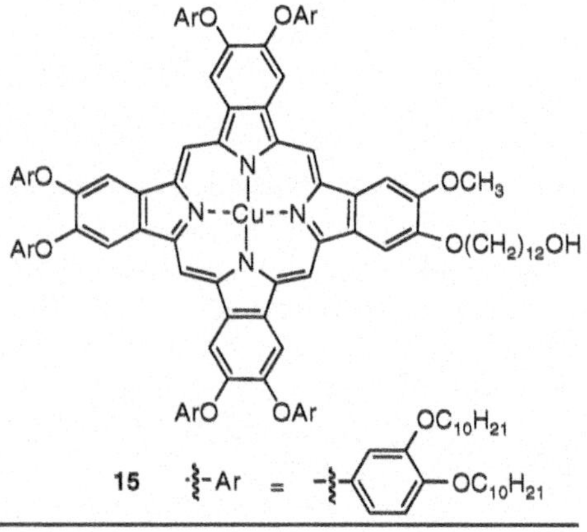

15 -Ar =

Table 3.5 Hexabenzocoronenes derivatives

Compound	Phase	Temperature (°C)	Mobility ($cm^2\ V^{-1}\ s^{-1} \times 10^3$)	Method	Ref.
16a, HBC-C_6	Cr	25	$\mu^+ = 0.95$	FET	[63]
16b, HBC-C_{10}	Cr	25	$\Sigma\mu_{1D} = 550$	PR-TRMC	[64, 65]
16b, HBC-C_{10}	Cr	115	$\Sigma\mu_{1D} = 550$	PR-TRMC	[66]
16b, HBC-C_{10}	Col_h	133	$\Sigma\mu_{1D} = 260$	PR-TRMC	[66]
16b, HBC-C_{10}	Cr	25	$\mu^+ = 50$	FET	[29]
16c, HBC-C_{12}	Cr	25	$\Sigma\mu_{1D} = 760$	PR-TRMC	[64, 65]
16c, HBC-C_{12}	Cr	98	$\Sigma\mu_{1D} = 960$	PR-TRMC	[66]
16c, HBC-C_{12}	Col_h	110	$\Sigma\mu_{1D} = 380$	PR-TRMC	[66]
16c, HBC-C_{12}	Cr	25	$\mu^+ = 0.05$	FET	[29]
16d, HBC-C_{14}	Cr	25	$\Sigma\mu_{1D} = 1{,}000$	PR-TRMC	[64, 65]
16d, HBC-C_{14}	Cr	98	$\Sigma\mu_{1D} = 1{,}130$	PR-TRMC	[66]
16d, HBC-C_{14}	Col_h	116	$\Sigma\mu_{1D} = 310$	PR-TRMC	[66]
16e, HBC-$C_{6,2}$	Cr	NR (0–97)	$\Sigma\mu_{1D} = 250$	PR-TRMC	[67]
16f, HBC-$C_{6,2}$	Col_h	NR (97–420)	$\Sigma\mu_{1D} = 290$	PR-TRMC	[67]
16f, HBC-$C_{8,2}$	Cr	25	$\Sigma\mu_{1D} = 430$	PR-TRMC	[64]
16f, HBC-$C_{8,2}$	Cr	71	$\Sigma\mu_{1D} = 620$	PR-TRMC	[64]
16f, HBC-$C_{8,2}$	Col_h	91	$\Sigma\mu_{1D} = 300$	PR-TRMC	[64]
16f, HBC-$C_{8,2}$	Cr	25	$\mu^+ = 1.00$	FET	[68]
16g, HBC-$C_{8,2*}$	Cr	25	$\Sigma\mu_{1D} = 460$	PR-TRMC	[64]
16g, HBC-$C_{8,2*}$	Cr	89	$\Sigma\mu_{1D} = 540$	PR-TRMC	[64]
16g, HBC-$C_{8,2*}$	Col_h	109	$\Sigma\mu_{1D} = 260$	PR-TRMC	[64]
16h, HBC-$C_{10,6}$	Col_p	0	$\Sigma\mu_{1D} = 730$	PR-TRMC	[67]
16h, HBC-$C_{10,6}$	Col_h	70	$\Sigma\mu_{1D} = 80$	PR-TRMC	[67]
16h, HBC-$C_{10,6}$	Cr	25	$\mu^+ = 1.4$	ToF	[69]
16h, HBC-$C_{10,6}$	Col_p	5	$\mu^+ = 0.4$	ToF	[69]
16h, HBC-$C_{10,6}$	Col_h	25	$\mu^+ = 0.7$	ToF	[69]
16h, HBC-$C_{10,6}$	Col_h	90	$\mu^+ = 1.01$	ToF	[69]
16i, HBC-$C_{14,10}$	Col_p	NR (0–46)	$\Sigma\mu_{1D} = 470$	PR-TRMC	[67]
16j, HBC-PhC_{12}	D_1	−78	$\Sigma\mu_{1D} = 170$	PR-TRMC	[66]
16j, HBC-PhC_{12}	D_2	22	$\Sigma\mu_{1D} = 220$	PR-TRMC	[66]
16j, HBC-PhC_{12}	D_3	192	$\Sigma\mu_{1D} = 460$	PR-TRMC	[66]

16a, HBC-C_6, R=C_6H_{13}
16b, HBC-C_{10}, R=$C_{10}H_{21}$
16c, HBC-C_{12}, R=$C_{12}H_{25}$
16d, HBC-C_{14}, R=$C_{14}H_{29}$
16e, HBC-$C_{6,2}$, R=$CH_2CH(C_2H_5)(C_4H_9)$
16f, HCB-$C_{8,2}$, R=$(CH_2)_2CH(CH_0)(CH_2)_3CH(CH_3)_2$
16g, HCB-$C_{8,2*}$, R=$(CH_2)_2CH(CH_3)(CH_2)_3CH(CH_3)_2$
16h, HCB-$C_{10,6}$, R=$CH_2CH(C_6H_{13})(C_8H_{17})$
16i, HCB-$C_{14,10}$,R=$CH_2CH(C_{10}H_{21})(C_{12}H_{25})$
16j, HCB-PhC_{12}, R=$C_6H_4C_{12}H_{25}$

* Chiral

Cr – Crystalline phase
D_{1-3} – Discotic mesophase not assigned

Table 3.6 Hexabenzocoronenes with reduced outer rings or extended polyaryl cores

Compound	Phase	Temperature (°C)	Mobility ($cm^2\ V^{-1}\ s^{-1} \times 10^3$)	Method	Ref.
17, HBC-H2-C_{12}	Cr	0	$\Sigma\mu_{1D} = 600$	PR-TRMC	[70]
17, HBC-H2-C_{12}	Col	50	$\Sigma\mu_{1D} = 750$	PR-TRMC	[70]
17, HBC-H2-C_{12}	Col	100	$\Sigma\mu_{1D} = 900$	PR-TRMC	[70]
18a, HBC60-C_{12}	Cr	25	$\Sigma\mu_{1D} = 520$	PR-TRMC	[71]
18a, HBC60-C_{12}	Cr	92	$\Sigma\mu_{1D} = 800$	PR-TRMC	[71]
18a, HBC60-C_{12}	Col	112	$\Sigma\mu_{1D} = 260$	PR-TRMC	[71]
18b, HBC60-$_{C8,2}$	Cr	25	$\Sigma\mu_{1D} = 450$	PR-TRMC	[71]
18b, HBC60-$_{C8,2}$	Cr	98	$\Sigma\mu_{1D} = 900$	PR-TRMC	[71]
18b, HBC60-$_{C8,2}$	Col	118	$\Sigma\mu_{1D} = 290$	PR-TRMC	[71]
19a, HBC96-C_7	Col	25	$\Sigma\mu_{1D} = 200$	PR-TRMC	[72]
19a, HBC96-C_7	Col	100	$\Sigma\mu_{1D} = 230$	PR-TRMC	[72]
19b, HBC96-C_{12}	Col	25	$\Sigma\mu_{1D} = 160$	PR-TRMC	[73]
19b, HBC96-C_{12}	Col	100	$\Sigma\mu_{1D} = 200$	PR-TRMC	[73]
19c, HBC96-$C_{16,4}$	Col	25	$\Sigma\mu_{1D} = 60$	PR-TRMC	[73]
19c, HBC96-$C_{16,4}$	Col	100	$\Sigma\mu_{1D} = 80$	PR-TRMC	[73]

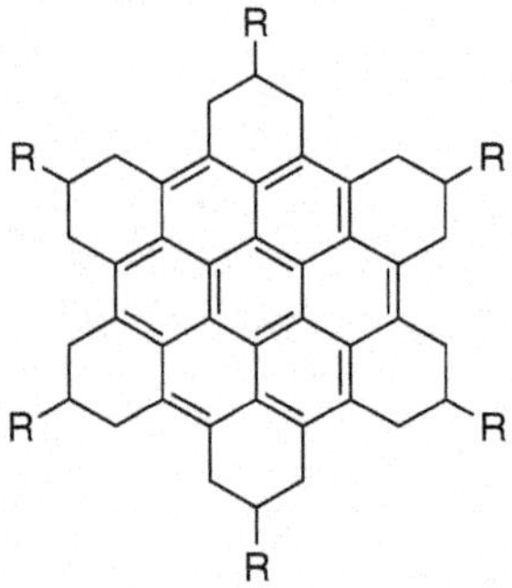

17,HBC-H_2-C_{12},R=$C_{12}H_{25}$

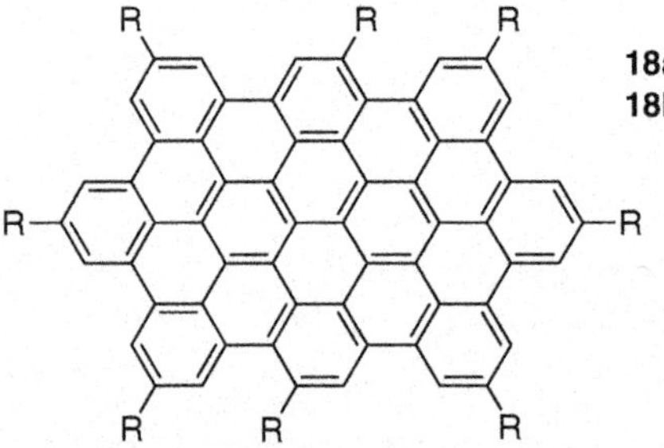

18a, HBC_{60}-C_{12}, R=$C_{12}H_{25}$
18b, HBC_{60}-$C_{8,2}$, R=$(CH_2)_2CH(CH_3)(CH_2)_3CH(CH_3)_2$

(continued)

3.5.2 *The Effect of Positional Order on the Mobility*

Generally, in molecular materials, there is a correlation between charge carrier mobility and positional order. The highest mobilities are associated with single

Table 3.6 (continued)

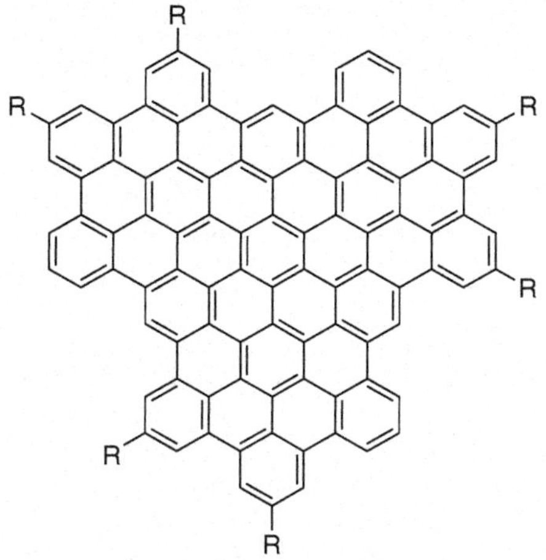

19a, HBC_{96}-C_7, R=C_7H_{15}
19b, HBC_{96}-C_{12}, R=$C_{12}H_{25}$
19c, HBC_{96}-$C_{16,4}$, =$(CH_2)_2CH(CH_3)(CH_2)_3CH(CH_3)(CH_2)_3CH(CH_3)(CH_2)_3CH(CH_3)_2$

crystal (grain boundary trap free) crystalline materials and the lowest with highly disordered amorphous materials. Liquid crystals occupy the 'middle ground'. In the specific case of discotic liquid crystals, positional order is also the most important single factor that determines the mobility. This is nicely illustrated by the HAT discotics (Fig. 3.7). In this series there is a marked dependence of the mobility on the chain-length with three orders of magnitude difference between HAT3 and HAT11. Along the series, the separation between the columns increases monotonically with side-chain length (16.4 Å for HAT3 to 27.0 Å for HAT11) [5, 39, 83, 84] and there is also a small increase in the disc-disc separation (3.52 Å for HAT4 to 3.62 Å for HAT11) [1, 5, 39, 83, 84]. However, this small differences cannot account for the differences in hole mobility. Rather, this is accounted for by the degree of disorder. HAT3 and HAT4 form Col_p phases in which the correlation length disc to disc along the column is very long and these clearly show the highest mobilities. HAT5-HAT11 form Col_h phases in which this correlation length decreases gradually from >12 molecules for HAT6 to only 2 or 3 molecules for HAT11 [84, 85] and this accounts for the decrease in mobility within the series. The same effect is seen in those systems that exhibit more than one columnar phase. The lower temperature phases (Col_r, Col_{ob}, H etc.) are always more ordered than the higher temperature Col_h phase and, for a given system, always show higher mobility. Perhaps the nicest examples of the effect of positional order on mobility are provided by those systems which have been deliberately engineered to increase the molecular coherence length along the column (Sect. 3.4.1.3). This can be done by adding a gelating agent to stiffen up the intercolumnar space [40, 43], by introducing hydrogen bonds between

Table 3.7 Other discogens

Compound	Phase	Temperature	Mobility × 10^3	Method	Ref.
Trisoxadiazole **20**	Col		$\mu^- = 0.1–1.0$	ToF	[74]
Tricyloquinazoline **21a** + 6 mol% K	Col_h	87	$\mu^- = 0.025$	Doping	[8]
Tricyloquinazoline **21b** + 10 mol% K	Col_h	150	$\mu^- < 0.45$	ToF	[75]
Pyrene **22**	Col_h	25–84	$\mu^+ = 1.1–3.2$	ToF	[76]
	Col_h	60	$\mu^- = 3.2$	ToF	[76]
Carbazole **23**	Col_h	120	$\mu^+ = 1.2$	ToF	[76]
	Col_h	120	$\mu^- = 0.0073$	ToF	[76]
Perylene **24a**	Col	50	$\Sigma\mu_{1D} \sim 20$	PR-TRMC	[77]
Perylene **24b**	Col	100	$\Sigma\mu_{1D} \sim 40$	PR-TRMC	[77]
Perylene **24c**	Col	50	$\Sigma\mu_{1D} \sim 70$	PR-TRMC	[77]
Perylene **24d**	Col	100	$\Sigma\mu_{1D} \sim 40$	PR-TRMC	[77]
Perylene **25**	Col[a]	177–190	$\Sigma\mu_{1D} = 110–80$	PR-TRMC	[78]
Dye **26**	Col[b]	25	$\mu_{sat}{}^+ = 0.0–1.0$	FET	[79]
Dye **26**	Col[b]	25	$\mu_{sat}{}^- = 0.2–1.5$	FET	[79]
Porphryn **27**	Col[c]	90	$\mu^+ = 0.24$	ToF	[80]
Porphryn **28a**	Col	87	$\mu^+ = 4.5$	ToF	[81]
Porphryn **28a**	Col	87	$\mu^- = 6.4$	ToF	[81]
Porphryn **28a**	Col	106	$\Sigma\mu_{1D} = 60$	PR-TRMC	[13, 82]
Porphryn **28b**	Col	106	$\Sigma\mu_{1D} = 56$	PR-TRMC	[13, 82]
Porphryn **28c**	Col	106	$\Sigma\mu_{1D} = 65$	PR-TRMC	[13, 82]
Porphryn **28d**	Col	106	$\Sigma\mu_{1D} = 53$	PR-TRMC	[13]
Porphryn **28e**	Col	106	$\Sigma\mu_{1D} = 61$	PR-TRMC	[13]
Porphryn **29**	Col_r	16	$\Sigma\mu_{1D} = 270$[d]	FP-TRMC	[18]

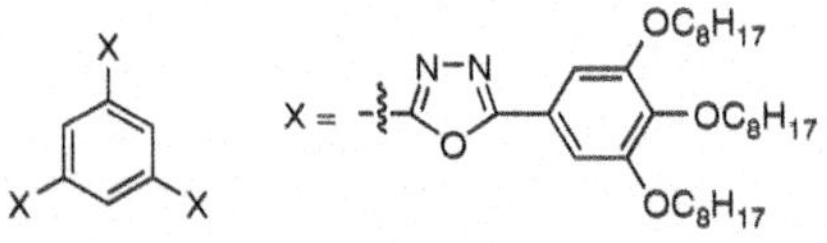

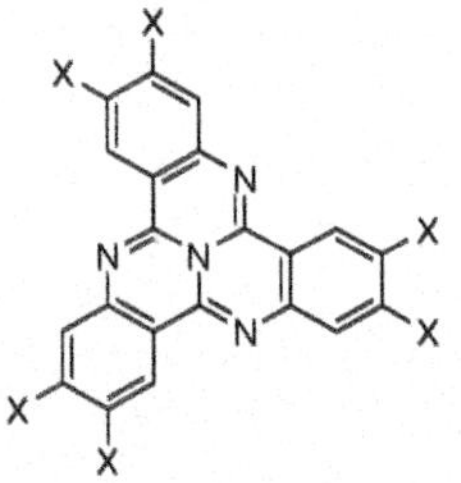

21a X = SC_6H_{13}

21b X = $O(CH_2CH_2O)_2CH_3$

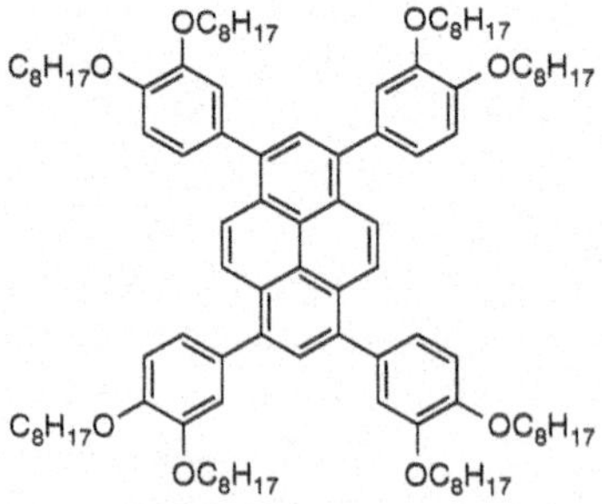

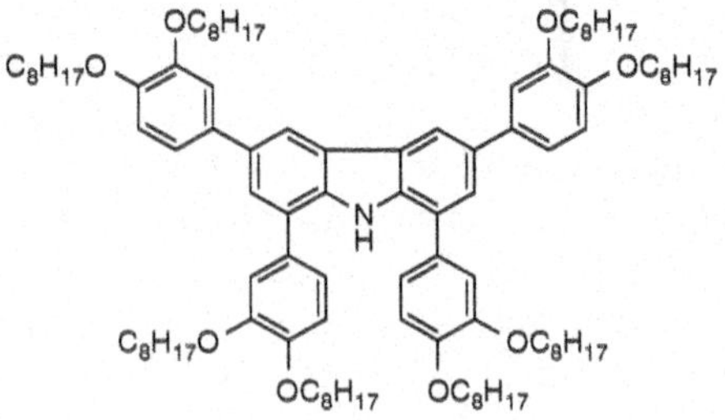

(continued)

Table 3.7 (continued)

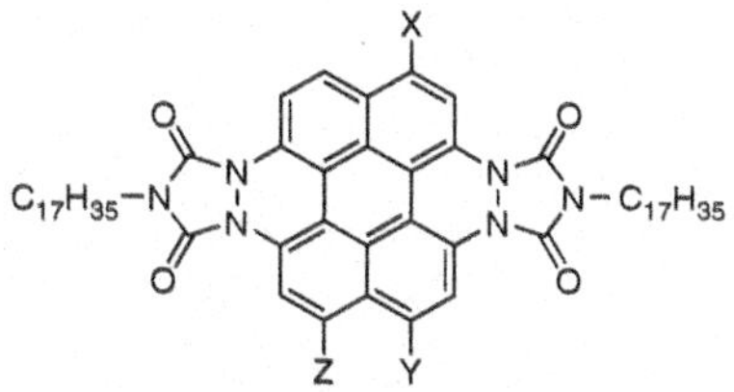

24a X = Y = Z = H

24b X = Y = $C_{18}H_{37}$, Z = H

24c X = Y = C_6H_{13}, Z = H

24d X = Z = C_6H_{13}, Y = H

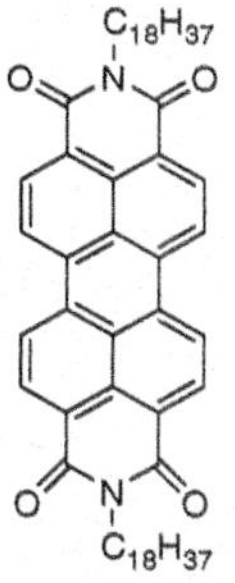

25

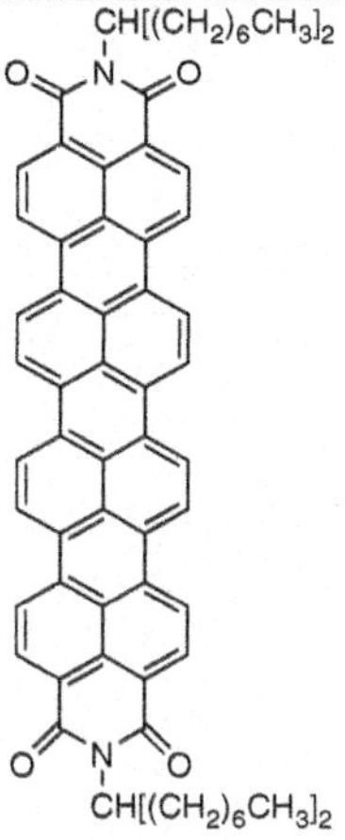

26

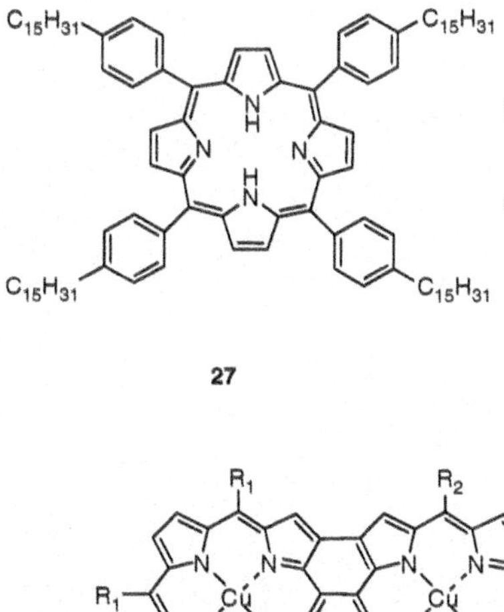

27

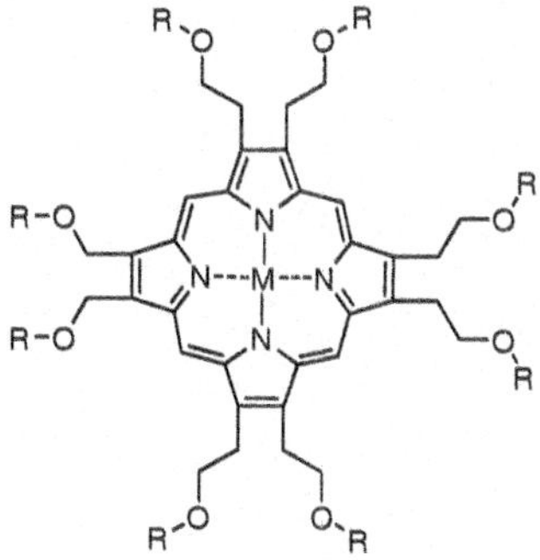

28a M = Zn, R = C_9H_{19}

28b M = Pd, R = $C_{10}H_{21}$

28c M = Cu, R = $C_{10}H_{21}$

28d M = Co, R = $C_{10}H_{21}$

28e M = Zn, R = $C_{10}H_{21}$

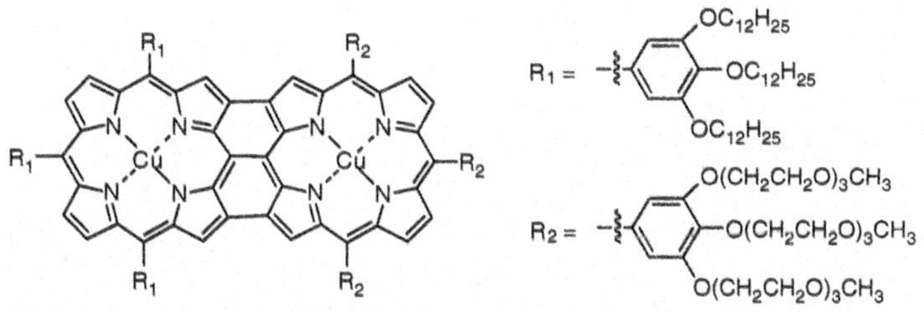

29

R_1 = 3,4,5-tris($OC_{12}H_{25}$)phenyl

R_2 = 3,4,5-tris($O(CH_2CH_2O)_3CH_3$)phenyl

[a]'A highly ordered liquid crystalline phase with characteristics of both columnar and smectic'
[b]'Discotic amorphous'
[c]'Columnar lamellar'
[d]The authors argue that this is measurement of μ^-

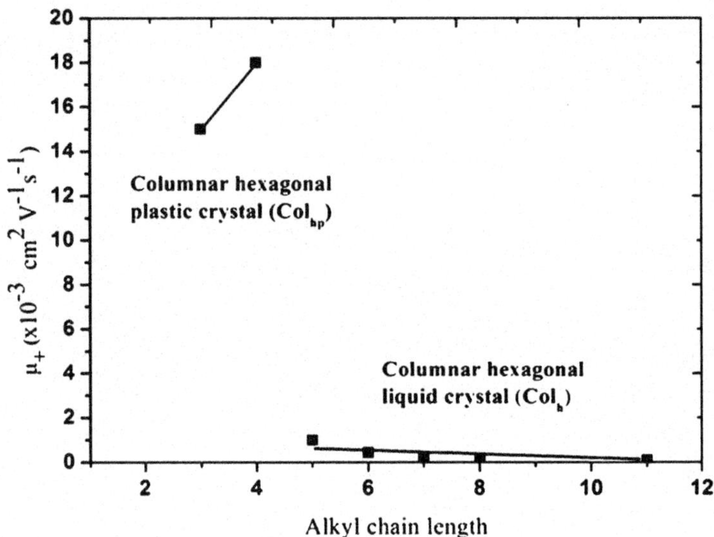

Fig. 3.7 Effect of chain-length on the ToF hole mobilities of HAT liquid crystals

the aryl cores [52–54], or, as in the CPI systems, by causing the aryl cores to lock together rather like a stack of 'Lego' bricks. In the case of the CPI discotics [86] the columns consist of alternating HAT and PTP or PDQ molecules [87] so that a charge carrier moving along the column must experience alternating 'well-depths'. However, the extra order inherent in the CPI structure means that these are some of the highest mobilities found [41].

3.5.3 *The Effect of Field Disorder on the Mobility*

According to most treatments of hopping between shallow traps, the mobility is expected to decrease with increasing dispersion of the energies of the traps [51]. This not only results in a decrease in mobility with increasing positional disorder but also (since the charge carriers are charged!) with increasing disorder in the field and hence disorder in the surrounding molecular dipoles. Hence, in a given series of amorphous materials mobility goes down with increasing molecular dipole [88]. In the discotics this is nicely illustrated by the comparison between the behavior of HAT4 and various ester derivatives [49] and the behavior of HAT6 and its nitro derivative HAT6-NO_2 [46]. It is probably also the reason for the relatively low mobilities of most of the unsymmetrically substituted triphenylenes (Sect. 3.4.1.2). However, more systematic work is needed on the effect of molecular dipole on mobility in discotic systems.

3.5.4 *The Effect of Orbital Overlap on the Mobility*

Regardless of what mechanism applies to charge migration in a particular discotic liquid crystal system (molecule to molecule hopping, narrow banded conduction or some intermediate mechanism), it is clear that the degree of overlap between the π-systems of neighboring rings is a key factor. This is a large part of the reason why conduction perpendicular to the columns (no π-π overlap) is so much less than conduction along the columns (significant π-π overlap). It is also one of the factors why positional order is so important, because the effective π-π overlap also increases as the system becomes more ordered. There are a number of other ways that rational molecular design can be used to manipulate the overlap. One approach that has proved effective for all sorts of organic semiconductors is to incorporate second and third row elements into the molecular structure: elements associated with more diffuse orbitals giving a greater orbital overlap. The most common approach is to replace oxygen with sulphur or selenium. This is probably the main reason why mobilities are higher in the Col phases of HHTT **2a** than they are in the Col_h phases of the HAT discotics **1**. The disadvantage of this approach is that such changes usually decrease the stability of the molecules and in particular substitution of oxygen for sulphur increases the tendency of the materials to undergo aerial oxidative degradation. It is also possible to manipulate the overlap integral by changing the nature of the side chain. Hence, in comparing the tri-indole compounds **30a** and **30b**: on replacing the nonylphenyl with the octylethynyl side chain the stacking distance the Col_h phase decreases from 4.4 to 3.9 Å and the mobility increases by four orders of magnitude [27]. One of the most interesting aspects of this work is that it suggests another general strategy by which the mobility in discotics can be 'improved' by the use of alkynyl side-chains.

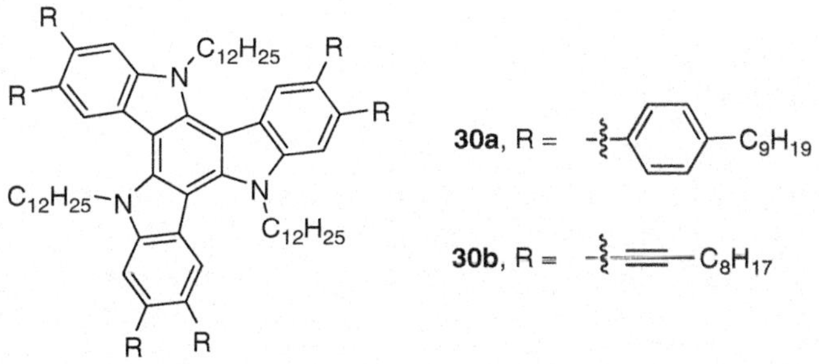

30a, R = $-C_6H_4-C_9H_{19}$

30b, R = $-C{\equiv}C-C_8H_{17}$

3.5.5 *The Importance of Lattice Reorganization Energy in Determining the Mobility*

Treatment of hopping conduction in terms of Marcus theory stresses the importance of the reorganization energy term as a major factor in determining the charge

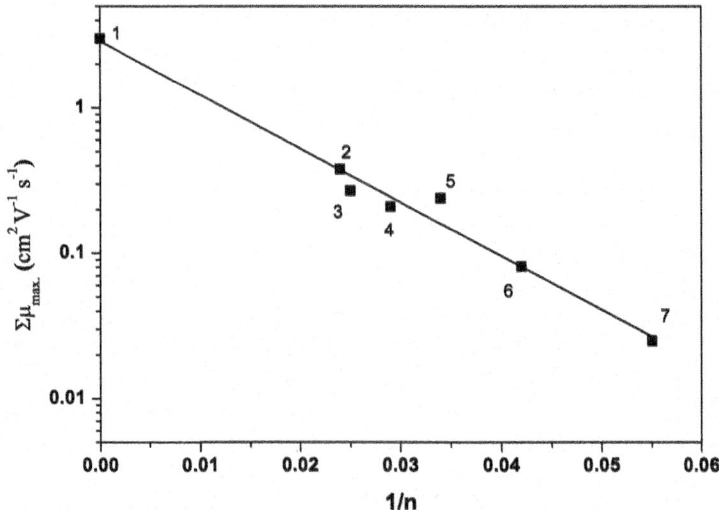

Fig. 3.8 Mobility of *1* graphite perpendicular to the plane and PR-TRMC maximum mobilities for the Col_h phases of discogens based on *2*, hexabenzocoronene, *3*, phthalocyanine, *4* axocarboxyldi-imidoperylene, *5* coronenemonoimide, *6* porphryn, and *7* triphenylene aromatic cores showing the effect of the size of the aromatic core

carrier mobility [89]. This is undoubtedly correct, but unfortunately this energy term is generally unknown and the methods used for calculating reorganisation energies – particularly the intermolecular components- invariably involve many approximations.

3.5.6 *The Possible Effect of the Size of the Aromatic Core on the Mobility*

In an interesting paper in 2001, van de Craats and Warman compared the PR-TRMC mobilities of the Col_h phases of many discogens [90]. They concluded that the maximum values (and perhaps the theoretical maximum attainable values) increased with the size of (the number of atoms, n, in) the aromatic core. Indeed, as shown in Fig. 3.8 there is a surprisingly good fit of the data to an equation of the form.

$$\sum \mu_{\text{max}} \approx 3 \exp\left[b/n\right]$$

Where $\Sigma\mu_{max}$ is the maximum observed for the sum of the intracolumnar hole and electron mobilities. It suggests that the maximum mobility achievable for any stack of pi systems 3.5 apart should be that found for conduction perpendicular to the plane in graphite (ca. 3 $cm^2\ V^{-1}\ s^{-1}$) and in their treatment would be the value for a well-ordered discogen with an 'infinite' core. The argument has since been criticized since the PR-TRMC mobilities of the 'benzocoronene' derivatives with super-large cores (Sect. 3.4.3) were smaller than the equation predicts [72]. However the amount of data on these discogens with very large cores is still very limited.

3.6 Problems and Challenges

Probably our main problems in this field arise from our very incomplete understanding of the columnar phase. Although the columnar phases of discotic liquid crystals have been known for 30 years they remain much less thoroughly studied and much less well understood than the smectic phases of calamitic liquid crystals. It is clear that, at least in principle, there can be difference between columnar phases not only in terms of the two-dimensional lattice, but also in terms of the tilt and order of the discs (positional correlation length) within the columns but this kind of information often remains hidden. Indeed, in some cases, there remain doubt even as to whether particular phases are genuinely liquid crystalline or if they are plastic crystals (a distinction which depends on our ability to see the often weak reflections in the low angle X-Ray region relating to 3D order) and if they are liquid crystal whether or not they are hexagonal (which depends on seeing the 110 reflection which can also be very weak). Sometimes only one reflection is seen in the wide angle region and it is then quite often assumed that the phase is hexagonal. Furthermore, although more is being learnt all the time from NMR and dielectric spectroscopy studies, we lack reliable data for the internal dynamics of most of these systems. These internal dynamical factors are critical for developing a convincing theoretical treatment of the conductivity across the whole spectrum of discotic liquid crystals.

There are still some interesting issues surrounding our understanding of the mechanism of conduction in columnar systems. There is no doubt this is best described overall by a hopping model but it is not always certain that this is molecule-to-molecule hopping. Within the sequence of aromatic molecular material types from low temperature crystalline to high temperature crystalline to plastic crystal to ordered columnar liquid crystal to disordered columnar liquid crystal to amorphous solid these must be a continuous evolution in the conduction mechanism from banded conduction to molecule-molecule hopping. Starting from the low temperature crystalline polynuclear aromatic end of the spectrum, as the temperature is raised or more disorder is introduced, the bands narrow, electron phonon coupling increases and band transport begins to break down. Presumably, this leads to domain-to-domain hopping or polaron to polaron hopping before the molecule-to-molecule regime is reached. Starting at the extreme disordered amorphous solid end of the spectrum conduction undoubtedly involves simple molecule-to molecule hopping but as we progress to ever more ordered aryl stacks this becomes less obviously the case. For the columnar phase of a high mobility discotic $\mu \sim 3.5 \times 10^{-1}$ cm^{2}V^{-1} s^{-1} with a separation between the discs of 3.5 Å and a field of 10^5 V cm^{-1} the molecule-to molecule hopping rate would be 10^{12} s^{-1} if it were unidirectional. However, since it is not unidirectional but a biased random walk the fundamental molecule-to-molecule hopping rate must be (considerably?) higher than this. Maybe it is in a domain where a molecule-to-molecule hopping pictures would not permit sufficient time for molecular and lattice relaxation to occur. We may already be in a region where there is an element of domain-to-domain or aggregate-to aggregate (polaronic) hopping. This is particularly an issue for these types of molecules given the unique ability of polynuclear aromatic molecular materials to form non-covalently-bonded charged clusters: charged clusters which are known to be stable even in solution and in the vapor phase [91–94]. There is still much to learn!

References

1. Shen, X., et al.: Orientational ordering and dynamics in the columnar phase of a discotic liquid crystal studied by deuteron NMR spectroscopy. J. Chem. Phys. **108**(10), 4324–4332 (1998). doi:10.1063/1.475833
2. Dvinskikh, S.V., et al.: Molecular self-diffusion in a columnar liquid crystalline phase determined by deuterium NMR. Phys. Rev. E Stat. Nonlinear Soft Matter Phys. **65**(5 Pt 1), 050702/1–050702/4 (2002). doi:10.1103/PhysRevE.65.050702
3. Van Keulen, J., et al.: Electrical conductivity in hexaalkoxytriphenylenes. Recueil des Travaux Chimiques des Pays-Bas. **106**(10), 534–536 (1987). doi:10.1002/recl.19871061004
4. Boden, N., et al.: One-dimensional electronic conductivity in discotic liquid crystals. Chem. Phys. Lett. **152**(1), 94–99 (1988). doi:10.1016/0009-2614(88)87334-2
5. Arikainen, E.O., et al.: Effects of side-chain length on the charge transport properties of discotic liquid crystals and their implications for the transport mechanism. J. Mater. Chem. **5**(12), 2161–2165 (1995). doi:10.1039/JM9950502161
6. Bushby, R.J.: Unpublished
7. Arikainen, E.O.: Spectroscopic Studies of the nature of charge carriers in one-dimensional electronically conducting discotic liquid crystals, p. 192. School of Chemistry, University of Leeds (1996)
8. Boden, N., et al.: First observation of a n-doped quasi-One-dimensional electronically-conducting discotic liquid crystal. J. Am. Chem. Soc. **116**(23), 10807–10808 (1994). doi:10.1021/ja00102a065
9. Boden, N., Bushby, R.J., Clements, J.: Mechanism of quasi-one-dimensional electronic conductivity in discotic liquid crystals. J. Chem. Phys. **98**(7), 5920–5231 (1993). doi:10.1063/1.464886
10. Boden, N., et al.: Characterization of the cationic species formed in p-doped discotic liquid crystals. J. Mater. Chem. **5**(10), 1741–1748 (1995). doi:10.1039/JM9950501741
11. Borner, R.C.: Electronically conducting discotic liquid crystals, p. 168. School of Chemistry, University of Leeds (1992)
12. Schouten, P.G., et al.: Radiation-induced conductivity in polymerized and nonpolymerized columnar aggregates of phthalocyanine. J. Am. Chem. Soc. **114**(23), 9028–9034 (1992). doi:10.1021/ja00049a039
13. Warman, J.M., Van De Craats, A.M.: Charge mobility in discotic materials studied by PR-TRMC. Mol. Cryst. Liq. Cryst. **396**, 41–72 (2003). doi:10.1080/15421400390213186
14. Van de Craats, A.M., et al.: Mechanism of charge transport along columnar stacks of a triphenylene dimer. J. Phys. Chem. B **102**(48), 9625–9634 (1998). doi:10.1021/jp9828989
15. Piris, J., et al.: Aligned thin films of discotic hexabenzocoronenes: anisotropy in the optical and charge transport properties. Adv. Funct. Mater. **14**(11), 1053–1061 (2004). doi:10.1002/adfm.200400182

16. Piris, J., Pisula, W., Warman, J.M.: Anisotropy of the optical absorption and photoconductivity of a zone-cast film of a discotic hexabenzocoronene. Synth. Met. **147**(1–3), 85–89 (2004). doi:10.1016/j.synthmet.2004.06.032
17. Saeki, A., et al.: Charge-carrier dynamics in polythiophene films studied by in-situ measurement of flash-photolysis time-resolved microwave conductivity (FP-TRMC) and transient optical spectroscopy (TOS). Philos. Mag. **86**(9), 1261–1276 (2006). doi:10.1080/14786430500380159
18. Sakurai, T., et al.: Prominent electron transport property observed for triply fused metalloporphyrin dimer: directed columnar liquid crystalline assembly by amphiphilic molecular design. J. Am. Chem. Soc. **130**(42), 13812–13813 (2008). doi:10.1021/ja8030714
19. Adam, D., et al.: Transient photoconductivity in a discotic liquid crystal. Phys. Rev. Lett. **70**(4), 457–460 (1993). doi:10.1103/PhysRevLett.70.457
20. Adam, D., et al.: Fast photoconduction in the highly ordered columnar phase of a discotic liquid crystal. Nature **371**(6493), 141–143 (1994). doi:10.1038/371141a0
21. Kepler, R.G.: Charge carrier production and mobility in anthracene crystals. Phys. Rev. **119**, 1226–1229 (1960). doi:10.1103/PhysRev.119.1226
22. Muller-Horsche, E., Haarer, D., Scher, H.: Transition from dispersive to nondispersive transport: photoconduction of polyvinylcarbarzole. Condens. Matter Mater. Phys. **35**(3), 1273–1280 (1987). doi:10.1103/PhysRevB.35.1273
23. Christ, T., Stuempflen, V., Wendorff, J.H.: Light-emitting diodes based on a discotic main chain polymer. Macromol. Rapid Commun. **18**(2), 93–98 (1997). doi:10.1002/marc.1997.030180204
24. Mott, N.F., Gurney, D.: Electronic Processes in Ionic Crystals. Academic Press, New York (1970)
25. Bushby, R.J., et al.: Enhanced charge conduction in discotic liquid crystals. J. Mater. Chem. **11**, 1982–1984 (2001). doi:10.1039/b104112f
26. McNeill, A., et al.: Discotic liquid crystals. In: 3D Nanoelectronic Computer Architecture and Implementation. Taylor & Francis, Philadelphia (2004)
27. Garcia-Frutos, E.M., et al.: High charge mobility in discotic liquid-crystalline triindoles: just a core business? Angew. Chem. **50**, 7399–7402 (2011). doi:10.1002/anie.201005820
28. Bjornholm, T., Hassenkam, T., Reitzel, N.: Supramolecular organization of highly conducting organic thin films by the Langmuir-Blodgett technique. J. Mater. Chem. **9**(9), 1975–1990 (1999). doi:10.1039/A903019K
29. Pisula, W., et al.: A zone-casting technique for device fabrication of field-effect transistors based on discotic hexa-peri-hexabenzocoronene. Adv. Mater. **17**(6), 684–689 (2005). doi:10.1002/adma.200401171
30. Pisula, W., et al.: Exceptionally long-range self-assembly of hexa-peri-hexabenzocoronene with dove-tailed alkyl substituents. J. Am. Chem. Soc. **126**(26), 8074–8075 (2004). doi:10.1021/ja048351r
31. Gearba, R.I., et al.: Homeotropic alignment of columnar liquid crystals in open films by means of surface nanopatterning. Adv. Mater. **19**(6), 815–820 (2007). doi:10.1002/adma.200602460
32. Shklyarevskiy, I.O., et al.: High anisotropy of the field-effect transistor mobility in magnetically aligned discotic liquid-crystalline semiconductors. J. Am. Chem. Soc. **127**(46), 16233–16237 (2005). doi:10.1021/ja054694t
33. Bramble, J.P., et al.: Planar alignment of columnar discotic liquid crystals by isotropic phase dewetting on chemically patterned surfaces. Adv. Funct. Mater. **20**(6), 914–920 (2010). doi:10.1002/adfm.200902140
34. de Leeuw, D.M., et al.: Stability of n-type doped conducting polymers and consequences for polymeric microelectronic devices. Synth. Met. **87**(1), 53–59 (1997). doi:10.1016/S0379-6779(97)80097-5
35. Iino, H., et al.: Fast ambipolar carrier transport and easy homeotropic alignment in a metal-free phthalocyanine derivative. Jpn. J. Appl. Phys. Part 2 Lett. Express Lett. **44**(42–45), L1310–L1312 (2005). doi:10.1143/JJAP.44.L1310
36. Iino, H., et al.: High electron mobility of 0.1 $cm^2\ V^{-1}\ s^{-1}$ in the highly ordered columnar phase of hexahexylthiotriphenylene. Appl. Phys. Lett. **87**(19), 192105/1–192105/3 (2005). doi:10.1063/1.2128066

37. Iino, H., et al.: Fast electron transport in discotic columnar phases of triphenylene derivatives. Jpn. J. Appl. Phys. Part 1 Regul. Pap. Br. Commun. Rev. Pap. **45**(1B), 430–433 (2006). doi:10.1143/JJAP.45.430
38. Boden, N., et al.: Enhanced conduction in the discotic mesophase. Mol. Cryst. Liq. Cryst. **410**, 541–549 (2004). doi:10.1080/15421400490434324
39. Simmerer, J., et al.: Transient photoconductivity in a discotic hexagonal plastic crystal. Adv. Mater. **8**(10), 815–819 (1996). doi:10.1002/adma.19960081010
40. Hirai, Y., et al.: Enhanced hole-transporting behavior of discotic liquid-crystalline physical gels. Adv. Funct. Mater. **18**(11), 1668–1675 (2008). doi:10.1002/adfm.200701313
41. Wegewijs, B.R., et al.: Charge-carrier mobilities in binary mixtures of discotic triphenylene derivatives as a function of temperature. Phys. Rev. B Condens. Matter Mater. Phys. **65**(24), 245112/1–245112/8 (2002). doi:10.1103/PhysRevB.65.245112
42. Nakayama, H., et al.: Measurements of carrier mobility and quantum yield of carrier generation in discotic liquid crystal hexahexyl-oxytriphenylene by time-of-flight method. Jpn. J. Appl. Phys. Part 2 Lett. **38**(9A/B), L1038–L1041 (1999). doi:10.1143/JJAP.38.L1038
43. Mizoshita, N., et al.: The positive effect on hole transport behaviour in anisotropic gels consisting of discotic liquid crystals and hydrogen-bonded fibres. Chem. Commun. **5**, 428–429 (2002). doi:10.1039/B111380C
44. Miyake, Y., et al.: Carrier mobility of a columnar mesophase formed by a perfluoroalkylated triphenylene. Synth. Met. **159**(9–10), 875–879 (2009). doi:10.1016/j.synthmet.2009.01.044
45. Van de Craats, A.M., et al.: The mobility of charge carriers in all four phases of the columnar discotic material hexakis(hexylthio)triphenylene. Combined TOF and PR-TRMC results. Adv. Mater. **8**(10), 823–826 (1996). doi:10.1002/adma.19960081012
46. Iino, H., et al.: Hopping conduction in the columnar liquid crystal phase of a dipolar discogen. J. Appl. Phys. **100**(4), 043716/1–043716/4 (2006). doi:10.1063/1.2219692
47. Bushby, R.J., et al.: Molecular engineering of triphenylene-based discotic liquid crystal conductors. Optoelectron. Rev. **13**(4), 269–279 (2005)
48. Tate, D.J.: Applications of discotic liquid crystals in organic electronics, p. 237. School of Chemistry, University of Leeds (2008)
49. Ochse, A., et al.: Transient photoconduction in discotic liquid crystals. Phys. Chem. Chem. Phys. **1**(8), 1757–1760 (1999). doi:10.1039/A808615J
50. Bleyl, I., et al.: Photopolymerization and transport properties of liquid crystalline triphenylenes. Mol. Cryst. Liq. Cryst. Sci. Technol. Section A Mol. Cryst. Liq. Cryst. **299**, 149–155 (1997). doi:10.1080/10587259708041987
51. Bleyl, I., et al.: One-dimensional hopping transport in a columnar discotic liquid-crystalline glass. Philos. Mag. B Phys. Condens. Matter Stat. Mech. Electron. Opt. Magn. Prop. **79**(3), 463–475 (1999). doi:10.1080/014186399257258
52. Paraschiv, I., et al.: H-bond-stabilized triphenylene-based columnar discotic liquid crystals. Chem. Mater. **18**(4), 968–974 (2006). doi:10.1021/cm052221f
53. Paraschiv, I., et al.: Hydrogen-bond stabilized columnar discotic benzenetrisamides with pendant triphenylene groups. J. Mater. Chem. **18**(45), 5475–5481 (2008). doi:10.1039/B805283B
54. Gearba, R.I., et al.: Tailoring discotic mesophases: columnar order enforced with hydrogen bonds. Adv. Mater. **15**(19), 1614–1618 (2003). doi:10.1002/adma.200305137
55. Kreouzis, T., et al.: Enhanced electronic transport properties in complementary binary discotic liquid crystal systems. Chem. Phys. **262**(2–3), 489–497 (2000). doi:10.1016/S0301-0104(00)00323-2
56. Donovan, K.J., et al.: Molecular engineering the phototransport properties of discotic liquid crystals. Mol. Cryst. Liq. Cryst. **396**, 91–112 (2003). doi:10.1080/15421400390213221
57. Tate, D.J., et al.: Improved syntheses of high hole mobility phthalocyanines: a case of steric assistance in the cyclo-oligomerisation of phthalonitriles. Beilstein J. Org. Chem. **8**(14), 120–128 (2012). doi:10.3762/bjoc.8.14
58. Schouten, P.G., et al.: The effect of structural modifications on charge migration in mesomorphic phthalocyanines. J. Am. Chem. Soc. **116**(15), 6880–6894 (1994). doi:10.1021/ja00094a048

59. van de Craats, A.M., Warman, J.M.: The influence of chain-to-core coupling on the charge transport and mesomorphic properties of discotic materials. Synth. Met. **121**(1–3), 1287–1288 (2001). doi:10.1016/S0379-6779(00)01219-4
60. van de Craats, A.M.: Charge transport in self-aggregating columnar systems such as phthalocyanines, triphenylenes and benzocoronenes. The formation, migration and recombination of charge carriers in various phases of the materials studied is investigated by making use of the time-resolved microwave conductivity technique, PR-TRMC. Opto-electronic Materials, Delft University of Technology (2000)
61. Ban, K., et al.: Discotic liquid crystals of transition metal complexes. 29. Mesomorphism and charge transport properties of alkylthio-substituted phthalocyanine rare-earth metal sandwich complexes. J. Mater. Chem. **11**(2), 321–331 (2001). doi:10.1039/B003984P
62. Fujikake, H., et al.: Time-of-flight analysis of charge mobility in a Cu-phthalocyanine-based discotic liquid crystal semiconductor. Appl. Phys. Lett. **85**(16), 3474–3476 (2004). doi:10.1063/1.1805178
63. Mori, T., Takeuchi, H., Fujikawa, H.: Field-effect transistors based on a polycyclic aromatic hydrocarbon core as a two-dimensional conductor. J. Appl. Phys. **97**(6), 066102/1–066102/3 (2005). doi:10.1063/1.1862757
64. Fechtenkotter, A., et al.: Discotic liquid crystalline hexabenzocoronenes carrying chiral and racemic branched alkyl chains: supramolecular engineering and improved synthetic methods. Tetrahedron **57**(17), 3769–3783 (2001). doi:10.1016/S0040-4020(01)00252-6
65. Ito, S., et al.: Synthesis and self-assembly of functionalized hexa-peri-hexabenzocoronenes. Chem. A Eur. J. **6**(23), 4327–4342 (2000). doi:10.1002/1521-3765(20001201)6:23<4327::AID-CHEM4327>3.0.CO;2-7
66. Van De Craats, A.M., et al.: Record charge carrier mobility in a room temperature discotic liquid-crystalline derivative of hexabenzocoronene. Adv. Mater. **11**(17), 1469–1472 (1999). doi:10.1002/(SICI)1521-4095(199912)11:17<1469::AID-ADMA1469>3.0.CO;2-K
67. Pisula, W., et al.: Relation between supramolecular order and charge carrier mobility of branched alkyl hexa-peri-hexabenzocoronenes. Chem. Mater. **18**(16), 3634–3640 (2006). doi:10.1021/cm0602343
68. van de Craats, A.M., et al.: Meso-epitaxial solution growth of self-organizing discotic liquid crystalline semiconductors. Adv. Funct. Mater. **15**(6), 495–499 (2003). doi:10.1002/adma.200390114
69. Kastler, M., et al.: Room-temperature nondispersive hole transport in a discotic liquid crystal. Appl. Phys. Lett. **89**(25), 252103/1–252103/3 (2006). doi:10.1063/1.2408654
70. Watson, M.D., et al.: Peralkylated coronenes via regiospecific hydrogenation of hexa-peri-hexabenzocoronenes. J. Am. Chem. Soc. **126**(3), 766–771 (2004). doi:10.1021/ja037522+
71. Iyer, V.S., et al.: A soluble C60 graphite segment. Angew. Chem. Int. Ed. **37**(19), 2696–2699 (1998). doi:10.1002/(SICI)1521-3773(19981016)37:19<2696::AID-ANIE2696>3.0.CO;2-E
72. Debije, M.G., et al.: The optical and charge transport properties of discotic materials with large aromatic hydrocarbon cores. J. Am. Chem. Soc. **126**(14), 4641–4645 (2004). doi:10.1021/ja0395994
73. Tomovic, Z., Watson, M.D., Muellen, K.: Superphenalene-based columnar liquid crystals. Angew. Chem. Int. Ed. **43**(6), 755–758 (2004). doi:10.1002/anie.200352855
74. Zhang, Y.-D., et al.: Columnar discotic liquid-crystalline oxadiazoles as electron-transport materials. Langmuir **19**(16), 6534–6536 (2003). doi:10.1021/la0341456
75. Boden, N., et al.: 2,3,7,8,12,13-Hexakis[2-(2-methoxyethoxy)ethoxy]tricycloquinazoline: a discogen which allows enhanced levels of n-doping. Liq. Cryst. **28**(12), 1739–1748 (2001). doi:10.1080/02678290110082383
76. Sienkowska, M.J., et al.: Photoconductivity of liquid crystalline derivatives of pyrene and carbazole. J. Mater. Chem. **17**(14), 1392–1398 (2007). doi:10.1039/B612253A
77. Van de Craats, A.M., et al.: Charge transport in mesomorphic derivatives of perylene. Synth. Met. **102**(1–3), 1550–1551 (1999). doi:10.1016/S0379-6779(98)00554-2
78. Struijk, C.W., et al.: Liquid crystalline perylene diimides: architecture and charge carrier mobilities. J. Am. Chem. Soc. **122**(45), 11057–11066 (2000). doi:10.1021/ja000991g

79. Tsao, H.N., et al.: From ambi- to unipolar behavior in discotic dye field-effect transistors. Adv. Mater. **20**(14), 2715–2719 (2008). doi:10.1002/adma.200702992
80. Monobe, H., Mima, S., Shimizu, Y.: Carrier mobility of discotic lamellar mesophases of 5,10,15,20-tetrakis(4-n-pentadecylphenyl)porphyrin. Chem. Lett. **9**, 1004–1005 (2000)
81. Yuan, Y., Gregg, B.A., Lawrence, M.F.: Time-of-flight study of electrical charge mobilities in liquid-crystalline zinc octakis(beta -octoxyethyl) porphyrin films. J. Mater. Res. **15**(11), 2494–2498 (2000). doi:10.1557/JMR.2000.0358
82. Schouten, P.G., et al.: Charge migration in supramolecular stacks of peripherally substituted porphyrins. Nature **353**(6346), 736–737 (1991). doi:10.1038/353736a0
83. Destrade, C., et al.: Disk-like mesogen polymorphism. Mol. Cryst. Liq. Cryst. **106**(1–2), 121–146 (1984). doi:10.1080/00268948408080183
84. Chiang, L.Y., et al.: Highly oriented fibers of discotic liquid crystal. J. Chem. Soc. Chem. Commun. **11**, 695–696 (1985). doi:10.1039/C39850000695
85. Safinya, C.R., et al.: Synchrotron x-ray scattering study of freely suspended discotic strands. Mol. Cryst. Liq. Cryst. **123**(1–4), 205–216 (1985). doi:10.1080/00268948508074778
86. Arikainen, E.O., et al.: Complimentary polytopic interactions. Angew. Chem. Int. Ed. **39**(13), 2333–2336 (2000). doi:10.1002/1521-3757(20000703)112:13<2423::AID-ANGE2423>3.0.CO;2-R
87. Bushby, R.J., et al.: The stability of columns comprising alternating triphenylene and hexaphenyltriphenylene molecules: variations in the structure of the hexaphenyltriphenylene component. Liq. Cryst. **33**(6), 653–664 (2006). doi:10.1080/02678290600682078
88. Borsenberger, P.M., O'Regan, M.B.: The role of dipole moments on hole transport in triphenylamine doped poly(styrene). Chem. Phys. **200**(1,2), 257–263 (1995). doi:10.1016/0301-0104(95)00195-T
89. Lemaur, V., et al.: Charge transport properties in discotic liquid crystals: a quantum-chemical insight into structure-property relationships. J. Am. Chem. Soc. **126**, 3271–3279 (2004). doi:10.1021/ja0390956. Copyright (C) 2011 American Chemical Society (ACS). All Rights Reserved
90. van de Craats, A.M., Warman, J.M.: The core-size effect on the mobility of charge in discotic liquid crystalline materials. Adv. Mater. **13**(2), 130–133 (2001). doi:10.1002/1521-4095(200101)13:2<130::AID-ADMA130>3.0.CO;2-L
91. Meot-Ner, M.: Dimer cations of polycyclic aromatics. Experimental bonding energies and resonance stabilization. J. Phys. Chem. **84**(21), 2724–2728 (1980)
92. Mautner, M.: Structurally complex organic ions: thermochemistry and noncovalent interactions. Acc. Chem. Res. **17**(5), 186–193 (1984). doi:10.1021/ar00101a006
93. Terahara, A., et al.: Transannular interactions in dimer cation radicals of naphthalene derivatives. Conformation anomaly and stabilization energy. J. Phys. Chem. **90**(8), 1564–1571 (1986). doi:10.1021/j100399a022
94. Ohya-Nishiguchi, H., Ide, H., Hirota, N.: Spin densities in the trimer cation radical of coronene. Chem. Phys. Lett. **66**(3), 581–583 (1979). doi:10.1016/0009-2614(79)80344-9

Chapter 4
Synthesis of Columnar Liquid Crystals

Sandeep Kumar

4.1 Introduction

Thermotropic liquid crystals consist primarily of two main types, i.e., rod-like, commonly known as calamitic liquid crystals and disk-like, described as discotic liquid crystals. Recently discovered banana-shaped liquid crystals are basically rod-like systems in which two calamitic units are connected to each other *via* a central bridge in such a way that the overall molecular shape is banana-like or bow-like. The synthetic strategies used to prepare these materials are essentially same as that used to generate rod-like liquid crystals. More than 100,000 calamitic liquid crystals have so far been synthesized using classical as well as modern synthetic methods. Most of the calamitic liquid crystalline compounds consist of two or more ring structures, bonded together directly or *via* linking groups. They usually have terminal hydrocarbon chains and sometimes lateral substituents as well. The typical chemical structure of these molecules can be represented by the general template as shown in Fig. 4.1, where A and B are core units such as, benzene, naphthalene, biphenyl, etc., R and R′ are flexible moieties such as, normal and/or branched alkyl chains, M and N are generally small lateral substituents e.g., –Cl, –Br, $–NO_2$, $–CH_3$, $–OCH_3$, –CN, etc. Y is a linking group to the core units and X & Z are linking groups of terminal chains and core units. Clearly, several permutations and combinations are possible to generate new calamitic liquid crystals and consequently over hundred thousand calamitic liquid crystals have so far been synthesized. In most cases, well-known classical organic chemistry reactions are applied to produce these materials. Basic synthetic strategies towards the synthesis of calamitic liquid crystals have been extensively covered in chapters of the "Handbook of Liquid Crystals" [1] and, therefore, the chemistry of these

S. Kumar (✉)
Soft Condensed Matter Group, Raman Research Institute, C.V. Raman Avenue, Sadashivanagar, Bangalore 560 080, India
e-mail: skumar@rri.res.in

R.J. Bushby et al. (eds.), *Liquid Crystalline Semiconductors*, Springer Series in Materials Science 169, DOI 10.1007/978-90-481-2873-0_4,

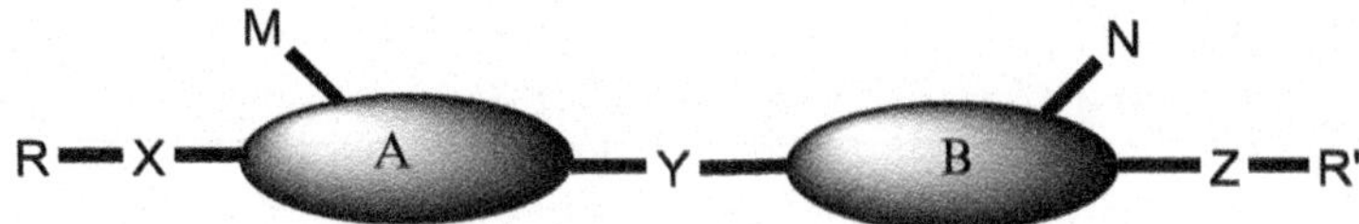

Fig. 4.1 General structural template for calamitic liquid crystals where A and B are core units R and R′ are commonly flexible chains, M and N are generally small lateral substituents and X, Y, Z are linking groups

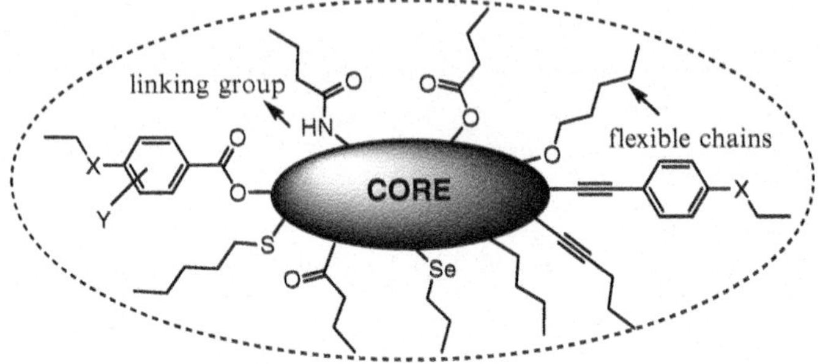

Fig. 4.2 General structural template for discotic liquid crystals

materials by giving specific examples is not covered in this chapter. However, some important organic reactions e.g., carbon-carbon bond formation reactions are presented in the synthesis of discotic liquid crystals section.

From a semiconducting properties point of view, discotic liquid crystals forming columnar phases are of primary importance, see Chaps. 3, 8 and 9 [2–15]. Molecules forming columnar mesophases are often made of a central discotic core substituted by a large number (usually 3–12) of saturated chains of three or more carbon atoms. A general template for the molecular shape of discotic mesogens is shown in Fig. 4.2. By tailoring the shape, size and nature of the central core as well as the type of the attached side chains, a variety of discotic liquid crystals with different mesophase morphologies can be generated. These materials often have two, three, four or sixfold rotational symmetry. The liquid crystallinity occurs due to the presence of two contrasting units in the molecule, i.e., the crystalline character is governed by the interaction between the conjugated cores while the liquid character originates from the melting of the saturated alkyl chains in the mesophase. The phase transition in discotic liquid crystals occurs due to the fact that at the melting point of the molecular crystals, the alkyl chains melt and provide fluidity to the mesophase while retaining the order due to core-core stacking. At the clearing point, the unstacking of the central cores occurs leading to an isotropic liquid state. Such discotic molecular architectures commonly organize spontaneously in the form of one-dimensional columns to create various columnar phases, such as, columnar hexagonal mesophase (Col_h); columnar rectangular mesophase (Col_r); columnar oblique phase (Col_{ob}); columnar square (tetragonal) phase (Col_{tet}); columnar lamellar phase (Col_L); columnar plastic phase (Col_p); and columnar helical phase (H). It is worth mentioning that the molecular fluctuations in the liquid crystalline phases support the self-healing of structural defects and hence improve various physical properties such as the charge carrier mobilities along the columnar stacks.

Mesophases formed by disc-shaped molecules are primarily two types: (i) nematic and, (ii) columnar. Additionally, a few examples are known to exhibit smectic and cubic phases. The columnar phase is ubiquitous in discotics (about 95%) followed by nematic phase whereas the other phases are rarely observed. A majority of the discotics exhibit only one type of mesophase, but a few examples are known to exhibit polymorphism. A single discotic molecule exhibiting all the mesophase has not been reported so far in the literature. It must be emphasized that though columnar phases are most characteristics for discotic (disc-shaped) mesogens, they are by no means unique to discotic molecules. Several other molecular architectures, such as, surfactants (forming lyotropic columnar phases), polycatenar molecules, dendrimers and bent-core mesogens are also known to exhibit thermotropic columnar phases and some of these have also been studied for charge migration properties, but this chapter covers the synthesis of only columnar phase forming discotic liquid crystals described in Chap. 3 and related materials.

4.2 Synthesis of Discotic Liquid Crystals

Although the synthesis of DLCs is, relatively, more difficult than the synthesis of calamitic liquid crystals, strategies to prepare discotic mesogens are fairly straightforward. If a polyfunctional aromatic core is commercially available, then it is quite convenient to generate DLCs by attaching aliphatic chains, e.g., esterification of naturally occurring scyllo-inositol yields hexaalkanyoloxy-scylloinositol discotics [16]. Unfortunately, this option is limited as only a few polyfunctional cores suitable to prepare DLCs are commercially available. The direct electrophilic aromatic substitution of polycyclic aromatic hydrocarbons, e.g., Friedel-Crafts acylation of decacyclene [17], has been used in some cases to generate liquid crystalline materials, but because of regioselectivity and reactivity problems, such examples are also not common. Another attractive option is the preparation of polyfunctional cores directly starting from unprotected functionalized precursor molecules, e.g., one-step preparation of hexahydroxy-rufigallol from gallic acid [18]. However, this option is also limited as many functional groups are sensitive towards the drastic reaction conditions used to build the core. The most common method used to prepare DLCs involves construction of the core with protected functional groups to avoid the formation of unwanted side products. The protecting groups e.g., methoxy groups, can be cleaved to generate reactive groups such as hydroxyl groups, which can be used to graft pendant chains necessary to induce mesomorphism. Most of the esters of discotic cores have been obtained this way. In recent times the protection and deprotection steps to prepare discotics with ether linkages were considered unnecessary and thus, long aliphatic chains required to induce mesogenity were attached in the starting molecules and then transformed them directly to liquid crystalline materials using classical or improved synthetic methods of today. This methodology has proved highly beneficial to prepare not only a number of alkoxy-substituted discotics, but also a number of alkyl-substituted DLCs. Often the preparation of DLCs is not that demanding, but their purification is usually quite difficult and tedious. This is primarily because of incomplete substitution of a polyfunctional molecule. Partial substitution of a polyfunctional core often gives a mixture of structurally similar products having almost identical R_f values on a chromatographic plate and, therefore, their separation is generally very difficult. Efforts should be made to find reaction conditions to push the reaction towards completion by using a large excess of the reagent and/or longer reaction time. However, in many cases, the excess of the reagent or longer reaction time cause side reactions and thus a complex mixture of products and impurities. Finding optimum reaction conditions is the crucial part of the synthesis. The synthesis and purification of many DLCs has been improved significantly over the last few years. Highly regioselective and high-yielding synthetic methods are now available for the synthesis of some DLCs. Chemists from all over the word have exploited about 60 different molecular cores to create more than 3,000 discotic liquid crystals [19]. A detailed description of the chemistry of all these materials is beyond the scope of this chapter and, therefore, only the chemistry of some DLCs and related materials used in the study of charge transport properties, see Chap. 3, is presented in the following sections.

4.3 Chemistry of Triphenylene Based DLCs

Triphenylene (TP) derivatives have been described as "the work horses" in the field of DLCs [11]. It is the most studied discotic core system [20, 21]. TP (**1**), see Scheme 4.1, belongs to the polycyclic aromatic hydrocarbon (PAH) group and has been known in the chemical literature for more than a century. This trimer of benzene was isolated from the pyrolytic products of benzene by Schultz who named it as "triphenylene" [22]. It can also be isolated from coal tar. Trimerization of cyclohexanone followed by dehydrogenation has been used to generate TP in the early twentieth century and its various chemical and physical properties were studied [22].

The French groups explored the potential of TP core to generate DLCs [23, 24] almost at the same time when the first examples of DLCs based on benzene core were reported by Chandrasekhar and co-workers. TP derivatives are thermally and chemically stable, their chemistry is quite developed and fairly accessible, they show a variety of mesophases and their one-dimensional charge and energy migration properties offer tremendous potential applications. As a consequence, numerous synthetic efforts have been made to prepare a variety of TP-based DLCs. More than 500 TP derivatives have been realized to evaluate their mesomorphic and other physical properties [20, 21]. Some important routes to prepare TP discotics are as follows.

4.3.1 Hexaalkylthio-TPs

One of the derivatives of this series, namely 2,3,6,7,10,11-hexahexylthio-TP (HHTT), displays a highly ordered helical phase [25]. The photo-induced charge carrier mobilities up to 0.1 $cm^2\ V\ s^{-1}$ were achieved in this phase [26]. With the exception of organic single crystals, these were the highest electronic mobility values reported at that time and, therefore, efforts have been made to prepare hexaalkylthio-TP derivatives in high yield and with high purity [25–32]. In order to prepare these materials, the 2,3,6,7,10,11-hexabromo-TP **2** was reacted with an excess of sodium alkylthiolate in a polar aprotic solvent DMEU at 100 °C for 1–2 h, see Scheme 4.1. The reaction yields about 40–55% hexaalkylthio-TP, **3**. An improvement in this process was reported by generating the thiolates with sodium hydride in dry ether, instead of sodium ethoxide in ethanol, and subsequently heating with hexabromo-TP in DMEU at 70 °C for 30 min. This improves the yield to 78–84% and the product is less contaminated with side products. We anticipated that the poor yield and contamination of the product could be due to the nucleophilic dealkylation reactions, where the initially formed hexaarylalkyl sulphide may be attacked by excess thiolate anion to produce dialkyl sulphide and thiophenolate. In order to re-alkylate these exposed thiophenolate groups, the reaction mixture was worked up with the appropriate l-bromoalkane. This modification not only improved the yield substantially (~95%) but also provided very high purity (~99.9%). Thus, thiolates were generated by reacting alkanethiol with potassium- *tert.*-butoxide in *N*-methyl-2-pyrrolidone and subsequently heated with hexabromo-TP at 70 °C for 25 min in the same solvent. Quenching the reaction with an appropriate 1-bromoalkane affords the desired products.

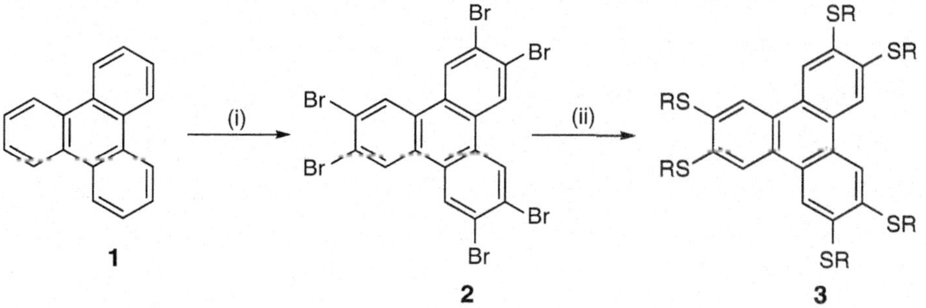

Scheme 4.1 Synthesis of hexaalkylthio-triphenylenes; (i) Br_2, Fe, $C_6H_5NO_2$, reflux; (ii) RSK, NMP, 70 °C

4.3.2 Hexaalkoxy-TPs

Hexaalkanoates and benzoates are the first liquid crystalline derivatives of TP core [24, 33, 34]. These discotics were prepared from hexahydroxy-TP as shown in Scheme 4.2. The trimerization of 1,2-methoxybenzene to hexamethoxy-TP using chloranil or iron(III)chloride as an oxidant was reported by Musgrave [35]. It is one of the unusual cases of Scholl reaction [36, 37] where more than one aryl–aryl bonds are formed. In this reaction, linear coupling products have not been realized. However, if one of the *para*-alkoxy positions is blocked, formation of biphenyl has been realized. The cyclization is enthalpically favored by the aromatic character of the central ring and the greater conjugation of the planar TP relative to nonplanar acyclic oligomers and polymers [38]. Hexaalkanoates and benzoates of TP have not been studied for their semiconducting properties as yet, probably because of their high clearing temperatures compared to those of hexaalkoxy-TPs, which are the most widely synthesized and studied discotic mesogens. Originally these materials were also prepared from hexahydroxy-TP in low yield, see Scheme 4.2. Unfortunately, the isolation of pure product was tedious. It may be noted that TP derivatives show mesomorphism only when the six peripheral alkoxy chains have a minimum of three carbon atoms, i.e., propyloxy chains. The trimerization using chloranil is limited to the preparation of hexamethoxy-TP; with higher homologues

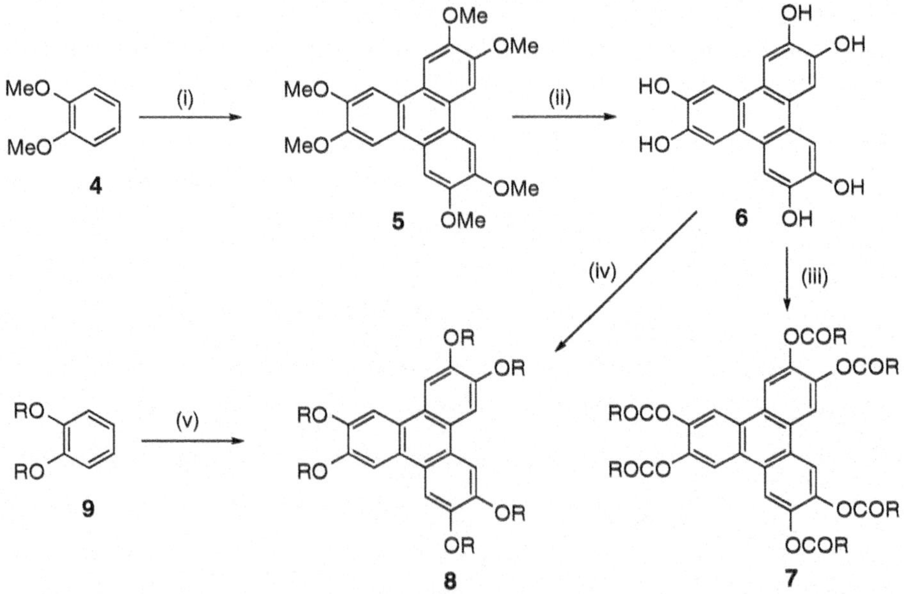

Scheme 4.2 Synthetic routes to hexaalkoxytriphenylenes; (i) chloranil, H_2SO_4; (ii) BBr_3 or HBr; (iii) RCOCl, pyridine; (iv) RBr, base; (v) $FeCl_3$ or $MoCl_5$ or $VOCl_3$, CH_2Cl_2, MeOH

such as 1,2-dihexyloxybenzene it gives only a poor yield of hexahexyloxy-TP with many side products. Therefore, in order to prepare these long chain derivatives, it was necessary to cleave the methoxy groups with boron tribromide or HBr, and realkylate the resulting hexaphenol with an appropriate alkyl halide to obtain different hexaalkoxy-TP discotics, see Scheme 4.2.

A major advancement in the synthesis of hexaalkoxy-TP discotics was achieved by Bushby and coworkers. They realized that the dealkylation (demethylation) followed by realkylation of the generated hexaphenol is a wasteful reaction and, therefore, can be avoided by taking the appropriate 1,2-dialkoxy benzene having the desired alkyl chain length for trimerization. Furthermore, they observed that only a catalytic amount of sulfuric acid is required to perform this reaction. Their method of cyclization of 1,2-dialkoxy benzene with $FeCl_3$ in dichloromethane having about 0.3% H_2SO_4 followed by a reductive work-up using methanol, see Scheme 4.2 became a milestone in TP discotics synthesis [39–42]. The reductive workup is a highly exothermic reaction and a large amount of formaldehyde is liberated. Therefore, care must be taken during the work-up particularly in the large scale synthesis. Subsequently two other reagents, molybdenum pentachloride [43] and vanadium oxytrichloride [44, 45], were reported to be highly efficient for the preparation of TP hexaethers. Furthermore, it was observed that even the catalytic amount of H_2SO_4 is also not always required for this trimerization. The liquid reagent $VOCl_3$ is particularly very attractive as it is miscible in various organic solvent and the trimerization occurs almost spontaneously. It is easy to handle the liquid reagent, workup of the reaction is convenient and the product yields are high, although it is a much more expensive reagent than $FeCl_3$. The choice of solvent is crucial in this reaction. It was observed that the reaction works well only in dichloromethane, all other solvents give only poor yield [45]. The reason for the poor performance of the reaction in various other solvents is still not clear. Although this oxidative coupling can be accomplished with a number of other reagents such as $SbCl_5$, VOF_3, $K_3Fe(CN)_6$, $CuCl_2$ or $Cu(OTf)_2$ and $AlCl_3$ in CS_2, $Pb(OAc)_4/BF_3$-Et_2O in MeCN, $Tl(OCOCF_3)_2$ in CF_3COOH, etc., the potential of these oxidants in the synthesis of TP-based DLCs has not yet been explored.

The above-mentioned trimerization methodology results only symmetrical hexaalkoxy-TPs. In order to modify the thermal and electronic properties of these discotics, several methods have been developed to prepare unsymmetrical TP discotics. In order to prepare hexaalkoxy-TPs having different peripheral chains, the most successful method is the biphenyl-phenyl oxidative coupling. Musgrave utilized this methodology in 1965 to couple tetramethoxybiphenyl with veratrole to prepare hexamethoxy-TP [35]. The oxidation of a mixture of 1,2-dialkoxybenzene and 3,3′,4,4′-tetraalkoxybiphenyl using $FeCl_3$, $MoCl_5$ or $VOCl_3$ oxidants, see Scheme 4.3 produces unsymmetrically substituted hexaalkoxy-TPs in good yield. In this reaction the cross coupling between the biphenyl and phenyl derivatives dominates over the oxidative dimerization of tetraalkoxybiphenyl or trimerization of dialkoxybenzene. In view of this efficient cross coupling, it has been suggested [35, 46] that tetraalkoxybiphenyl is probably the intermediate product during the

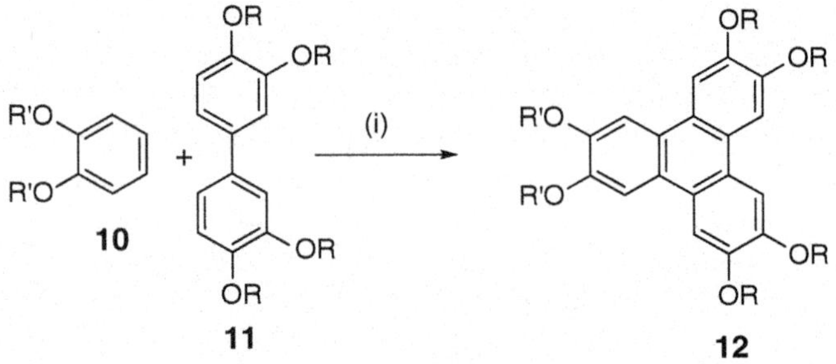

Scheme 4.3 Phenyl-biphenyl coupling route to the synthesis of unsymmetrical hexaalkoxy-triphenylenes; (i) $FeCl_3$ or $MoCl_5$ or $VOCl_3$, CH_2Cl_2

oxidative trimerization of 1,2-dialkoxybenzene to hexaalkoxy-TP. However, any tetraalkoxybiphenyl in the oxidative trimerization of 1,2-dialkoxybenzene could not be isolated.

The above-mentioned phenyl-biphenyl coupling route is extremely important to generate a number of unsymmetrical TPs, however, it works well only for the coupling of a di- or trialkoxy-benene with a tetraalkoxy-biphenyl. Moreover, the reaction is not regiospecific. TP derivatives with any degree of substitution with full regiocontrol can be obtained with the help of modern organometallic chemistry [47–56]. The methodology essentially involves assembling a terphenyl, which can be oxidatively cyclized to TP. Palladium-catalyzed Suzuki or Nigishi types of reactions are commonly used to prepare terphenyls. An example of TP synthesis using Suzuki coupling is shown in Scheme 4.4. It involves the coupling of a phenylboronic acid with a halide-functionalized biphenyl. The resultant terphenyl **15** is cyclized to the tetraalkoxy-TP **16** using oxidative coupling. As TP derivatives with less than six substitutions are generally not liquid crystalline, further substitutions of the compound **16** are required to induce mesomorphism [57].

Although, Suzuki coupling is very common in unsymmetrical TP synthesis, Nigishi coupling, which involves coupling of an organozinc compound with aryl halide, has also been applied to the synthesis of some unsymmetrical TPs, see Scheme 4.5.

4.3.3 α-Functionalized Hexaalkoxy-TPs

As the TP nucleus is a polycyclic hydrocarbon, electrophilic aromatic substitutions can be easily carried out on it. Electrophilic aromatic substitution in parent hydrocarbon **1** is directed by steric and electronic effects. Because of steric effect, substitution at β-position or 2-position is favored compared to α- or 1-position [46]. Bromination of TP is reported to yield mainly 2,3,5,6,10,11-hexabromo-TP [58], however, the electronic effect plays a major role in the nitration of TP.

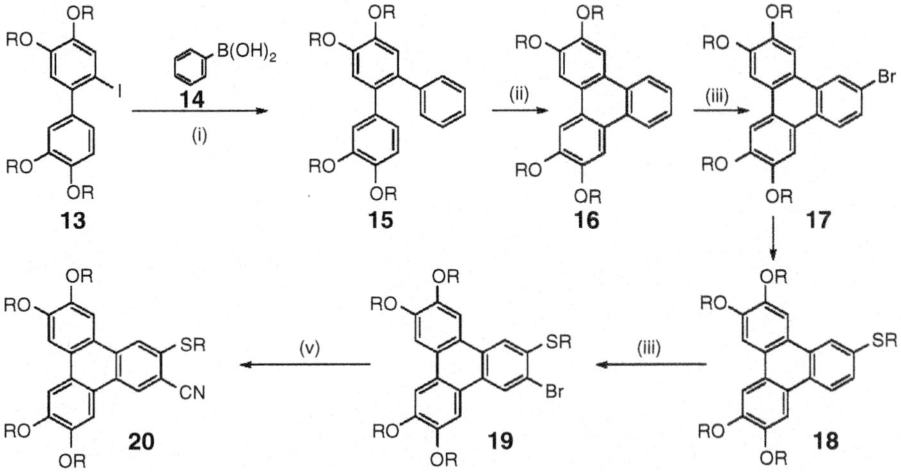

Scheme 4.4 Synthesis of unsymmetrical triphenylenes where the Suzuki coupling is used to construct the key precursor; (i) $[(C_6H_5)_3P]_4Pd$, Na_2CO_3, THF, reflux; (ii) CH_2Cl_2, $VOCl_3$; (iii) Br_2, CH_2Cl_2; (iv) $C_5H_{11}SK$, NMP, $[(C_6H_5)_3P]_4Pd$; (v) CuCN, NMP, reflux

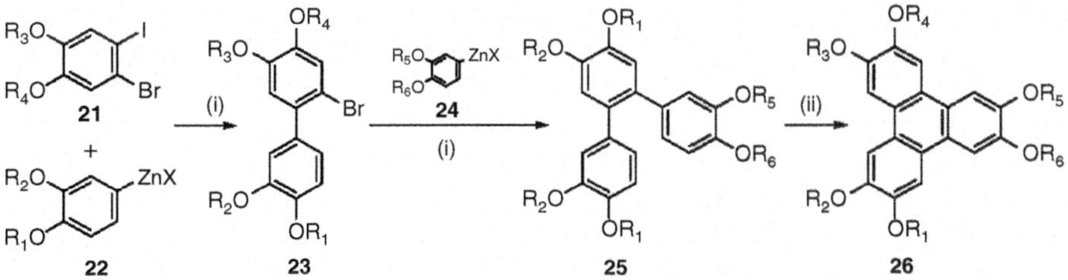

Scheme 4.5 Synthesis of unsymmetrical triphenylenes where the Nigishi coupling is used to construct the key precursor; (i) Pd(0), THF, reflux; (ii) $FeCl_3$

Nitration of the unsubstituted TP is reported to yield a trinitro derivative, but under controlled conditions a mixture of 1-nitro- and 2-nitro-TPs is formed [46]. In a 2,3,6,7,10-pentaalkoxy-TP, where both α and β positions are available, classical electrophilic aromatic substitution reactions such as Friedel–Crafts acylation and halogenation give only β-substitution product but nitration yields the α-nitro product [59]. Nitration of 2,3,6,7,10,11-hexaalkoxy-TP, where only the α-positions are free for further substitution, yields the α-nitro product and thus 1-nitro-2,3,6,7,10,11-hexaalkoxy-TP is formed [60–64]. Solvents play an important role in the nitration of hexaalkoxy-TPs. In a mixture of ether–acetic acid, almost exclusively the α-nitro product is formed. Even under exhaustive conditions, only a small amount of trinitrohexaalkoxy-TP is formed in this solvent system. Changing the solvent system from ether–acetic acid to dichloromethane–nitromethane imparts a dramatic effect and all three rings of TP can be successively nitrated under very mild conditions [63, 64]. The trinitration proceeds with high regioselectivity to give exclusively

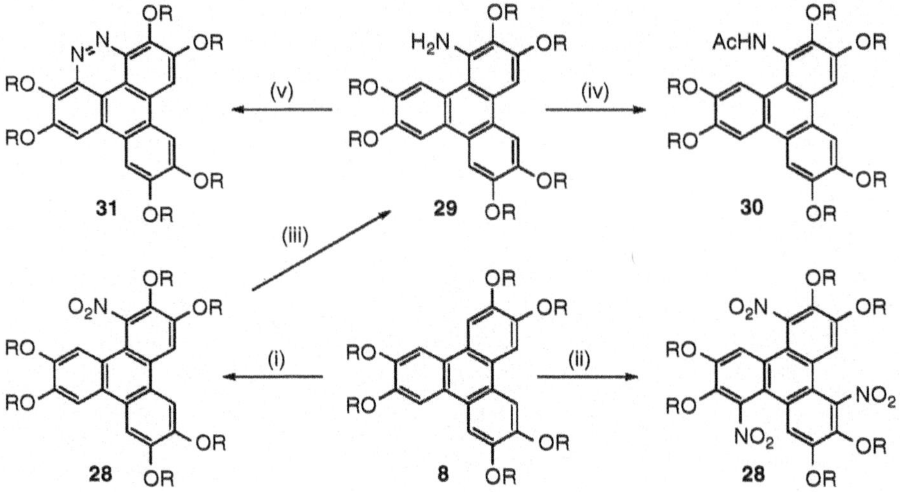

Scheme 4.6 Nitration of hexaalkoxy-TP; (i) HNO_3/Et_2O-AcOH; (ii) HNO_3/CH_3NO_2; (iii) NH_2NH_2/Pd or $NiCl_2/NaBH_4$; (iv) Ac_2O/Py; (v) $NaNO_2$/AcOH

one isomer having C_3 symmetry, i.e. 1,5,9-trinitro-2,3,6,7,10,11-hexaalkoxy-TP. Alkoxynitro-TPs are valuable precursors to several other derivatives, such as amino, mono- and di-alkylamino, acylamino and azo derivatives, and thus a number of new TP derivatives, see Scheme 4.6 can be prepared [60–66].

Similarly, hexaalkoxy-TPs can be halogenated in the α-position, but usually a mixture of polyhalogenated products is formed. Therefore, it is more convenient to prepare these materials *via* a phenyl-biphenyl coupling route. Thus, for example, oxidative coupling of 3,3′,4,4′-tetraahexyloxybiphenyl **11** and 1-fluoro-2,3-dihexyloxybenzene **32** provides 1-fluoro-2,3,6,7,10,11-hexahexyloxy-TP **33** in good yield, see Scheme 4.7 [67]. As the halogen-substitution is in the plane of the core, it does not affect the molecular breadth, but fills the space around the core and enhances the intermolecular forces. Consequently, the columnar phase stability increases significantly. Furthermore, the halogen atoms can be replaced by cyano group in order to further enhance the stability of the columnar phase.

4.3.4 Hydroxyl-Alkoxy-TPs

The hydroxy-functionalized TPs are very valuable precursors for the synthesis of many other discotics such as TP dimers, oligomers, polymers, networks, mixed tail derivatives and lower and higher degree substituted TP derivatives. Although a number of hydroxyl-alkokxy-TPs are possible and many of them have also been prepared, monohydroxy-pentaalkoxy-TPs are of particular importance. Accordingly, several methods have been developed to prepare these materials [68–74].

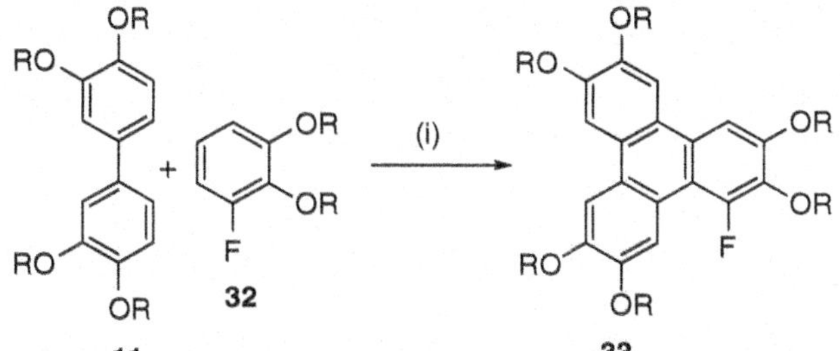

Scheme 4.7 Synthesis of α-halogenated-hexaalkoxy-TP (i) $FeCl_3$, CH_2Cl_2

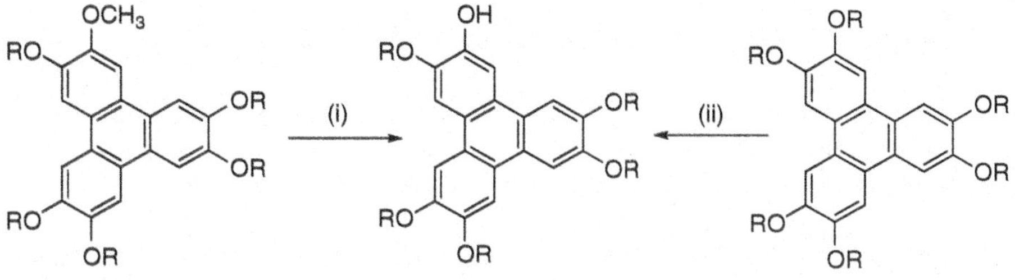

Scheme 4.8 The most common routes to prepare a monohydroxy-pentaalkoxy-TP; (i) lithium diphenylphosphide; (ii) bromocatecholborane

However, only two methods are commonly used to produce these compounds. Both are based on the cleavage of one alkoxy chain of the hexaalkoxy-TPs. As the lithium diphenylphosphide is known to cleave aryl-methyl ether selectively in presence of aryl-alkyl ethers, it is used to prepare monohydroxy-TPs by cleaving the methoxy group of monomethoxy-pentaalkoxy-TPs, see Scheme 4.8 which can be prepared easily *via* the phenyl-biphenyl coupling route. The other attractive route to prepare monohydroxy-pentaalkoxy-TPs is *via* the cleavage of one alkyl chains of hexaalkoxy-TP using bromocatecholborane (Cat-B-Br). The reagent can also be used to prepare di- and trihydroxy-TPs. Thus, treatment of hexapentyloxy-TP with 1.2 equivalent of Cat-B-Br gives almost 70% of monohydroxy-pentaalkoxy TP, see Scheme 4.8 and treatment with about 2.5 eq of Cat-B-Br yields a mixture of dihydroxy-tetraalkoxy-TP. On the other hand, the use of 3.6 equivalents of Cat-B-Br produces exclusively two products, the symmetrical 2,6,10-trihydroxy-3,7,11-tris(pentyloxy)-TP (61%) and the non-symmetrical 2,7,10-trihydroxy-3,6,11-tris(pentyloxy)-TP (38%) [73].

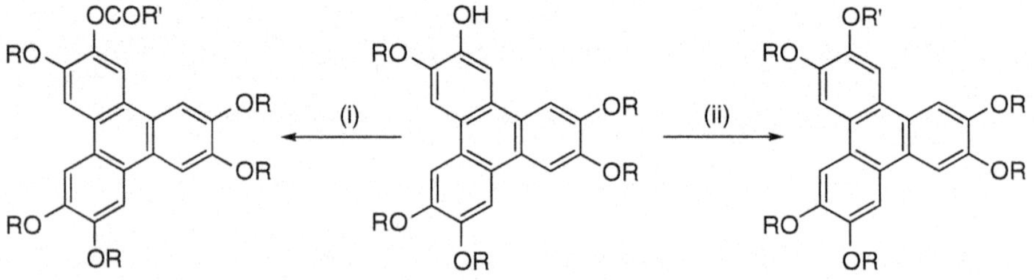

Scheme 4.9 Esterification and alkylation of monohydroxy-pentaalkoxytriphenylenes to prepare unsymmetrical triphenylene discotics; (i) R′COCl, pyridine; (ii) R′Br, base

4.3.5 Discotics Derived from Hydroxy-Alkoxy-TPs

A large number of DLCs have been derived from hydroxy-alkoxy-TPs. However, only a few derivatives have been studied for charge transport properties. Unsymmetrical TP derivatives such as **36** can be easily prepared by esterifying monohydroxy-pentaalkoxy-TP with the desired acid derivative, see Scheme 4.9. Similarly, alkylation of monohydroxy-pentaalkoxy-TPs with various alkyl halides or functionalized alkyl halides yields hexaalkoxy-TPs, **37**, having one alkyl chain different than other five, see Scheme 4.9. Such unsymmetrical TP derivatives have been extensively studied for charge transport properties, see Chap. 3 [75, 76].

4.3.6 Discotic Dimers and Oligomers

A discotic dimer is composed of two discotic units linked, commonly, *via* a flexible spacer and rarely *via* a rigid spacer. Liquid crystalline dimers show interesting mesomorphic behavior depending on the length of the spacer and structure of the linking group. The most commonly prepared triphenylene based dimers contain two identical triphenylene moieties connected *via* a spacer. Often the spacer is a polymethylene chain, but in some cases ester or amide linkage in the middle of the spacer or at the terminal positions have also been used [69, 77–85]. In most of flexible discotic dimers the columnar phase freezes into glassy state on cooling. The stability of the glassy state depends on the spacer length as well as on the symmetry of the molecule. The non-symmetrical dimers (with two discotic units different) usually give longer-lived glasses. As the length of linking chain increases, in general, the lifetime of the glassy state decreases [69].

The symmetrical dimers can be synthesized either in a single step by reacting the monohydroxy-TP with 0.5 equivalent of the appropriate α, ω-dibromoalkane under classical etherification reaction conditions or in two steps. In the two-step procedure, the monohydroxy-pentaalkoxy-TP is first reacted with an excess of the appropriate

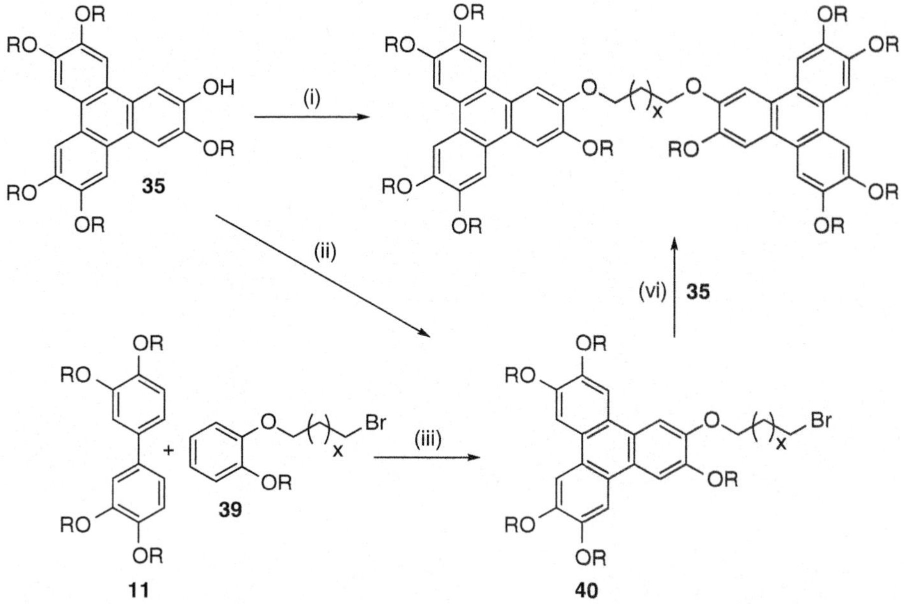

Scheme 4.10 Synthetic routes to symmetrical triphenylene dimers; (i) $Br(CH_2)_nBr$ (0.5 equiv.), K_2CO_3; (ii) $Br(CH_2)_nBr$ (excess), K_2CO_3; (iii) $FeCl_3$, CH_2Cl_2; (iv) K_2CO_3

α, ω-dibromoalkane to obtain the ω-brominated product **40**, see Scheme 4.10 which may be reacted further with monohydroxy-TP to obtain the desired dimer. This two-step process is particularly important for preparing non-symmetrical dimers. The intermediate ω-brominated product **40** can also be prepared using the biphenyl-phenyl oxidative coupling route. Though a large number of TP dimers have been synthesized [21], only simple symmetric dimers have been studied for their charge transport behavior.

Paraschiv et al. reported an interesting class of intramolecular H-bond-stabilized columnar liquid crystalline materials by synthesizing a series of 1,3,5-benzenetrisamide derivatives **43** with three hexaalkoxy-TP pendent groups [86, 87]. In these materials, the columnar phase is stabilized *via* intermolecular hydrogen bonding. This way, a high directionality of interaction could be achieved. These materials forms ordered columnar phases in which the charge carrier mobility was several times higher compared to monomeric TP discotics and close to the mobility observed in large core hexabenzocoronene derivatives. These trimers were prepared by reacting an amino-terminated hexaalkoxy-TP **41** with trimesoyl chloride **42**, see Scheme 4.11. The amino-functionalized TP derivative can be prepared from mono-functionalized TP. Alkylation of monohydroxy-pentaalkoxy-TP with α, ω-dibromoalkane gives the bromo-terminated TP which can be converted to the azide by treatment with sodium azide in ethanol. Reduction of the azide with lithium aluminum hydride furnished the required amino-terminated TP which can be coupled with 1,3,5-benzenetricarbonyl trichloride in the presence of triethylamine.

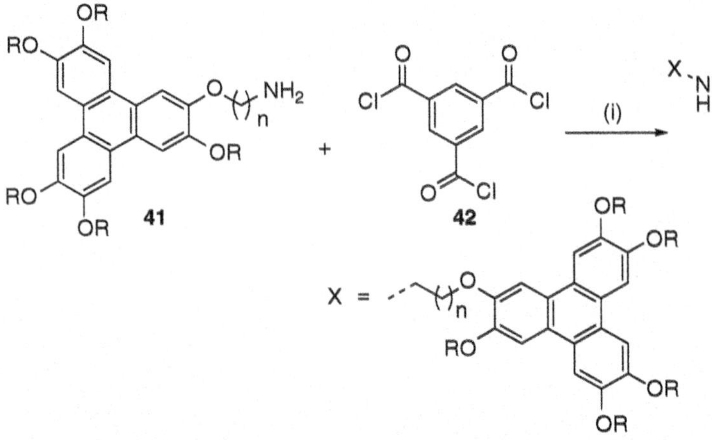

Scheme 4.11 Synthesis of star-shaped triphenylene trimers; (i) CH_2Cl_2, Et_3N

It may be noted that a large number of other discotic oligomers and polymers have been realized but their conducting properties have not been evaluated and, therefore, their synthesis has also not been covered here. Interested readers may look Ref. [21].

4.4 Hexaazatriphenylenes

1,4,5,8,9,12-Hexaazatriphenylene (HAT) is an electron-deficient polycyclic heteroaromatic discotic core. Hexaazatriphenylene derivatives have received a considerable attention because of their easy synthetic accessibility, electron deficiency, coordination properties and π-complexation ability. Praefcke and coworkers prepared several hexaalkyl and hexaalkoxy derivatives of hexaazatriphenylene but none of these compounds were found to be liquid crystalline in nature [88]. Hexaazatriphenylene derivatives can be prepared *via* a one-step cyclocondensation of benzene hexamine **44** and various symmetrical 1,2-diketones **45** as shown in Scheme 4.12. An improved synthesis of the parent core by the condensation of glyoxal with freshly prepared hexaaminobenzene was reported by Rogers [89].

The first liquid crystalline derivatives of HAT were reported by Bushby and coworkers [90–92]. They prepared several hexaphenyl-substituted HAT derivatives **48** with different peripheral chains as shown in Scheme 4.13. It is worth noting that some non-mesomorphic HAT derivatives when mixed with other discotic liquid crystals, exhibit stable columnar phases due to complimentary polytopic interactions [90, 91].

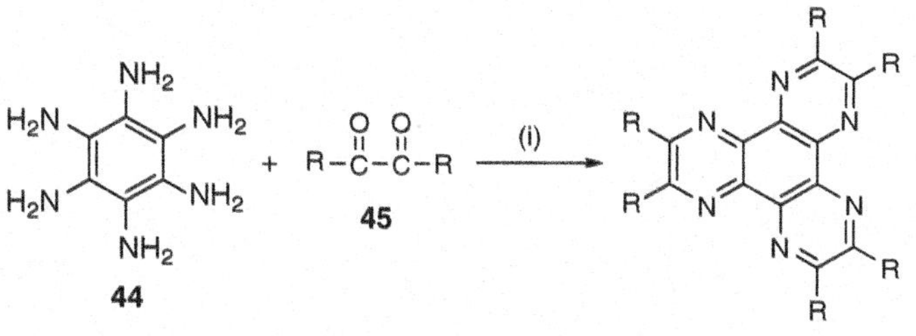

Scheme 4.12 Synthesis of hexaazatriphenylene derivatives; (i) AcOH, reflux

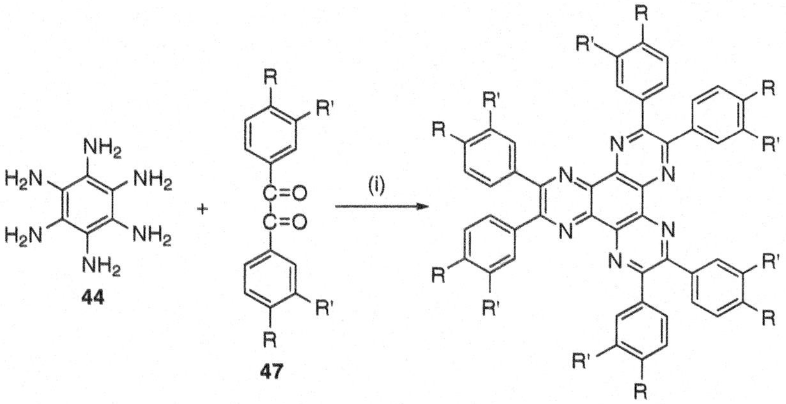

Scheme 4.13 Synthesis of liquid crystalline HAT derivatives; (i) AcOH, reflux

Czarnik and co-workers reported the synthesis of hydrogen-free HAT-hexacarbonitrile **51**, see Scheme 4.14 which acts as a potential precursor for the preparation of various liquid crystalline derivatives of HAT [93]. The regioselective displacement of the cyano groups of HAT-hexacarbonitrile **51** with alkoxy groups yields the symmetrical difunctional HATs **55** having alternative donor and acceptor substituents, see Scheme 4.14. On the other hand, displacement of the cyano groups of **51** with alkylthio groups produces hexathioethers of HAT **56**. The hexacarboxamido-HAT derivatives **57** can be easily prepared by reacting compound **53** with appropriate amines [93, 94]. Due to strong H-bonding, one of the compounds of this series exhibits very small intracolumnar inter-disc distance (3.18 Å) and quite large intracolumnar correlation length (120–180 Å). Owing to the small intracolumnar distance and high correlation length, it was anticipated that these materials should display high charge carrier mobility. The charge carrier mobility in this compound was measured by pulse-radiolysis time-resolved microwave conductivity and found to vary from 0.04 to 0.08 cm^2/Vs.

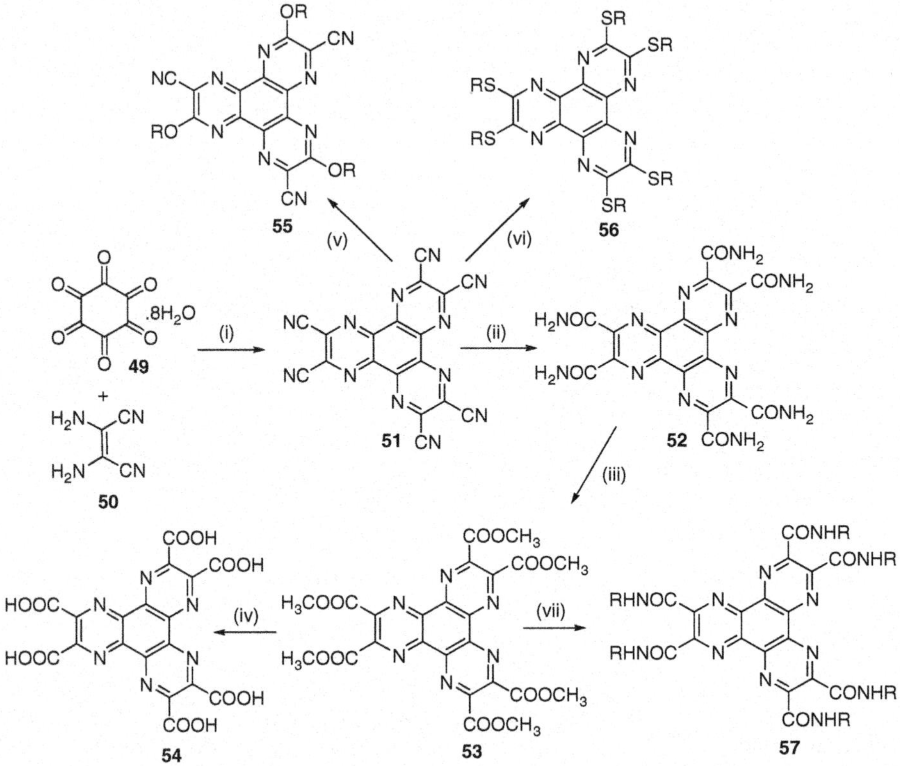

Scheme 4.14 Synthesis of hexazatriphenylenehexacarbonitrile and derived materials; (i) AcOH; (ii) Conc. H_2SO_4; (iii) MeOH, H_2SO_4; (iv) Et_3N/H_2O; (v) ROH, CH_3CN; (vi) RSH, K_2CO_3, DMF; (vii) RNH_2

4.5 Pyrene Core

Pyrene is an important molecule in organoelectronics due to its high fluorescence quantum yield. Its derivatives have been extensively studied as fluorophores in various areas of chemical biology. Various pyrene derivatives have found applications as sensors, organic light emitting diodes, photoconductors, fluorescent polymers, genetic probes, etc. [95–105]. The parent hydrocarbon can be easily substituted at 1,3,6 and 8-positions to generate liquid crystals. Thus, 1,3,6,8-tetrahalopyrene **59** may be obtained in excellent yield *via* halogenation (bromination or chlorination) of pyrene **58**, see Scheme 4.15 [106, 107]. Reaction of **59** with fuming sulphuric acid followed by hydrolysis yields oxidized product 3,6-dihydroxypyrene-1,6-quinone **60**. Reductive esterification of this quinone in the presence of different acid chlorides generates various 1,3,6,8-tetraalkanoyloxypyrene derivatives **61**. These tetraesters with long linear or branched alkyl chains were found to be non-mesomorphic but columnar phases can be induced *via* charge-transfer complexation

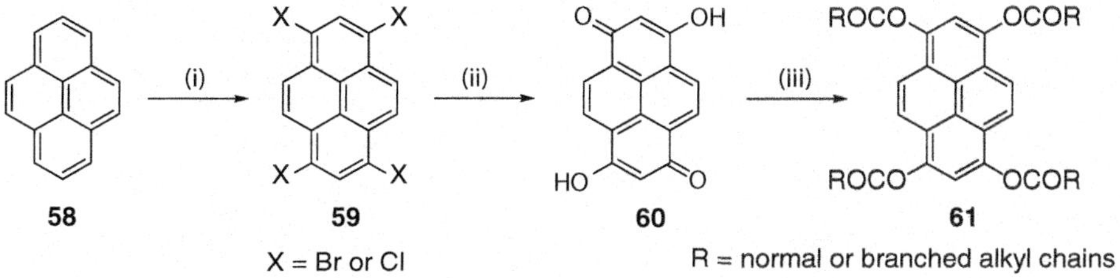

58 59 60 61

X = Br or Cl

R = normal or branched alkyl chains

Scheme 4.15 Synthesis of pyrene tetraesters; (i) Br_2, nitrobenzene or Cl_2, $C_2H_2Cl_4$; (ii) 25% H_2SO_4-SO_3; 40% H_2SO_4, H_2O; (iii) RCOCl, Zn, DMAP, THF, pyridine

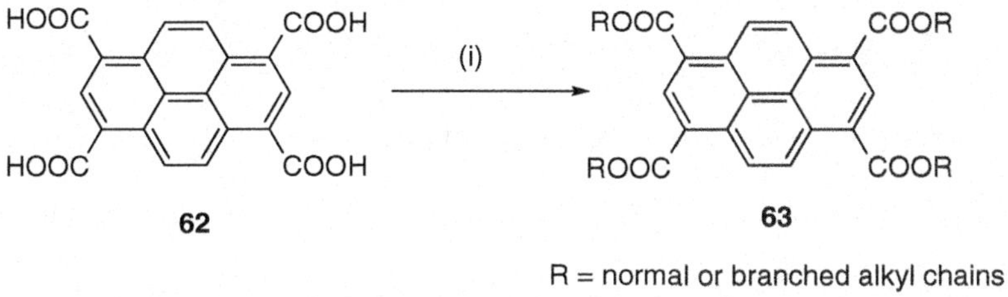

62 63

R = normal or branched alkyl chains

Scheme 4.16 Synthesis of pyrene 1,3,6,8-tetracarboxylic acid esters; (i) ROH, DCC, DMAP

with an electron acceptor, trinitroflourenone. On the other hand, some chiral tetraester exhibits a monotropic columnar phase [106]. These materials have been used to demonstrate ferroelectric switching in discotic liquid crystals.

Surprisingly, short chain normal alkyl esters **63** (R = methyl or ethyl group) and racemic 2-ethylhexyl ester, **63** (R = 2-ethylhexyl group), of pyrene 1,3,6,8-tetracarboxylic acid are reported to be liquid crystalline [108–110]. These tetraesters have been used to fabricate an organic light emitting diode device [109]. The synthesis of such tetraesters is straight forward *via* the esterification of commercial pyrene 1,3,6,8-tetracarboxylic acid **62**, see Scheme 4.16.

A variety of discotics can be prepared from the tetrabromopyrene **59** as shown in Scheme 4.17. Thus, the Suzuki coupling between **59** and commercially available methoxyphenylboronic acid or 1-fluorophenylboronic acid yields tetraphenyl derivatives **64** and **67**, respectively. Demethylation of **64** with HBr yields 1,3,6,8-tetrakis(4-hydroxyphenyl)-pyrene **65** which can be alkylated with different alkyl halides to tetraethers **66**. Similarly, esterification of the tetraphenol with acid chloride or benzoic acid affords tetraesters **66** (R = COR) or benzoate **66** (R = COPhR), respectively [111]. Replacement of fluoro atoms with alkylthiolates generates nonmesomorphic tetra-thioethers of pyrene **68**.

The Suzuki coupling reaction of tetrabromopyrene **59** with appropriate boronic acid ester yields tetraaryl-pyrenes **69**, see Scheme 4.18 [112, 113] while the Sonogashira-Hagihara reaction in between **59** and 3,4,5-tridodecyloxyphenylacetylene produces 1,3,6,8-tetrakis(3,4,5-trisdodecyloxyphenylethynyl)pyrenes **70** [114].

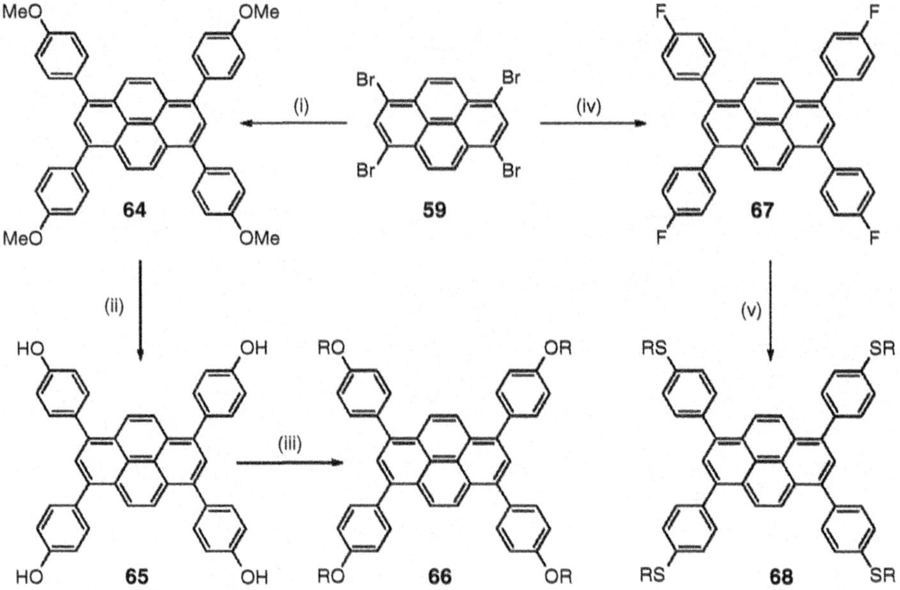

Scheme 4.17 Synthesis of tetraarylpyrene derivatives; (i) 1-methoxyphenylboronic acid, K_2CO_3, $Pd(PPh_3)_4$, dioxane; (ii) HBr, 130 °C, 14 h; (iii) RBr, K_2CO_3, DMF, 85 °C, 16 h; (iv) 1-fluorophenylboronic acid, K_2CO_3, $Pd(PPh_3)_4$, dioxane; (v) RSH, NaH, DMF

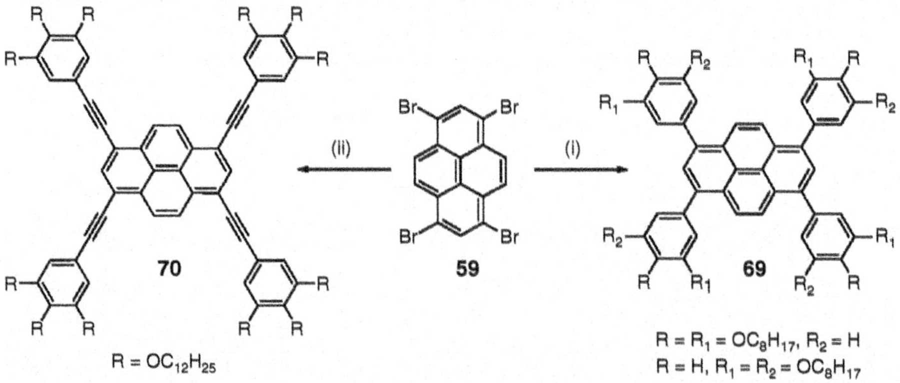

Scheme 4.18 Synthesis of Synthesis of tetraaryl and alkynylpyrene derivatives; derivatives; (i) arylboronic acid, K_2CO_3, $Pd(PPh_3)_4$, toluene; (ii) 3,4,5-tridodecyloxyphenylacetylene, $Pd(PPh_3)_4$,CuI, NEt_3, toluene

4.6 Perylene Discotics

In a similar way to pyrene derivatives, the excellent electronic absorption and fluorescence emission properties of perylene derivatives lead to scientists working in various areas of materials science to study their physical properties. Depending

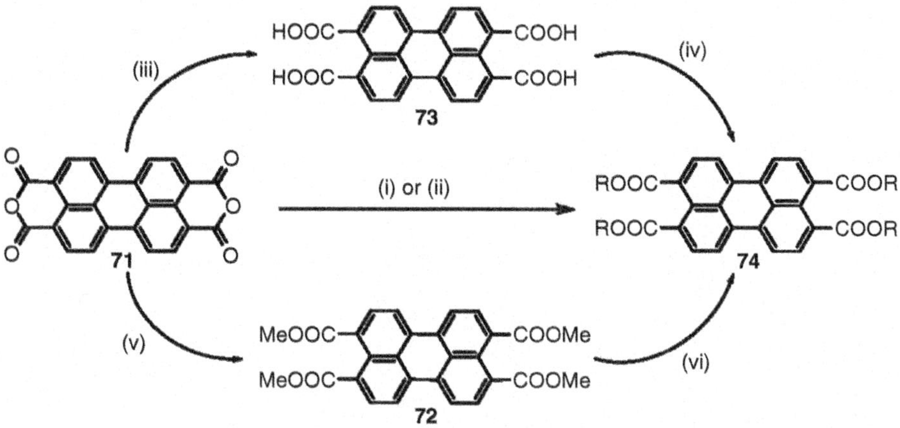

Scheme 4.19 Synthesis of perylene tetraesters; (i) ROH/RBr (1:1), K_2CO_3, 7 days; (ii) ROH/RBr (1:1), DBU, CH_3CN, 12–16 h; (iii) aq KOH, reflux, aq H_2SO_4 or HCl; (iv) $(C_8H_{17})_4NBr$ [TOAB], RBr, reflux, 2 h; (v) $(CH_3)_2SO_4$, aq. NaOH, reflux, 24 h, aq HCl; (vi) ROH, RONa

on the degree and nature of the substitution patterns, perylene derivatives can fluoresce anywhere across yellow-to-red parts of the visible spectrum. Therefore, perylene based-compounds are interesting materials for various applications, such as organic photovoltaic (OPV) cells [115, 116], see Chap. 8 and organic field effect transistors (OFETs) [117, 118], see Chap. 9 and organic light emitting diodes (OLEDs) [119, 120], see Chap. 7, etc. Perylene-based liquid crystalline materials are primarily two types, i.e., ester-based and imide-based. Only discotics derived from perylene imides have been studied for their charge transport properties so far. However, before describing the chemistry of perylene bisimides, the chemistry of liquid crystalline tetraalkyl esters of perylene is presented briefly first.

Though the synthesis of perylene tetraalkyl esters had been reported much earlier [121, 122], the liquid crystalline property of these materials were first identified by Bennings et al. in 2000 [123]. Various synthetic methods to prepare perylene tetracarboxylate have been reported in literature. Commercially available perylene-3, 4, 9, 10-tetracarboxylic acid dianhydride **71** on refluxing with dimethyl sulphate in aqueous NaOH for a day followed by neutralization furnished 3, 4, 9, 10-tetrakis(carbimethoxy)perylene **72**, see Scheme 4.19 [121]. Transesterification of perylenetetracarboxylate tetramethyl ester **72** with alkanol in the presence of sodium alkanoate yields 3, 4, 9, 10-tetra-(*n*-alkoxycarbonyl)-perylene **74** in good yield [122]. These tetraester **74** can also be prepared by reacting perylene dianhydride with equimolar amount of alkyl bromide or iodide and alkanol in presence of a base. Vigorous heating of perylene-3,4,9,10-tetracarboxylic acid dianhydride **71** with K_2CO_3 in a 1:1 mixture of alkyl halide and alkanol for 2–7 days produces 3,4,9,10-tetra-(*n*-alkoxycarbonyl)-perylene **74** in a single step [108, 123, 124]. The use of strong base 1,8-diazabicyclo[5.4.0]undec-7-ene (DBU) instead of potassium carbonate furnished the product in a short period [125]. Efficient product

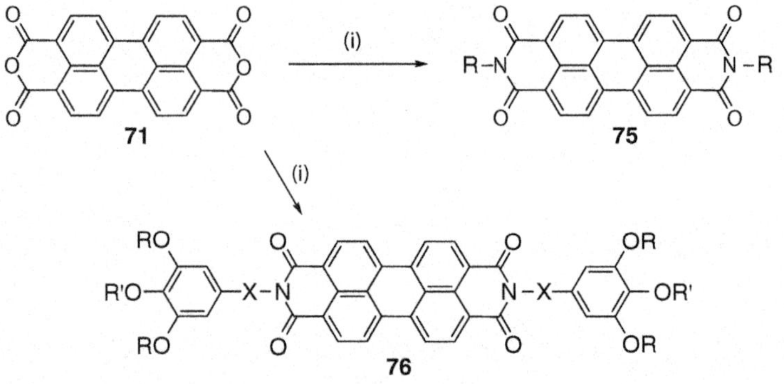

Scheme 4.20 Synthesis of perylene bisimides; (i) $RNH_2/ArNH_2$, $Zn(OAc)_2$, quinoline, 130–180 °C, 2–3 h or RNH_2, pyridine, 100–125 °C

formation was achieved by heating to reflux of a mixture of perylene-3,4,9,10-tetracarboxylic acid dianhydride **71,** alkyl bromide, alkanol [1:1 molar ratio] and DBU in acetonitrile for overnight. These esters can also be obtained from classical esterification of perylene-3,4,9,10-tetracarboxylic acid **73** which is prepared *via* hydrolysis of perylene-3, 4, 9, 10-tetracarboxylic acid dianhydride **71** by refluxing in aqueous potassium hydroxide followed by acidification [126–128]. The use of a phase transfer catalyst, tetra-octylammonium bromide (TOAB), is a modified way of esterification to avoid complicated and time consuming mode of conventional transesterification [122]. Refluxing vigorously a mixture of **71** and alkyl bromide in water and TOAB for about 2 h produces tetraalkylester **74** in good yield [127, 128].

Perylene tetracarboxylic acid bisimides are commonly known as perylene bisimides (PBIs). They are readily available, inexpensive and thermally very stable materials. Perylene bisimides have been used in the dye and pigment industry for a very long time. More recently, their applications in organic electronics have been recognised [129–131]. PBIs have been used as *n*-type organic semiconductor in solar cells [115, 132, 133], organic light emitting diodes [134–136] and organic thin film transistors [118, 137–139]. During the past two decades, a number of PBI derivatives have been synthesized and explored for excellent electronic and liquid crystalline properties [140–153]. Cormier and Gregg reported the existence of mesomorphic behavior in PBIs [140, 141]. These materials are conveniently prepared by condensing perylene-3,4,9,10-tetracarboxylic dianhydride **71** with suitable primary amines in presence of pyridine at 100–125 °C, see Scheme 4.20. At present, PBI derivatives are commonly prepared *via* zinc acetate catalyzed condensation of appropriate alkyl or aryl amine with perylene-3, 4, 9, 10-tetracarboxylic acid dianhydride **71**, see Scheme 4.20. Some PBI derivatives having branched alkyl chains or polyoxyethylene chains directly connected to the perylene nucleus are reported to be liquid crystalline, but the exact nature of the mesophase was not reported. Most likely, they form smectic phases as the shape of such molecules is more-or-less rod-like or lathe-like. In order to

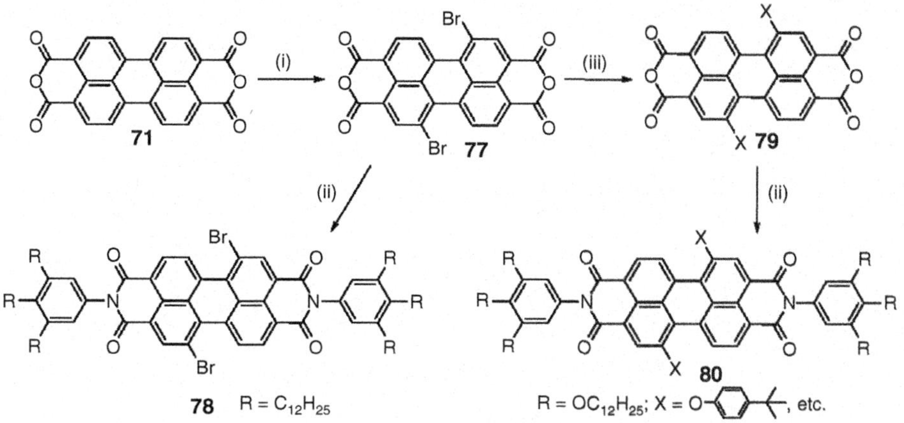

Scheme 4.21 Synthesis of bay-positions substituted perylene bisimides; (i) Br_2, I_2, H_2SO_4, 85 °C, 10 h; (ii) $ArNH_2$, $Zn(OAc)_2$, quinoline; (iii) ArOH, NMP, K_2CO_3, 80 °C

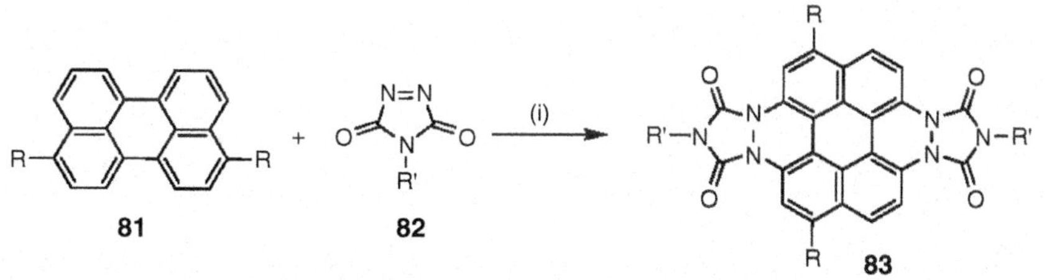

Scheme 4.22 Synthesis of 2,9-dialkyl-5,6,11,12-tetrahydrocoronene-5,6,11,12-tetracarboxylic acid bisalkylimide derivatives; (i) xylene, reflux

induce columnar phases, bulky groups were attached on the imide nitrogen atoms directly or *via* other groups, see Scheme 4.20, as well as in the bay positions, see Scheme 4.21. Bromination of perylene-3,4,9,10-tetracarboxylic acid dianhydride **71** yields 1,7-dibrominated bisimide **77** [154]. The brominated bisimide compound can be converted to the liquid crystalline imide **78** by treating with amines. Furthermore, halogen substituents in the bay regions are reactive towards substitution with nucleophiles. Thus, the reaction of halogenated bisanhydride with phenol in presence of potassium carbonate and NMP at 80 °C affords phenoxy substituted bisanhydrides **79**, which can be condensed with various amines to produce liquid crystalline imines **80**.

Müllen and co-workers reported the synthesis and mesomorphic behavior of 2,9-dialkyl-5,6,11,12-tetrahydrocoronene-5,6,11,12-tetracarboxylic acid derivatives **83** [155, 156]. The synthesis of these compounds is presented in Scheme 4.22. Only a mono-cycloaddition product of dienophiles to perylene was observed under normal conditions. However, the bisaddition product could be produced under drastic conditions, but only in low yield, which is attributable to the lower solubility

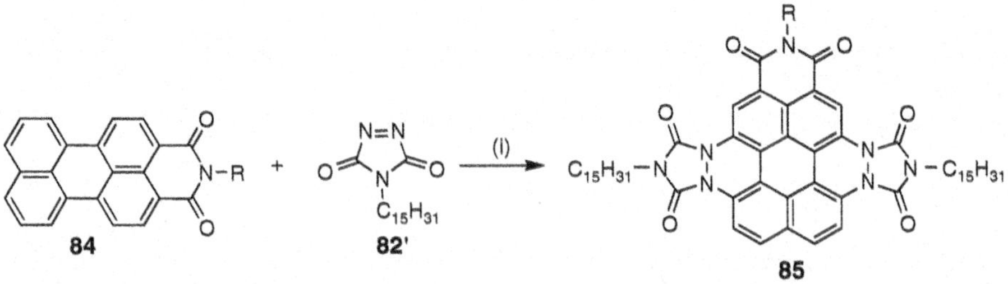

Scheme 4.23 Synthesis of T-shaped perylene mesogens; (i) *m*-Xylene, reflux

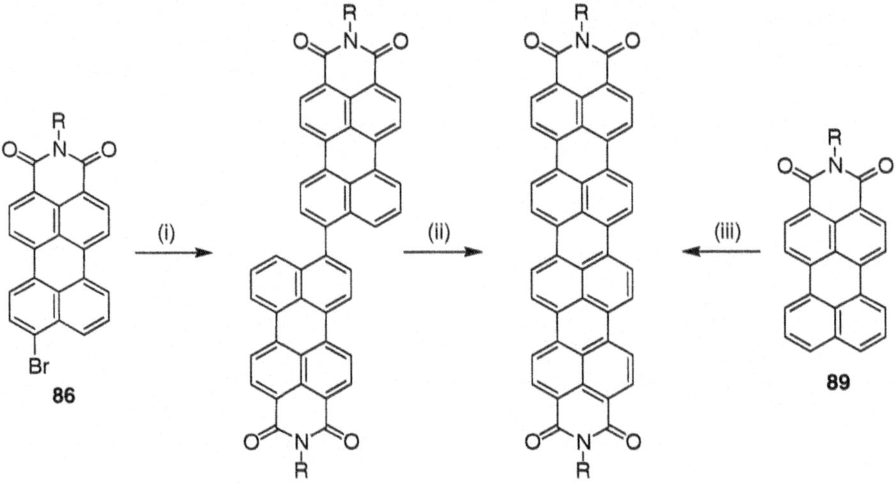

Scheme 4.24 Synthesis of quaterrylene diimides; (i) Ni^0; (ii) K_2CO_3; (iii) *tert.*-BuONa

and reduced reactivity of the enophile produced after the monoaddition reaction. Reasonable reaction yields for **83** were achieved by refluxing 3,10-dialkyl perylene **81** with an excess amount of 4-*n*-alkyl-3,5-dioxatriazole **82** in xylene. Periodical addition of **82** in small amounts to the reaction mixture until complete disappearance of the yellow starting material as well as the red monoaddition product improves the yield of the di-adduct.

Similarly, the Diels-Alder reaction of *N*-alkylated perylene monoimide **84** with *N*-pentadecyl-1,2,4-triazoline-3,5-dione **82** in boiling xylene yields the T-shaped, thermally as well as photochemically stable, blue perylene imide mesogens **85**, see Scheme 4.23 [157]. The low reactivity of **84** and poor stability of **82** require the addition of a large excess (ten equivalents) of **82** in several portions to a solution of **84** in refluxing *m*-xylene to achieve **85** in good yield.

Müllen and co-workers extended the synthesis of perylene discotics to various other core-enlarged perylene dyes, such as terrylene and quaterrylene diimides [158]. An example of quaterrylene diimides synthesis is shown in Scheme 4.24.

These compounds can be prepared either by a two-step synthesis involving homocoupling of perylene monobromide **86** to bis-perylene derivative **87** followed by cyclodehydrogenation or by a one-pot, direct coupling of two perylene units **89** under strong basic conditions.

4.7 Tricycloquinazoline Discotics

As a consequence of the many interesting physical properties, such as, trigonal symmetry, extended conjugation, colour, extraordinary thermal and chemical stability, low ionization potential and electronic properties, of tricycloquinazoline (TCQ) based DLCs, several examples of this class of organic semiconductor have been prepared recently. TCQ was synthesized by Cooper and Partridge during investigations on cyclic amidines [159]. Cyclotrimerization of *o*-aminobenzaldehyde or *o*-aminonitrile or anthranil can easily yield TCQ nucleus [159–161]. Trimerization of a disubstituted anthranil in presence of ammonium acetate in sulfolane-acetic acid has been used to prepare various TCQ based DLCs [162–171]. In general this trimerization is a poor-yielding process. Addition of ammonium acetate periodically during the reaction improves the yield marginally.

Two series of TCQ discotics have been reported. The first series was concerned with the 2,3,7,8,12,13-hexaalkylthio-TCQ derivatives with alkyl side chain length varying from 3 to 18 carbon atoms [162–166]. The synthesis of these derivatives is depicted in Scheme 4.25. Nitration of 3,4-dichlorotoluene, **90**, yielded

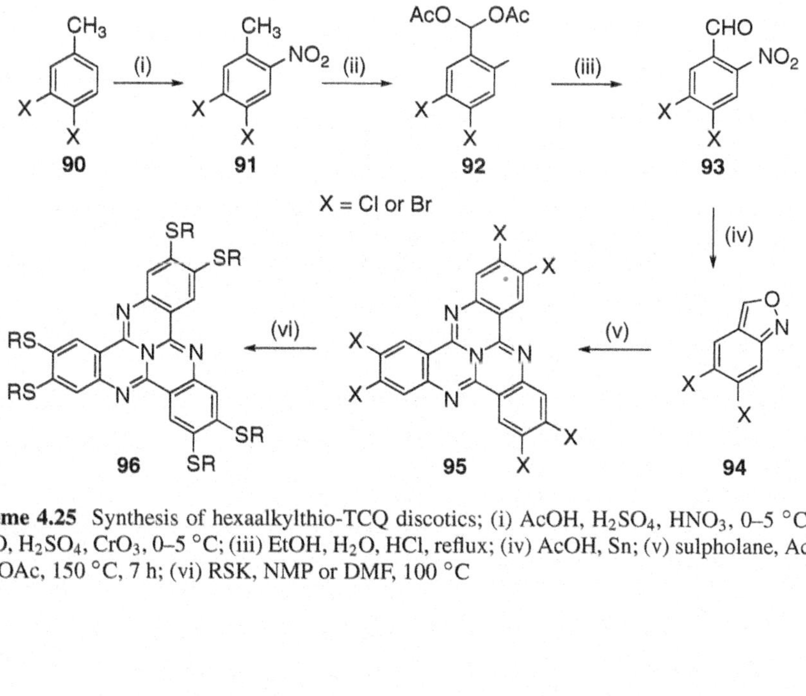

Scheme 4.25 Synthesis of hexaalkylthio-TCQ discotics; (i) AcOH, H_2SO_4, HNO_3, 0–5 °C; (ii) Ac_2O, H_2SO_4, CrO_3, 0–5 °C; (iii) EtOH, H_2O, HCl, reflux; (iv) AcOH, Sn; (v) sulpholane, AcOH, NH_4OAc, 150 °C, 7 h; (vi) RSK, NMP or DMF, 100 °C

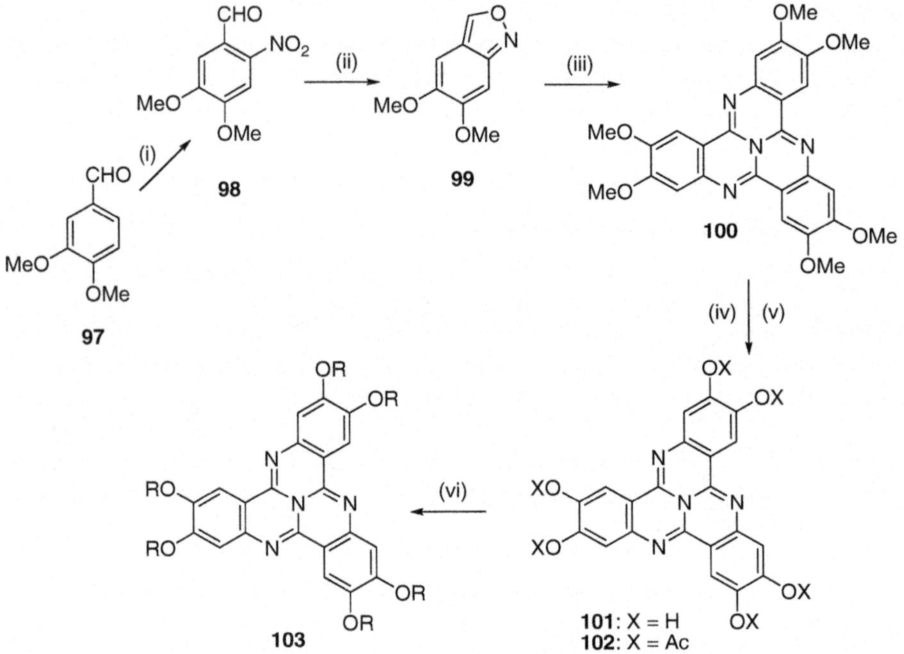

Scheme 4.26 Synthesis of hexaalkoxy-TCQ discotics; (i) HNO_3; (ii) AcOH, Sn; (iii) sulpholane, AcOH, NH_4OAc, reflux; (iv) pyridine.HCl, 220 °C; (v) pyridine, Ac_2O; (vi) KOH, DMSO, RBr

3,4-dichloro-2-nitrotoluene,**91**, which on oxidation with chromium trioxide in acetic anhydride afforded 3,4-dichloro-2-nitro-α-α-diacetoxytoluene, **92**. Hydrolysis of **92** furnished 3,4-dichloro-2-trinitrobenzaldehyde, **93**. Compound **93** can also be prepared directly from **91** in a single step using ceric ammonium nitrate. Partial reduction of **93** yielded anthranil **94**, which can be trimerized to hexachloro-TCQ, **95**. Hexabromo-TCQ can also be prepared in the same manner, but in lower yield. Treatment of hexachloro- or hexabromo-TCQ with alkylthiolate in DMF or NMP at 100 °C afforded hexathioethers of TCQ, **96**. It should be noted that, although both hexachloro- and hexabromo-TCQ are insoluble in either DMF or NMP, they react readily with alkylthiolates and give rise to soluble products.

The second series is based on 2,3,7,8,12,13-hexaalkoxy-TCQ derivatives [167–171]. Nucleophilic substitution of hexahalo-TCQ with alkoxide could produce TCQ hexaethers, however, this reaction with hexachloro-TCQ failed to produce hexaalkoxy-TCQ derivativs. Therefore, these derivatives were prepared from hexahydroxy-TCQ as shown in Scheme 4.26. Veratraldehyde, **97,** was nitrated to 2-nitroveratraldehyde, **98**, at low temperature in high yield. Partial reduction of **98** using tin foil and acetic acid produced 5,6-dimethoxyanthranil, **99**, which was successfully trimerized in the presence of ammonium acetate in refluxing sulfolane-acetic acid to produce hexamethoxy-TCQ, **100,** in low yield. Demethylation of all six methoxy groups of **100** was carried out in molten pyridinium hydrochloride at elevated temperatures, affording hexahydroxy-TCQ, **101,** in about **50%** yield. Hexaalkylation of **101** was achieved in low to moderate yields using excess of the appropriate n-bromoalkane in KOH/DMSO. The yield of final product **103** depends upon the purity of hexaphenol **101** which is sensitive to air oxidation and, therefore, a stored hexahydroxy-TCQ generally resulted in poor yield of **103**. The in situ conversion of hexaphenol **101** into its hexaacetate **102** followed by direct alkylation of the hexaacetate improved the yield of hexaalkoxy-TCQ significantly [168, 169]. Efforts have also been made to introduce desired alkoxy chains prior to trimerization step similar to hexaalkoxy-triphenylene synthesis but not much improvement was observed [170].

4.8 Prophyrin Discotics

Porphyrins are natural products which are important not only in biological sciences but also in materials science. Porphyrins and their metal complexes have also stirred interdisciplinary interest due to their intriguing physical, chemical and biological properties [172–177]. Because of their biological and materials science importance, almost all the metals have been incorporated in the porphyrin nucleus (central cavity). The interest in porphyrin derivatives for organic electronic applications originates from their photo-stability, photo-absorption over a broad range of wavelengths, interesting photophysical properties and convenient chemical synthesis. These attractive properties make them suitable for many applications, such as, organic photovoltaic, electrophtographic, photoelectrochemical applications, etc. A large number of compounds based on porphyrins are known to display thermotropic mesophases. About 300 discotic liquid crystalline derivatives derived from porphyrin nucleus have already been reported. Most of these fall into two categories; β-substituted porphyrin derivatives and *meso*-substituted porphyrin derivatives.

The porphyrin macrocycle, consists of four pyrrole rings joined by four interpyrrolic methine bridges to give a highly conjugated macrocycle. Porphyrins can be prepared *via* several methods, such as, tetramerization of monopyrroles, dimerization of dipyrromethanes and from open chain pyrrolic derivatives. However, tetramerization of β-substituted pyrroles to generate symmetrical octasubstituted-porphyrins is one of the most common methods to generate porphyrin based DLCs. Gregg et al. prepared a series of porphyrin based octaether discotic liquid crystals [178] as shown in Scheme 4.27. The pyrrole trimethyl ester **104** on selective hydrolysis using sodium methoxide in methanol yields the diacid monomethylester **105**. This compound on reduction with BH_3/THF followed by tosylation yields the ditosylated derivative **107**. The ditosylate derivative on heating with the appropriate alcohol forms the corresponding diether **108** which is converted to compound **109***via* oxidation. The pyrrole **109** was transformed into porphyrin **110** in one pot synthesis using KOH to hydrolyze the ester, hydrobromic acid for decarboxylation and cyclization, and chloranil for oxidation of the macrocycle to the porphyrin discotics. Alternatively, these porphyrin ethers have also been prepared from the

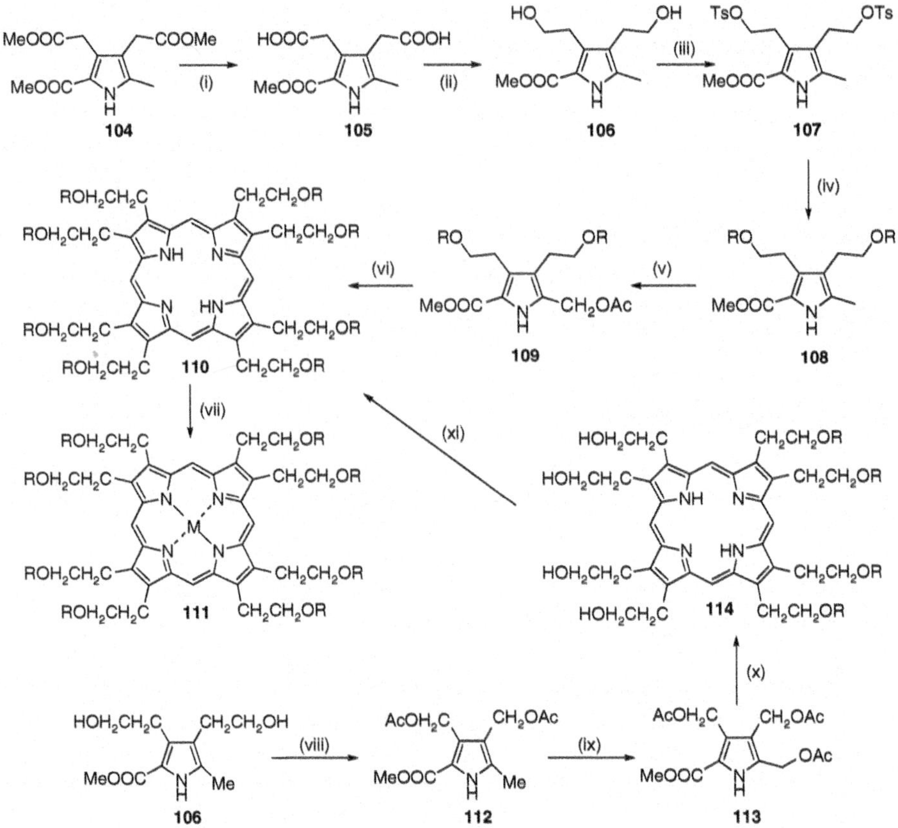

Scheme 4.27 Synthesis of Octaethanol porphyrin and its octaethers; (i) NaOMe/MeOH; (ii) BH_3/THF; (iii) TsCl/pyridine; (iv) ROH/toluene; (v) $Pb(OAc)_4$; (vi) (a) KOH/EtOH; (b) HBr/EtOH; (c) chloranil; (vii) $M(OAc)_2$, CH_2Cl_2/MeOH, reflux; (viii) AcCl/NEt_3; (ix) $Pb(OAc)_4$; (x) (a) KOH/EtOH; (b) HBr/EtOH; (c) O_2; (xi) NaH/DMSO, RX

porphyrin octaethanol **114**, which in turn can be generated from the pyrrole **106**. Metallated derivatives **111** were prepared by heating the metal-free compound with the metal salts. Porphyrin octaesters have also been prepared similarly [179].

The *meso*-substituted porphyrins, though not naturally occurring, are widely preferred candidates in various fields such as biomimetic models, materials chemistry, photodynamic therapy, catalysis, electron transfer, etc. [180]. The synthesis of these derivatives is much simpler compared to their β-substituted counterparts. Rothenmund et al. prepared the first *meso*-tetramethyl porphyrin derivative by the condensation of acetaldehyde and pyrrole [181]. A side-product "chlorin", a porphyrin-related macrocycle, which is largely, but not completely aromatic in nature, is usually formed in this reaction. However, chlorins can be easily oxidized to form the corresponding porphyrins. Alder et al. developed a new method in the 1960s to synthesize tetraphenyl derivatives by the condensation reaction of benzaldehyde and pyrrole in the presence of acidic solvents under refluxing

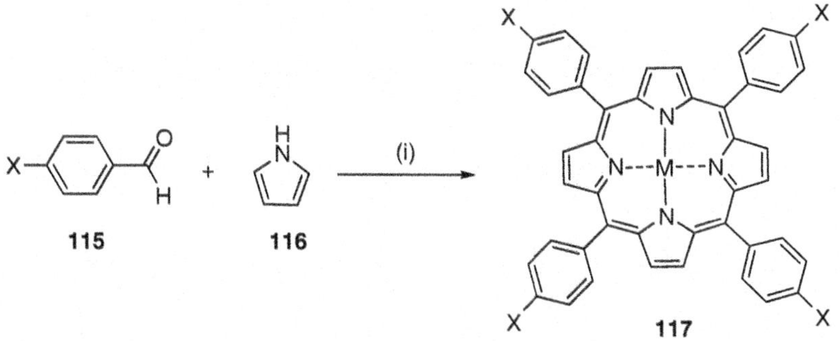

Scheme 4.28 Synthetic route to prepare *meso*-tetra(4-alkylphenyl)porphyrins; (i) propionic acid, reflux; DDQ; $M(OAc)_2$, DMF

conditions. This method gives much higher yield compared to that achieved for the tetramethyl derivative synthesized by Rothenmund et al. This reaction also produces chlorins which was oxidized by the use of DDQ. The methodology was modified slightly by Longo et al. and now commonly known as the Alder-Longo method [182]. Thus, the condensation of pyrrole and benzaldehyde in propionic acid under reflux for 30 min yields tetraphenylporphyrin in about 20% yield, see Scheme 4.28. Several other methods have been developed to improve the yield [183, 184]. The yield can be achieved up to 50% depending on the choice of aldehyde and acid [183].

A large number of *meso*-tetra-substituted porphyrins have been prepared to explore their mesomorphic properties [185]. While simple tetraalkyl-substituted porphyrins are generally non-mesomorphic, many tetra-(4-alkylphenyl), tetra-(4-alkoxyphenyl), tetra-(3,4-dialkylphenyl), tetra-(3,4-dialkoxyphenyl), tetra-(4-carboxyphenyl), tetra-(alkanoyloxyphernyl), etc., display mesomorphism. Although the importance of porphyrins in photosynthesis is well known, it is surprising that only a few discotic porphyrin derivatives have been studied for their charge transport properties. Recently an interesting highly conjugated metalloporphyrin dimeric DLC **126** was prepared by Sakurai et al. in order to investigate its electron transport properties [186]. It was consequently found that it exhibits a very large value for electron mobility. The synthesis of this material is shown in Scheme 4.29. Condensation of trialkoxybenzaldehyde **119** with the dipyrromethane **118** under standard porphyrin synthesis reaction conditions yields the disubstituted porphyrin nucleus **120**. It was converted to a metallated trisubstituted porphyrin nucleus **121** *via* a sequence of reactions. Bromination of **121** gives the mono-brominated product **122**, that can be converted to a boronic acid derivative **123**. Compound **124** can be prepared following the same set of reaction conditions, but having different alkyl chains in the starting benzaldehyde. Suzuki coupling of compounds **123** and **124** yields the dimer **125**, which can be cyclized to form the desired fused metalloporphyrin dimer **126**. Symmetrical dimers having identical peripheral chains have also been prepared in the same manner, but they do not display any observable mesomorphism.

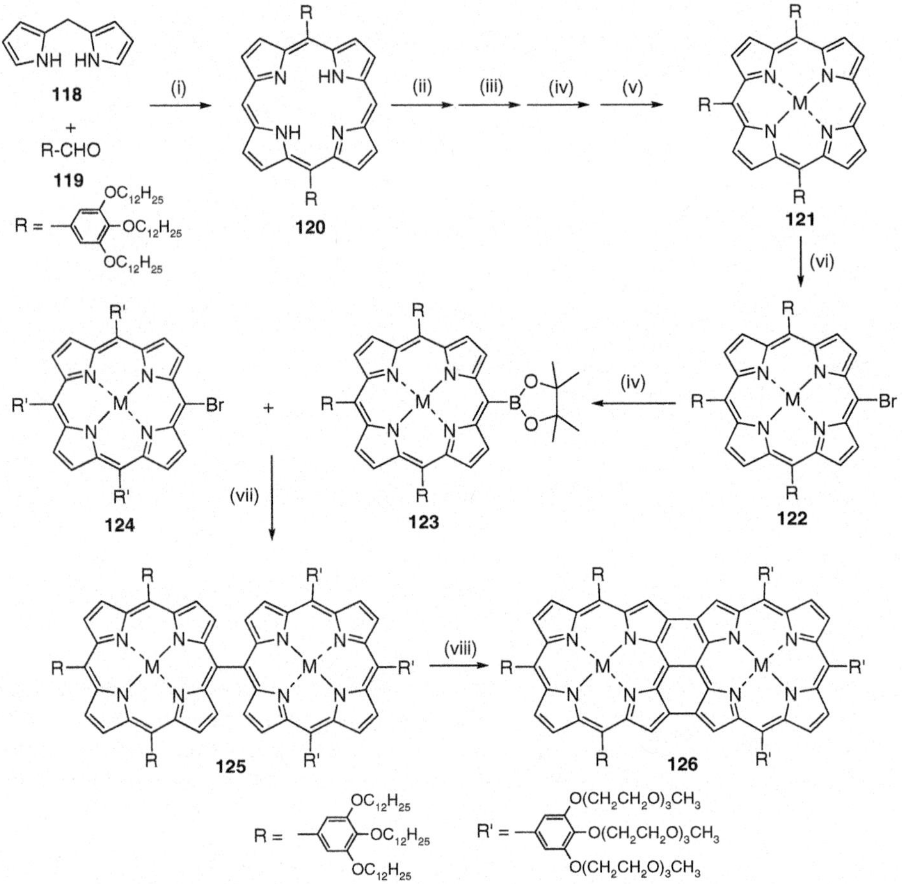

Scheme 4.29 Synthesis of a triply fused metalloporphyrin dimer; (i) $BF_3.OEt_2$, DDQ, CH_2Cl_2; (ii) NBS, $CHCl_3$; (iii) $Zn(OAc)_2$, $CHCl_3$/MeOH; (iv) 4,4,5,5-tetramethyl-1,3,2-dioxaborolane, $PdCl_2(dppf)_2$, Et_3N, $ClCH_2CH_2Cl$; (v) 3,4,5-triiododecyloxtbromobenzene, $Pd(PPh_3)_4$, Cs_2CO_3, DMF, toluene; (vi) NBS, pyridine, $CHCl_3$; (vii) $Pd(PPh_3)_4$, Cs_2CO_3, DMF, toluene; (viii) DDQ, $Sc(OTf)_3$, toluene

4.9 Phthalocyanine Discotics

The phthalocyanine (Pc) nucleus is closely related to the porphyrin nucleus and, therefore, compounds based on this molecular building block are often referred to as tetrabenzo-tetraazaporphyrins. Pcs and their metallo-derivatives are not only well known in the dyes and pigment industries, but they have also been recognized as building blocks for the construction of new molecular materials for electronics and optoelectronics and they have also attracted increasing interest for biological and biomedical applications. Peripheral substitution of the Pc nucleus with long

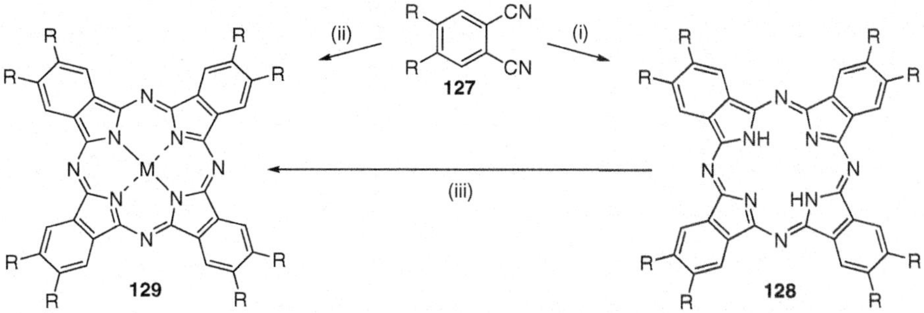

Scheme 4.30 Tetramerization of a phthalonitril derivative to phthalocyanine; (i) reflux in a high boiling solvent; (ii) reflux in a high boiling solvent with metal salt; (iii) reflux in EtOH with metal salt

hydrocarbon chains not only makes them processable, due to enhanced solubility in various organic solvents, but also induces liquid crystallinity in many cases. Thermotropic mesomorphism in these materials was recognised in 1982 by Piechocki et al. [187]. After that many examples of mesogenic Pcs have been prepared with variations in the nature, number, length and position of the flexible aliphatic side-chains. The linking group attaching the side-chains to the Pc core and the central metal ion exert a substantial influence on the nature and transition temperatures of the mesophases exhibited by these derivatives. A large number of symmetrical as well as nonsymmetrical Pcs with substitutions at the α or/and β sites have been prepared to study various aspects of their physical properties.

Although Pcs can be prepared from many different precursors, tetramerization of a disubstituted phthalonitrile, see Scheme 4.30, is the most common method to prepare octa-substituted Pc discotics. The cyclization is commonly carried out in a high-boiling solvent in the presence of a non-nucleophilic strong base. A metal salt can be added in the same reaction to obtain the metallated phthalocyanine. Alternatively, free phthalocyanine can be first prepared and then reacted with appropriate metal salt. However, the isolation of metal-free phthalocyanine is often tedious and low yielding and, therefore, it is convenient to prepare metallated phthalocyanines directly, unless the metal-free phthalocyanine is absolutely required. This methodology has been utilized to prepare a variety of Pc discotics such as, peripherally octaalkoxymethyl substituted Pcs, octaalkoxy substituted Pcs, octaalkyl substituted Pcs, octathioalkyl substituted Pcs, octathiaalkylmethyl substituted Pcs, octaalkoxyphenyl- and alkoxyphenoxy substituted Pcs, nonperipherally substituted octaalkyl and octaalkoxymethyl Pcs, nonsymmetrical octa-, hepta-, hexa-, and penta-substituted Pcs; crown-ether substituted Pcs and core-extended macrodiscotic Pcs, etc.

The synthesis of the desired phthalonitrile precursor is of primary importance in the preparation of Pc discotics. These phthalonitriles are generally prepared using classical chemical reactions. A typical method of synthesis of some phthalonitriles

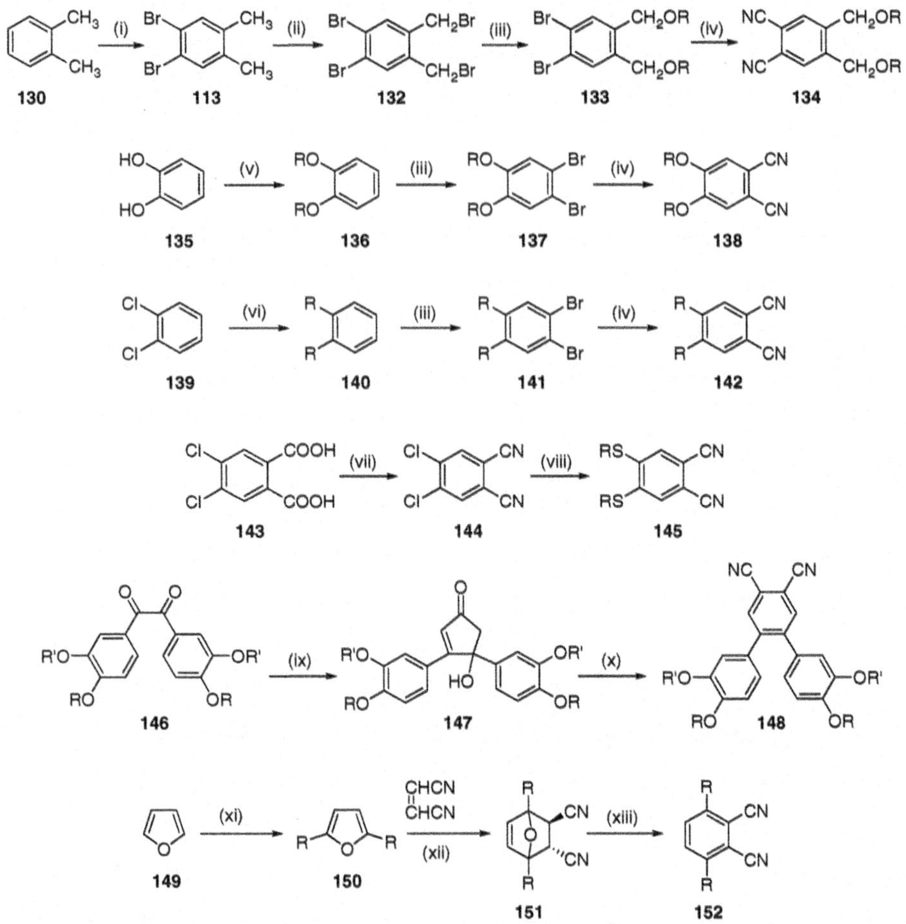

Scheme 4.31 Synthetic routes to prepare some phthalonitriles used in the preparation of phthalocyanine discotics; (i) Br_2; (ii) NBS; (iii) ROH, K_2CO_3; (iv) CuCN, DMF; (v) RBr, base; (vi) RMgBr, Ni catalyst; (vii) Ac_2O; $HCONH_2$, reflux; NH_3; $SOCl_2$; (viii) RSH, K_2CO_3/DMSO; (ix) potassium *tert.*-butanolate, acetone, ethanol; (x) dicyanoacetylene, p-toluenesulfonic acid, chlorobenzene; (xi) (i) BuLi; RBr; (xii) $LiN(SiMe_3)_2$, THF; (xiii) H_2O

used in producing Pc discotics is shown in Scheme 4.31. Thus, for instance, dialkoxymethyl substituted phthalonitrile, used to prepare octaalkoxymethyl substituted phthalocyanines, are prepared from *o*-xylene. Bromination of *o*-xylene **130** yields the dibromo-derivative of xylene, **131**. Bromination of the methyl groups of **131** is accomplished by treating with NBS (*N*-bromosuccinimide) to afford the tetrabrominated compound **132**. Treatment of **132** with appropriate alkoxide displaces the bromine atom attached to methylene group and furnishes compound **133**, which can be converted to phthalonitrile **134** by reacting with Cu(I)CN in DMF. Similarly, the octa-alkoxy-substituted phthalocyanines, which are the most widely studied phthalocyanine discotics, are prepared from cyclization of 1,2-dicyano-4,5-bis(alkoxy)benzene **138**. Catechol **135** can be alkylated to 1,2-dialkoxybenzene **136**, which on bromination easily furnishes the dibromo-derivative **137**. Replacement of the bromo-groups by a cyano-moiety can be accomplished as described above by reacting with Cu(I)CN in DMF or NMP. Alkylation of *o*-dichloro benzene **139** with a Grignard reagent in the presence of a nickel catalyst gives dialkylbenzenes **140**. Bromination with molecular bromine followed by reaction with copper cyanide in DMF yields the dialkyl phthalonitrile **141**, which is used to prepare octaalkyl phthalocyanines. On the other hand, 1, 2-dichloro-4,5-dicyanobenzene **144** can be prepared from 4, 5-dichlorophthalic acid **143**. Replacement of chlorine atoms by alkylthio-groups can be achieved by reacting **144** with the potassium salt of an alkanethiol in DMSO to yield the thioalkyl-substituted phthalonitrile **145**. Thus octathioalkyl-substituted Pcs can be obtained from this phthalonitrile. The synthesis of peripheral octaalkoxyphenyl-substituted Pcs needs 3,3′,4,4′-tetraalkoxy-*o*-terphenyl-4′,5′-dicarbonitrile precursor **148**. It can be prepared from 3,3′,4,4′-tetra-*n*-alkyloxybenzil **146** by treating with acetone and potassium *tert.*-butanoate to give 3,4-bis(3,4-dialkyloxyphenyl)-4-hydroxy-2-cyclopenten-1-one **147**. This product on reaction with dicyanoacetylene gives the phthalonitrile **148.** In order to prepare non-peripherally substituted Pcs, alkyl, alkoxy, etc., substitutions are required at the α-positions. Thus, the 3,6-dialkylphthalonitrile **152** is prepared from 2,5-dialkylfuran **150**, which is synthesized by lithiation and alkylation of the furan **149**. A Diels-Alder reaction between 2,5-dialkylfuran and fumaronitrile furnishes **151**. The reaction is reversible and requires several weeks at low temperature to reach the equilibrium. The use of excess of fumaronitrile to push the reaction towards adduct formation causes problems as the product cannot be isolated by conventional purification methods and it is used as such, without further purification, to prepare phthalonitrile by reacting the adduct with lithium bis(trimethylsilyl)amide, which is a non-nucleophilic base. Unreacted fumaronitrile reacts with the base and creates problems in isolating the pure material. It is also generally difficult to isolate pure phthalonitriles and, therefore, in most cases it was used as prepared for conversion into the phthalocyanine. Similar strategies have been used to prepare a number of different phthalonitriles used in the synthesis of various Pc based discotics.

4.10 Hexabenzocoronene and Other Large Discotic Cores

4.10.1 Hexabenzocoronene Discotics

It is expected that enlarging the aromatic macrocycle would enhance the columnar order due to intense, intermolecular π-π interactions and, thus, increase the charge carrier mobility. Hexa-peri-hexabenzocoronene (HBC) is one of the largest and most symmetrical of all-benzenoid polycyclic aromatic hydrocarbons, that function as a core fragment for DLCs. Hexa-*peri*-hexabenzocoronene contains 42 carbon atoms and 13 phenyl rings, so it can be considered as a nano-graphene. Müllen

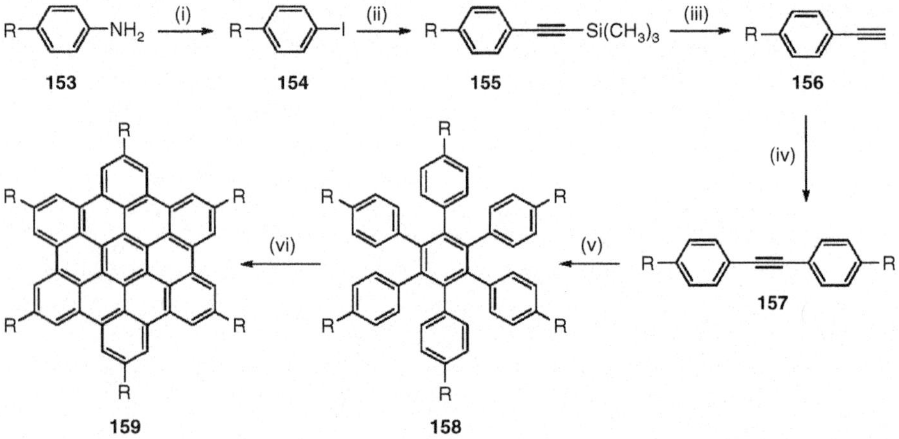

Scheme 4.32 Synthesis of HBC discotics; (i) $C_5H_{11}NO_2$, KI; (ii) TMSA, $[PPh_3]_2PdCl_2$, PPh_3, CuI, piperidine; (iii) KF, DMF; (iv) **154**, $[PPh_3]_4Pd$, CuI, piperidine; (v) $Co_2(CO)_8$; (vi) $AlCl_3$, $Cu(CF_3SO_3)_2$, CS_2

group exploited HBC and other large size polyaromatic hydrocarbons extensively to prepare DLCs and to study their charge transport properties. Some of these discotics have been successfully demonstrated to be useful for the fabrication of devices such as field effect transistors, photovoltaic solar cells, etc. Although the synthesis of parent HBC was reported by Clar et al. [188], Halleux et al. [189] and Hendel et al. [190], it was the Müllen's group, who first developed an efficient route to prepare the parent HBC, its derivatives and related PAH-structures by a Scholl-type intramolecular oxidative cyclodehydrogenation of oligophenylenes with Cu(II) salts such as $CuCl_2$ and $Cu(OTf)_2$ catalyzed by Lewis acid $AlCl_3$ [8, 191–196]. Subsequently, the reaction was performed using anhydrous $FeCl_3$. Iron(III) chloride acts both as a fairly strong Lewis acid and a mild oxidizing agent. It possesses an oxidation potential sufficient for the C-C bond formation. The general synthetic route of HBC and its sixfold symmetric derivatives is shown in Scheme 4.32. It starts with commercially available 4-alkyl aniline **153** which is transformed into 4-alkyliodobenzene **154** *via* the Sandmayer reaction. Sonogashira coupling in between **154** and trimethylsilylacetylene yields compound **155**. Deprotection of the trimethylsilyl group gives the free 4-alkylphenylacetylene **156**, which is coupled with 1-alkyl-4-iodobenzene under Sonogashira coupling reaction conditions to yield the 4,4′-di-*n*-alkyltolane **157**, the crucial building block of HBC derivatives. Cobalt octacarbonyl catalyzed cyclotrimerization of **157** yields the oligophenylene **158**. The oxidative cyclodehydrogenation (intramolecular Scholl reaction) of this hexaalkylphenyl benzenes **158** in the presence of $AlCl_3$ and $Cu(CF_3SO_3)_2$ or $FeCl_3$ results the desired HBC derivative **159**. Introduction of a phenyl group between the HBC core and the alkyl chains results in the formation of a liquid crystalline HBC derivative, which exhibits mesomorphic behavior at room temperature [197]. This compound also displays very high charge carrier mobility in the mesophase.

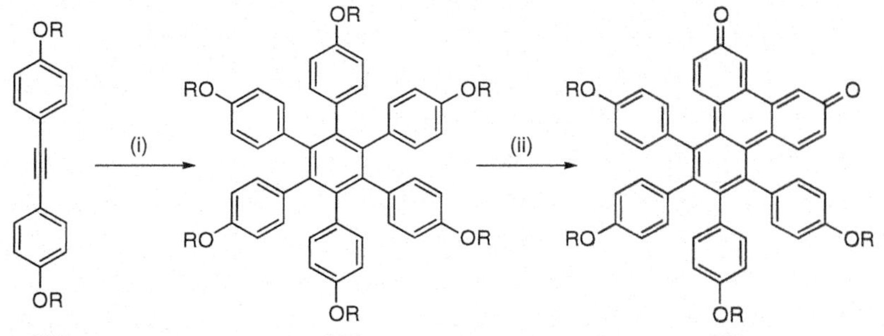

Scheme 4.33 Attempted synthesis of hexaalkoxy-HBC discotics; (i) $[Co_2(CO)_8]$, dioxane; (ii) $FeCl_3$, $MeNO_2$, CH_2Cl_2

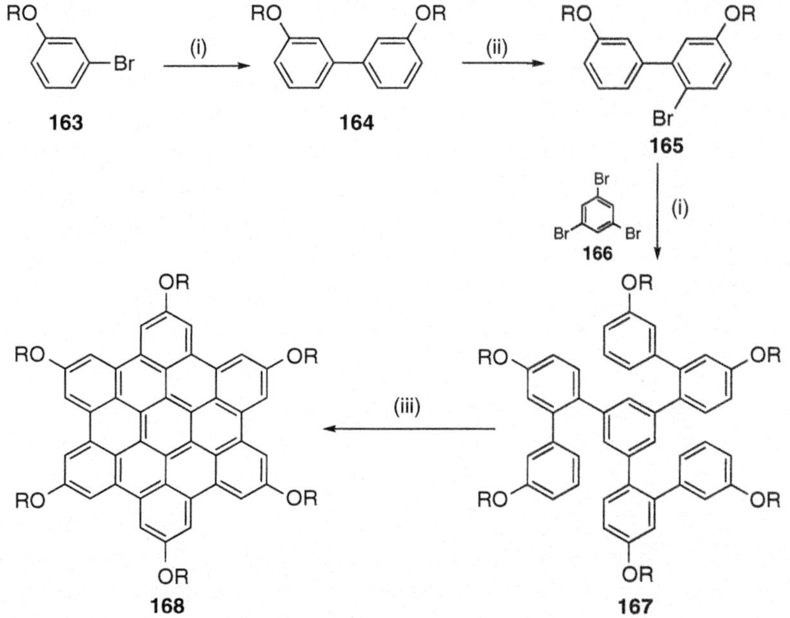

Scheme 4.34 Synthesis of hexaalkoxy-HBC discotics; (i) THF, -78 C, BuLi, $B(OMe)_3$; toluene/EtOH/H_2O/Na_2CO_3/$Pd(PPh_3)_4$; (ii) NBS, MeCN; (iii) $FeCl_3$, CH_3NO_2, CH_2Cl_2

Unlike the alkoxy-derivatives of many other discotic cores, the synthesis of alkoxy-substituted HBC derivatives could not be achieved for a long time, since the oxidative cyclodehydrogenation step results in dealkylation of the alkoxy-substituted hexaphenylbenzene **161** leading to a quinone product **162**, see Scheme 4.33 [198]. Recently, an alternative method, as shown in Scheme 4.34, has been developed to prepare this kind of materials successfully [199].

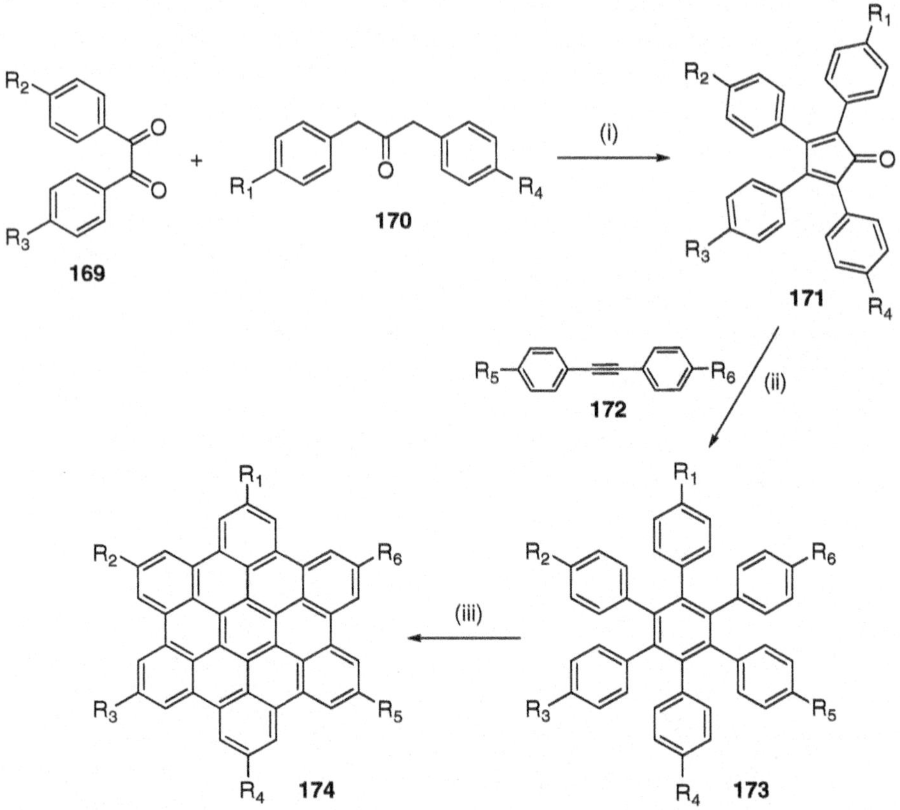

Scheme 4.35 Synthesis of unsymmetrically substituted HBC derivatives; (i) Bu_4NOH, MeOH, *tert.*-BuOH or KOH, EtOH, reflux; (ii) Ph_2O, 260 °C; (iii) $FeCl_3$, CH_2Cl_2, $MeNO_2$

The methodologies described could not be applied to the synthesis of unsymmetrical and some fluoro-substituted HBCs. Therefore an alternative method was developed as shown in Scheme 4.35 [200, 201]. It involves [4 + 2] Diels-Alder cycloaddition of a suitably substituted diphenyl acetylene **172** and 2,3,4,5-tetrarylcyclopenta-2,4-dien-1-one **171**. The cyclopentadienone is synthesized *via* double Knovenagel condensation between a 4,4′-substituted benzyl **169** and 1,3-diarylacetone **170**, see Scheme 4.35. Oxidative cyclodehydrogenation of **173** with iron (III) chloride yields HBC derivatives **174**. Thus both unsymmetrical and symmetrical HBC derivatives can be obtained *via* this versatile route.

It is remarkable that catalytic hydrogenation of HBC derivatives under moderate hydrogen pressure in the presence of palladium on activated carbon is reported to yield quantitatively regiospecific conversion to peralkylated coronenes **175**, see

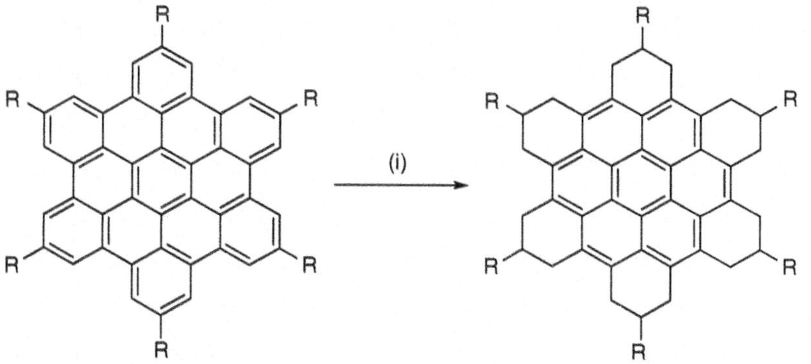

Scheme 4.36 Synthesis of peralkylated coronenes; (i) Pd, H_2

Scheme 4.36 [202]. Some of these coronene derivatives, e.g., having dodecyl and 3,7-dimethyloctyl peripheral chains, exhibit mesomorphic behavior, just like the precursor HBCs from which they were derived. These materials also exhibit high charge carrier mobilities. It is extremely intriguing that on the one hand, HBC exists in the atmosphere of stars, threby indicating its remarkable stability, but on the other hand, it undergoes facile catalytic hydrogenation, despite being an all-benzenoid polycyclic hydrocarbon.

4.10.2 Larger Discotic Cores

Discotic cores larger than hexabenzocoronene were designed and synthesized to move towards a graphene structure for better electronic properties. These cores were prepared using similar synthetic strategies. Thus, the $FeCl_3$- or $Cu(OTf)_2$-$AlCl_3$-mediated oxidative cyclodehydrogenation of branched hexaphenylbenzene derivative was applied to the synthesis of giant graphene molecules with different sizes and shapes [203–210]. Some of these are shown in Scheme 4.37. Diels-Alder reactions were applied to first prepare appropriate branched oligophenylenes and then these oligophenylenes were subjected to oxidative cyclodehydrogenation to generate planar graphene discs. Thus, polycyclic hydrocarbon cores with carbon atoms ranging from C44 to C132 have been realized. As the degree of conjugation increases, the colour of the compounds deepens. These compounds possess better optical and charge transport properties. Some of these discotics are excellent candidates for the fabrication of optoelectronic devices.

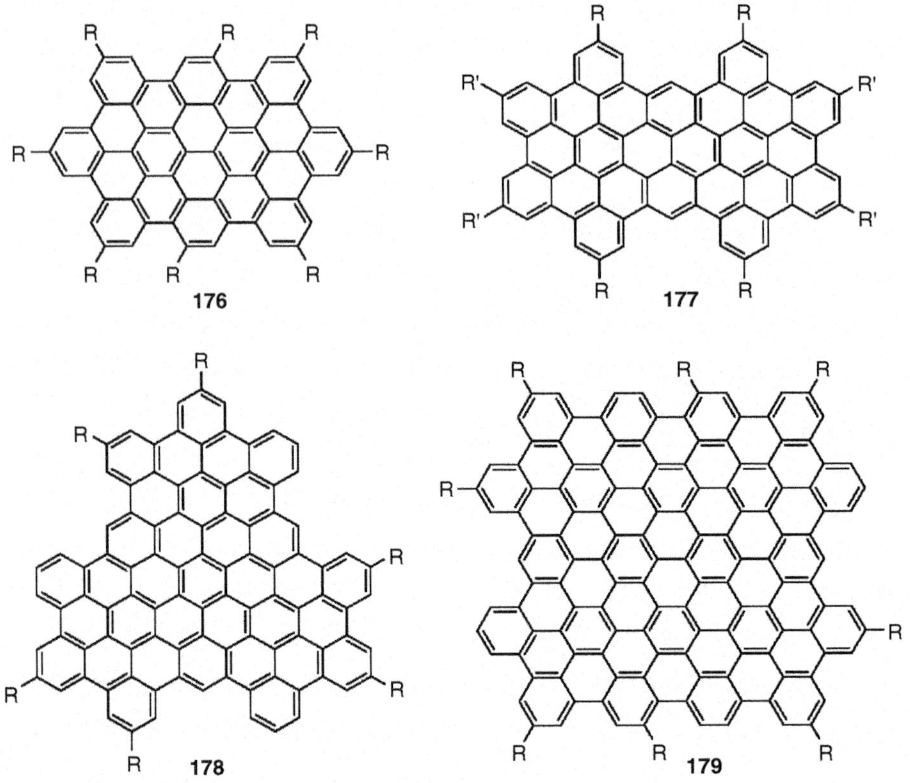

Scheme 4.37 Structures of some large core discotics

4.11 Miscellaneous Semiconducting Discotics

4.11.1 Oxadiazole Discotics

The time-of-flight electron mobility in a oxadiazole derivative namely, 1,3,5-tris{5-[3,4,5-tris(octyloxy)phenyl]-1,3,4-oxadiazole-2-yl}benzene, has recently been studied by Zhan et al. [211]. The synthesis of this material is straightforward as shown in Scheme 4.38. Condensation of 1,3,5-benzenetricarbonyl trichloride with the hydrazide **182** yields the precursor **183**, which can be converted to discotic oxadiazole derivative **184** by heating in $POCl_3$ at 80 °C [211, 212].

4.11.2 Carbazole Discotics

Carbazole derivatives are well known for their photoconducting properties. Incorporation of the carbazole moiety in the supramolecular structure of DLCs may enhance

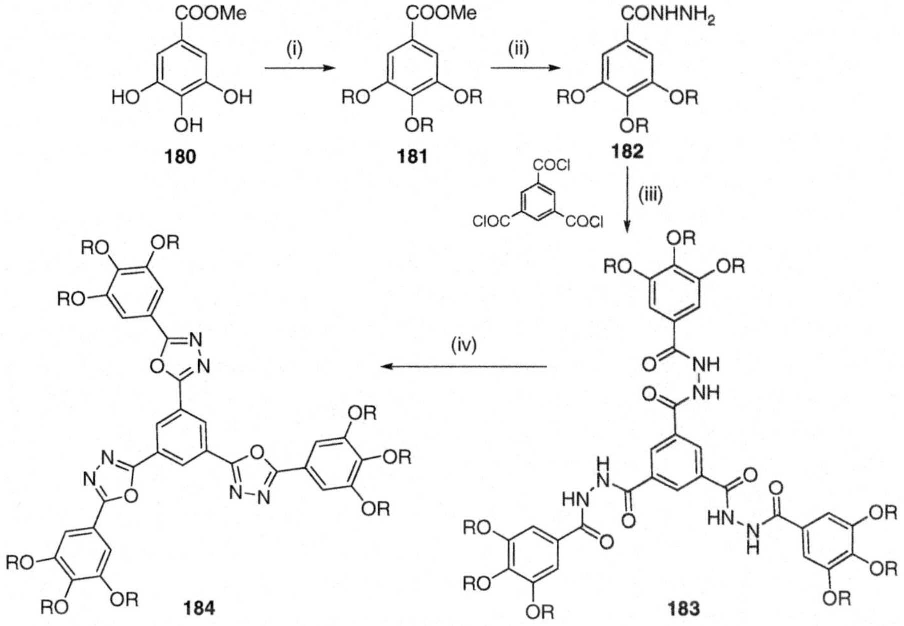

Scheme 4.38 Synthesis of oxadiazole-based DLCs; (i) RBr, K_2CO_3, DMF; (ii) NH_2NH_2, EtOH; (iii) THF, pyridine; (iv) $POCl_3$, 80 °C

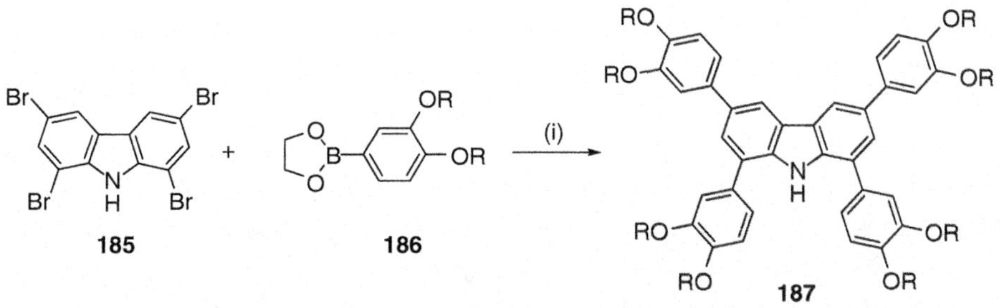

Scheme 4.39 Synthesis of carbazole discotics; (i) $Pd(PPH_3)_4$, Na_2CO_3, toluene

their physical properties. Accordingly, a number of discotics with pendent carbazole groups have been prepared, but none of them show mesomorphic properties in the pure state [213]. Recently, Sienkowska et al. converted the carbazole molecule into liquid crystalline materials by attaching four dialkoxyphenyl units around the periphery [112]. Thus, tetraarylcarbazoles **187** were prepared *via* Suzuki coupling between the 1,3,6,8-tetrabromocarbazole **185**, which can be prepared by direct bromination of carbazole, and the appropriate boronic ester **186**, see Scheme 4.39. The biphenyls are formed as byproduct in this reaction due to homocoupling of the corresponding boronic esters. Additionally, partially arylated arenes also form in the reaction, but the desired product **187** can still be isolated by column chromatography.

4.12 Summary

The field of DLCs is relatively a young research area. However, many significant advances have been achieved in a relatively short span of time. The subject area is growing very fast and several excellent reviews and a dedicated book on DLCs is now available. These molecules are of immense interest due to their intriguing supramolecular architectures. Their strong π-π interactions within a column lead to high charge carrier mobilities, a property that is essential in the development of eco-friendly and commercially viable organic electrooptic and electronic devices like photovoltaic solar cells, light emitting diodes, thin film transistors, etc. More than 3,000 DLCs derived from about 60 different cores have been realized but, unfortunately, the physical properties of only a few of these materials have studied in any detail so far. Efforts have been made to derive some structure–property relationships, but with limited success to date, and it is difficult to draw any concrete general conclusions. While it is imperative to prepare more new discotic liquid crystalline materials, it is also of paramount importance to evaluate the physical properties of these intriguing materials and translate the basic knowledge gained into real devices.

References

1. Demus, D., Goodby, J., Gray, G.W., Spiess, H.W., Vill, V. (eds.): Handbook of Liquid Crystals. Wiley-VCH, Weinhiem (1998)
2. Kaafarani, B.R.: Discotic liquid crystals for opto-electronic applications. Chem. Mater. **23**, 378–396 (2011)
3. Bisoyi, H.K., Kumar, S.: Liquid-crystal nanoscience: an emerging avenue of soft self-assembly. Chem. Soc. Rev. **40**, 306–319 (2011)
4. Bisoyi, H.K., Kumar, S.: Discotic nematic liquid crystals: science and technology. Chem. Soc. Rev. **39**, 264–286 (2010)
5. Kumar, S.: Self-organization of disc-like molecules: chemical aspects. Chem. Soc. Rev. **35**, 83–109 (2006)
6. Laschat, S., Baro, A., Steinke, N., Giesselmann, F., Hagele, C., Scalia, G., Judele, R., Kapatsina, E., Sauer, S., Schreivogel, A., Tosoni, M.: Discotic liquid crystals: from tailor-made synthesis to plastic electronics. Angew. Chem. Int. Ed. **46**, 4832–4887 (2007)
7. Sergeyev, S., Pisula, W., Geerts, Y.H.: Discotic liquid crystals: a new generation of organic semiconductors. Chem. Soc. Rev. **36**, 1902–1929 (2007)
8. Wu, J., Pisula, W., Mullen, K.: Graphenes as potential material for electronics. Chem. Rev. **107**, 718–747 (2007)
9. Boden, N., Bushby, R.J., Clements, J., Movaghar, B.: Device applications of charge transport in discotic liquid crystals. J. Mater. Chem. **9**, 2081–2086 (1999)

10. Bushby, R.J., Lozman, O.R.: Photoconducting liquid crystals. Curr. Opin. Solid State Mater. Sci. **6**, 569–578 (2002)
11. Bushby, R.J., Lozman, O.R.: Discotic liquid crystals 25 years on. Curr. Opin. Coll. Interface Sci. **7**, 343–354 (2002)
12. Ohta, K., Hatsusaka, K., Sugibayashi, M., Ariyoshi, M., Ban, K., Maeda, F., Naito, R., Nishizawa, K., van de Craats, A.M., Warman, J.M.: Discotic liquid crystalline semiconductors. Mol. Cryst. Liq. Cryst. **397**, 25–45 (2003)
13. Cammidge, A.N., Bushby, R.J.: Discotic liquid crystals- synthesis and structural features. In: Demus, D., Goodby, J., Gray, G.W., Spiess, H.W., Vill, V. (eds.) Handbook of Liquid Crystals, vol. 2B, pp. 693–748. Wiley-VCH, Weinheim (1998). Chapter VII
14. Chandrasekhar, S.: Discotic liquid crystals. A brief review. Liq. Cryst. **14**, 3–14 (1993)
15. Donnio, B., Guillon, D., Deschenaux, R., Bruce, D.W.: Metallomesogens. In: McCleverty, J.A., Meyer, T.J. (eds.) Comprehensive Coordination Chemistry II, vol. 7, pp. 357–627. Elsevier, Oxford (2003)
16. Kohne, B., Praefcke, K.: Hexa-*O*-alkanoyl-*scyllo*-inositols, the first Alicyclic, saturated, discotic liquid crystals. Angew. Chem. Int. Ed. **23**, 82–83 (1984)
17. Keinan, E., Kumar, S., Moshenberg, R., Ghirlando, R., Wachtel, E.J.: Trisubstituted decacyclene derivatives; bridging the gap between the carbonaceous mesophase and discotic liquid crystals. Adv. Mater. **3**, 251–254 (1991)
18. Kumar, S.: Rufigallol-based self-assembled supramolecular architectures. Ph. Transit. **81**, 113–128 (2008)
19. Kumar, S.: Chemistry of Discotic Liquid Crystals: From Monomers to Polymers. CRC Press, Boca Raton (2011)
20. Kumar, S.: Recent developments in the chemistry of triphenylene-based discotic liquid crystals. Liq. Cryst. **31**, 1037–1059 (2004)
21. Kumar, S.: Triphenylene-based discotic liquid crystal dimers, oligomers and polymers. Liq. Cryst. **32**, 1089–1113 (2005)
22. Buess, C.M., Lawson, D.D.: The preparation, reactions, and properties of triphenylenes. Chem. Rev. **60**, 313 (1960)
23. Billard, J., Dubois, J.C., Tinh, N.H., Zann, A.: Une mesophase disquotique. J. de. Chimie. **2**, 535–540 (1978)
24. Destrade, C., Mondon, M.C., Malthete, J.: Hexasubstituted triphenylenes: a new mesomorphic order. J. Phys. Colloq. **40**, C3-17–21 (1979)
25. Fontes, E., Heiney, P.A., de Jeu, W.H.: Liquid-crystalline and helical order in a discotic mesophase. Phys. Rev. Lett. **61**, 1202–1205 (1988)
26. Adam, D., Schuhmacher, P., Simmerer, J., Haussling, L., Siemensmeyer, K., Etzbach, K.H., Ringsdorf, H., Haarer, D.: Fast photoconduction in the highly ordered columnar phase of a discotic liquid crystal. Nature **371**, 141–143 (1994)
27. Kohne, V.B., Poules, W., Praefcke, K.: Erste flussigkristalline Hexakis-(alkylthio)-triphenylene. Chem. Z. **108**, 113 (1984)
28. Gramsbergen, E.F., Hoving, H.J., de Jeu, W.H., Praefcke, K., Kohne, B.: X-ray investigation of discotic mesophases of alkylthio substituted triphenylenes. Liq. Cryst. **1**, 397–400 (1986)
29. Marguet, S., Markovitsi, D., Millie, P., Sigal, H., Kumar, S.: Influence of disorder on electronic excited states: an experimental and numerical study. J. Phys. Chem. B. **102**, 4697–4710 (1998)
30. Khone, R.B., Praefcke, K., Derz, T., Frischmuth, W., Gansau, C.: Uber Selen-substitution des Hexakis-(alkylseleno)-triphenylene, erste Selen-haltige diskotische Flussigkristallklasse. Chem. Z. **108**, 408 (1984)
31. Lee, W.K., Heiney, P.A., Mccauley Jr., J.P., Smith III, A.B.: Fourier transform infrared absorption study of hexa(hexylthio)triphenylene: a discotic liquid crystal. Mol. Cryst. Liq. Cryst. **198**, 273–284 (1991)
32. Idziak, S.H.J., Heiney, P.A., Mccauley Jr., J.P., Carroll, P., Smith III, A.B.: Phase diagram of hexa-*N*-alkylthiotriphenylenes. Mol. Cryst. Liq. Cryst. **237**, 271–275 (1993)

33. Vauchier, C., Zann, A., Le Barny, P., Dubois, J.C., Billard, J.: Orientation of discotic mesophases. Mol. Cryst. Liq. Cryst. **66**, 103–114 (1981)
34. Destrade, C., Tinh, N.H., Gasparoux, H., Malthete, J., Levelut, A.M.: Disc-like mesogens: a classification. Mol. Cryst. Liq. Cryst. **71**, 111–135 (1981)
35. Matheson, I.M., Musgrave, O.C., Webster, C.J.: Oxidation of Veratrole by quinines. Chem. Commun. (Lond) **1965**, 278–279 (1965)
36. Scholl, R., Mansfeld, J.: *Meso*-Benzdianthron (Helianthron), *meso*-Naphthodianthron, und ein neuer Weg zum Flavanthren. Ber. Dtsch. Chem. Ges. **43**, 1734–1746 (1910)
37. King, B.T., Kroulik, J., Robertson, C.R., Rempala, P., Hilton, C.L., Korinek, J.D., Gortari, L.M.: Controlling the Scholl reaction. J. Org. Chem. **72**, 2279–2288 (2007)
38. Voisin, E., Williams, V.E.: Do catechol derivatives electropolymerize? Macromolecules **41**, 2994–2997 (2008)
39. Boden, N., Borner, R.C., Bushby, R.J., Cammidge, A.N., Jesudason, M.V.: The synthesis of triphenylene-based discotic mesogens new and improved routes. Liq. Cryst. **15**, 851–858 (1993)
40. Boden, N., Bushby, R.J., Cammidge, A.N.: A quick-and-easy route to unsymmetrically substituted derivatives of triphenylene: preparation of polymeric discotic liquid crystals. J. Chem. Soc. Chem. Commun. **1994**, 465–466 (1994)
41. Boden, N., Borner, R.C., Bushby, R.J., Cammidge, A.N., Jesudason, M.V.: The synthesis of triphenylene-based discotic mesogens new and improved routes. Liq. Cryst. **33**, 1443–1448 (2006)
42. Borner, R.C., Bushby, R.J., Cammidge, A.N., Boden, N., Jesudason, M.V.: Ferric chloride/methanol in the preparation of triphenylene-based discotic liquid crystals. Liq. Cryst. **33**, 1439–1442 (2006)
43. Kumar, S., Manickam, M.: Oxidative trimerization of o-dialkoxybenzenes to hexaalkoxytriphenylenes: molybdenum(v) chloride as a novel reagent. Chem. Commun. **1997**, 1615–1666 (1997)
44. Kumar, S., Varshney, S.K.: Vanadium oxytrichloride, a novel reagent for the oxidative trimerization of o-dialkoxybenzenes to hexaalkoxytriphenylenes. Liq. Cryst. **26**, 1841–1843 (1999)
45. Kumar, S., Varshney, S.K.: Synthesis of triphenylene and dibenzopyrene derivatives; vanadium oxytrichloride a novel reagent. Synthesis **2001**, 305–311 (2001)
46. Musgrave, O.C.: Oxidation of alkyl aryl ethers. Chem. Rev. **69**, 499–531 (1969)
47. Borner, R.C., Jackson, R.F.W.: A flexible and rational synthesis of substituted triphenylenes by palladium-catalysed cross-coupling of arylzinc halides. J. Chem. Soc. Chem. Commun. **1994**, 845–846 (1994)
48. Perez, D., Guitian, E.: Selected strategies for the synthesis of triphenylenes. Chem. Soc. Rev. **33**, 274–283 (2004)
49. Kumar, S., Naidu, J.J.: First rational synthesis of dibenzo[*fg, op*]naphthacene discotics. Liq. Cryst. **28**, 1435–1437 (2001)
50. Rose, B., Meier, H.: Liquid crystals in the series of 2,3,6,7-tetraalkoxytriphenylenes. Z. Naturforsch. **53B**, 1031–1034 (1998)
51. Meier, H., Rose, B.: Dehydrotriphenylenes for the generation of bent molecular ribbons. J. Prakt. Chem. **340**, 536–543 (1998)
52. Brenna, E., Fuganti, C., Serra, S.: New route to *o*-terphenyls: application to the synthesis of 6,7,10,11-tetramethoxy-2-(methoxycarbonyl)triphenylene. J. Chem. Soc. Perkin Trans. 1 **1998**, 901–904 (1998)
53. Cammidge, A.N., Gopee, H.: Structural factors controlling the transition between columnar-hexagonal and helical mesophase in triphenylene liquid crystals. J. Mater. Chem. **11**, 2773–2783 (2001)
54. Cammidge, A.N., Gopee, H.: Mixed alkyl-alkoxy triphenylenes. Mol. Cryst. Liq. Cryst. **397**, 117–128 (2003)
55. Cammidge, A.N.: The effect of size and shape variation in discotic liquid crystals based on triphenylene cores. Phil. Trans. Soc. A **364**, 2697–2708 (2006)

56. Cammidge, A.N., Chausson, C., Gopee, H., Li, J., Hughes, D.L.: Probing the structural factors influencing columnar mesophase formation and stability in triphenylene discotics. Chem. Commun. **2009**, 7375–7377 (2009)
57. Kumar, S., Naidu, J.J.: Novel hexasubstituted triphenylene discotic liquid crystals having three different types of peripheral substituent. Liq. Cryst. **29**, 899–906 (2002)
58. Breslow, R., Jaun, B., Kluttz, R.Q., Xia, C.Z.: Ground state Pi-electron triplet molecules of potential use in the synthesis of organic ferromagnets. Tetrahedron **38**, 863–867 (1982)
59. Rego, J.A., Kumar, S., Ringsdorf, H.: Synthesis and characterization of fluorescent, low-symmetry triphenylene discotic liquid crystals: tailoring of mesomorphic and optical properties. Chem. Mater. **8**, 1402–1409 (1996)
60. Boden, N., Bushby, R.J., Cammidge, A.N., Headdock, G.: Novel discotic liquid crystals created by electrophilic aromatic substitution. J. Mater. Chem. **5**, 2275–2281 (1995)
61. Boden, N., Bushby, R.J., Cammidge, A.N.: Preliminary communications functionalization of discotic liquid crystals by direct substitution into the discogen ring α-nitration of triphenylene-based discogens. Liq. Cryst. **18**, 673–676 (1995)
62. Boden, N., Bushby, R.J., Cammidge, A.N.: Functionalisation of triphenylene based discotic liquid crystals. Mol. Cryst. Liq. Cryst. **260**, 307–313 (1995)
63. Kumar, S., Manickam, M.: Nitration of triphenylene discotics: synthesis of mononitro-, dinitro- and trinitro-hexaalkoxy triphenylenes. Mol. Cryst. Liq. Cryst. **309**, 291–295 (1998)
64. Kumar, S., Manickam, M., Balagurusamy, V.S.K., Schonherr, H.: Electrophilic aromatic substitution in triphenylene discotics: synthesis of alkoxynitrotriphenylenes. Liq. Cryst. **26**, 1455–1466 (1999)
65. Bushby, R.J., Boden, N., Kilner, C.A., Lozman, O.R., Lu, Z., Liu, Q., Thornton-Pett, M.A.: Helical geometry and liquid crystalline properties of 2,3,6,7,10,11-hexaalkoxy-1-nitrotriphenylenes. J. Mater. Chem. **13**, 470–474 (2003)
66. Schonherr, H., Manickam, M., Kumar, S.: Surface morphology and molecular ordering in thin films of polymerizable triphenylene discotic liquid crystals on HOPG revealed by atomic force microscopy. Langmuir **18**, 7082–7085 (2002)
67. Boden, N., Bushby, R.J., Cammidge, A.N., Duckworth, S., Headdock, G.: α-halogenation of triphenylene-based discotic liquid crystals: towards a chiral nucleus. J. Mater. Chem. **7**, 601–605 (1997)
68. Henderson, P., Ringsdorf, H., Schuhmacher, P.: Synthesis of functionalized triphenylenes and dibenzopyrenes precursor molecules for polymeric discotic liquid crystals. Liq. Cryst. **18**, 191–195 (1995)
69. Boden, N., Bushby, R.J., Cammidge, A.N., El-Mansoury, A., Martin, P.S., Lu, Z.: The creation of long-lasting glassy columnar discotic liquid crystals using 'dimeric' discogens. J. Mater. Chem. **9**, 1391–1402 (1999)
70. Rose, A., Lugmair, C.G., Swager, T.M.: Excited-state lifetime modulation in triphenylene-based conjugated polymers. J. Am. Chem. Soc. **123**, 11298–11299 (2001)
71. Bushby, R.J., Lu, Z.: Isopropoxy as a masked hydroxy group in aryl oxidative coupling reactions. Synthesis **2001**, 763–767 (2001)
72. Kumar, S., Lakshmi, B.: A convenient and economic method for the synthesis of monohydroxy-pentaalkoxy- and hexaalkoxytriphenylene discotics. Tetrahedron Lett. **46**, 2603–2605 (2005)
73. Kumar, S., Manickam, M.: Synthesis of functionalized triphenylenes by selective ether cleavages with *B*-bromocatecholboron. Synthesis **1998**, 1119–1122 (1998)
74. Pal, S.K., Bisoyi, H.K., Kumar, S.: Synthesis of monohydroxy-functionalized triphenylene discotics: green chemistry approach. Tetrahedron **63**, 6874–6878 (2007)
75. Ochse, A., Kettner, A., Kopitzke, J., Wendorff, J.H., Bassler, H.: Transient photoconduction in discotic liquid crystals. Phys. Chem. Chem. Phys. **1**, 1757–1760 (1999)
76. Bleyl, I., Erdelen, C., Etzbach, K.H., Paulus, W., Schmidt, H.W., Siemensmeyer, K., Haarer, D.: Photopolymerization and transport properties of liquid crystalline triphenylenes. Mol. Cryst. Liq. Cryst. **299**, 149–155 (1997)

77. Zamir, S., Poupko, R., Luz, Z., Hueser, B., Boeffel, C., Zimmermann, H.: Molecular ordering and dynamics in the columnar mesophase of a new dimeric discotic liquid crystal as studied by X-ray diffraction and deuterium NMR. J. Am. Chem. Soc. **116**, 1973–1980 (1994)
78. Adam, D., Schuhmacher, P., Simmerer, J., Haussling, L., Paulus, W., Siemensmeyer, K., Etzbach, K.H., Ringsdorf, H., Haarer, D.: Photoconductivity in the columnar phases of a glassy discotic twin. Adv. Mater. **7**, 276–280 (1995)
79. Boden, N., Bushby, R.J., Cammidge, A.N., Martin, P.S.: Glass-forming discotic liquid-crystalline oligomers. J. Mater. Chem. **5**, 1857–1860 (1995)
80. Kumar, S., Schuhmacher, P., Henderson, P., Rego, J., Ringsdorf, H.: Synthesis of new functionalized discotic liquid crystals for photoconducting applications. Mol. Cryst. Liq. Cryst. **288**, 211–222 (1996)
81. Bacher, A., Bleyl, I., Erdelen, C.H., Haarer, D., Paulus, W., Schmidt, H.W.: Low molecular weight and polymeric triphenylenes as hole transport materials in organic two-layer LEDs. Adv. Mater. **9**, 1031–1035 (1997)
82. van de Craats, A.M., Siebbeles, L.D.A., Bleyl, I., Haarer, D., Berlin, Y.A., Zharikov, A.A., Warman, J.M.: Mechanism of charge transport along columnar stacks of a triphenylene Dimer. J. Phys. Chem. B. **102**, 9625–9634 (1998)
83. Kumar, S., Manickam, M., Schonherr, H.: First examples of functionalized triphenylene discotic dimers: molecular engineering of advanced materials. Liq. Cryst. **26**, 1567–1571 (1999)
84. Manickam, M., Smith, A., Belloni, M., Shelley, E.J., Ashton, P.R., Spencer, N., Preece, J.A.: Introduction of bis-discotic and bis-calamitic mesogenic addends to C 60. Liq. Cryst. **29**, 497–504 (2002)
85. Kranig, W., Huser, B., Spiess, H.W., Kreuder, W., Ringsdorf, H., Zimmermann, H.: Phase behavior of discotic liquid crystalline polymers and related model compounds. Adv. Mater. **2**, 36–40 (1990)
86. Paraschiv, I., Giesbers, M., van Lagen, B., Grozema, F.C., Abellon, R.D., Siebbeles, L.D.A., Marcelis, A.T.M., Zuilhof, H., Sudholter, E.J.R.: H-bond-stabilized triphenylene-based columnar discotic liquid crystals. Chem. Mater. **18**, 968–974 (2006)
87. Paraschiv, I., de Lange, K., Giesbers, M., van Lagen, B., Grozema, F.C., Abellon, R.D., Siebbeles, L.D.A., Sudholter, E.J.R., Zuilhof, H., Marcelis, A.T.M.: Hydrogen-bond stabilized columnar discotic benzenetrisamides with pendant triphenylene groups. J. Mater. Chem. **18**, 5475–5481 (2008)
88. Kohne, B., Praefcke, K.: Eine neue und einfache synthese des dipyrazino[2,3-*f*:2′,3′-*h*]-chinoxalin-ringsystems. Leibigs. Ann. Chem. **1985**, 522–528 (1985)
89. Rogers, D.Z.: Improved synthesis of 1,4,5,8,9,12-hexaazatriphenylene. J. Org. Chem. **51**, 3904–3905 (1986)
90. Arikainen, E.O., Boden, N., Bushby, R.J., Lozman, O.R., Vinter, J.G., Wood, A.: Complimentary polytopic interactions. Angew. Chem. Int. Ed. **39**, 2333–2336 (2000)
91. Lozman, O.R., Bushby, R.J., Vinter, J.G.: Complementary polytopic interactions (CPI) as revealed by molecular modelling using the XED force field. J. Chem. Soc. Perkin Trans. 2 **2**, 1446–1452 (2001)
92. Boden, N., Bushby, R.J., Headdock, G., Lozman, O.R., Wood, A.: Syntheses of new 'large core' discogens based on the triphenylene, azatriphenylene and hexabenztrinaphthylene nuclei. Liq. Cryst. **28**, 139–144 (2001)
93. Kanakarajan, K., Czarnik, A.W.: Syntheses of some hexacarboxylic acid derivatives of hexaazatriphenylene. J. Heterocycl. Chem. **25**, 1869–1872 (1988)
94. Gearba, R.I., Lehmann, M., Levin, J., Ivanov, D.A., Koch, M.H.J., Barbera, J., Debije, M.G., Piris, J., Geerts, Y.H.: Tailoring discotic mesophases: columnar order enforced with hydrogen bonds. Adv. Mater. **15**, 1614–1618 (2003)
95. Pu, L.: Fluorescence of organic molecules in chiral recognition. Chem. Rev. **104**, 1687–1716 (2004)
96. Martinez-Manez, R., Sancenon, F.: Fluorogenic and chromogenic chemosensors and reagents for anions. Chem. Rev. **103**, 4419–4476 (2003)

97. Daub, J., Engl, R., Kurzawa, J., Miller, S.E., Schneider, S., Stockmann, A., Wasielewski, M.R.: Competition between conformational relaxation and Intramolecular electron transfer within phenothiazine—pyrene dyads. J. Phys. Chem. A **105**, 5655–5665 (2001)
98. Baker, L.A., Crooks, R.M.: Photophysical properties of pyrene-functionalized poly(propylene imine) dendrimers. Macromolecules **33**, 9034–9039 (2000)
99. Modrakowski, C., Flores, S.C., Beinhoff, M., Schluter, A.D.: Synthesis of pyrene containing building blocks for dendrimer synthesis. Synthesis **2001**, 2143–2155 (2001)
100. Chaiken, R.F., Kearns, D.R.: Intrinsic photoconduction in pyrene crystals. J. Chem. Phys. **49**, 2846–2850 (1968)
101. Holroyd, R.A., Preses, J.M., Boettcher, E.H., Schmidt, W.F.: Photoconductivity induced by single-photon excitation of aromatic molecules in liquid hydrocarbons. J. Phys. Chem. **88**, 744–749 (1984)
102. Jones II, G., Vullev, V.I.: Photoinduced electron transfer between non-native donor—acceptor moieties incorporated in synthetic polypeptide aggregates. Org. Lett. **4**, 4001–4004 (2002)
103. Yamana, K., Fukunaga, Y., Ohtani, Y., Sato, S., Nakamura, M., Kim, W.J., Akaike, T., Maruyama, A.: DNA mismatch detection using a pyrene–excimer-forming probe. Chem. Commun. **2005**, 2509–2511 (2005)
104. Hwang, G.T., Seo, Y.J., Kim, B.H.: A highly discriminating quencher-free molecular beacon for probing DNA. J. Am. Chem. Soc. **126**, 6528–6529 (2004)
105. Fujimoto, K., Shimizu, H., Inouye, M.: Unambiguous detection of target DNAs by excimer—monomer switching molecular beacons. J. Org. Chem. **69**, 3271–3275 (2004)
106. Bock, H., Helfrich, W.: Field dependent switching angle of a columnar pyrene. Liq. Cryst. **18**, 707–713 (1995)
107. Hirose, T., Kawakami, O., Yasutake, M.: Induction and control of columnar mesophase by charge transfer interaction and side chain structures of tetrasubstituted pyrenes. Mol. Cryst. Liq. Cryst. **451**, 65–74 (2006)
108. Hassheider, T., Benning, S.A., Kitzerow, H.S., Achard, M.F., Bock, H.: Color-tuned electroluminescence from columnar liquid crystalline alkyl arenecarboxylates. Angew. Chem. Int. Ed. **40**, 2060–2063 (2001)
109. Keuker-Baumann, S., Bock, H., Sala, F.D., Benning, S.A., Hassheider, T., Frauenheim, T., Kitzerow, H.S.: Absorption and luminescence spectra of electroluminescent liquid crystals with triphenylene, pyrene and perylene units. Liq. Cryst. **28**, 1105–1113 (2001)
110. Dantras, E., Dandurand, J., Lacabanne, C., Laffont, L., Tarascon, J.M., Archambeau, S., Seguy, I., Destruel, P., Bock, H., Fouet, S.: HRTEM, TSC and broadband dielectric spectroscopy of a discotic liquid crystal. Phys. Chem. Chem. Phys. **6**, 4167–4173 (2004)
111. de Halleux, V., Calbert, J.P., Brocorens, P., Cornil, J., Declercq, J.P., Bredas, J.L., Geerts, Y.: 1,3,6,8-tetraphenylpyrene derivatives: towards fluorescent liquid-crystalline columns? Adv. Fun. Mater. **14**, 649–659 (2004)
112. Sienkowska, M.J., Monobe, H., Kaszynski, P., Shimizu, Y.: Photoconductivity of liquid crystalline derivatives of pyrene and carbazole. J. Mater. Chem. **17**, 1392–1398 (2007)
113. Sienkowska, M.J., Farrar, J.M., Zhang, F., Kusuma, S., Heiney, P.A., Kaszynski, P.: Liquid crystalline behavior of tetraaryl derivatives of benzo[*c*]cinnoline, tetraazapyrene, phenanthrene, and pyrene: the effect of heteroatom and substitution pattern on phase stability. J. Mater. Chem. **17**, 1399–1411 (2007)
114. Hayer, A., de Halleux, V., Kohler, A., El-Garoughy, A., Meijer, E.W., Barbera, J., Tant, J., Levin, J., Lehmann, M., Gierschner, J., Cornil, J., Geerts, Y.H.: Highly fluorescent crystalline and liquid crystalline columnar phases of pyrene-based structures. J. Phys. Chem. B **110**, 7653–7659 (2006)
115. Tang, C.W.: Two-layer organic photovoltaic cell. Appl. Phys. Lett. **48**, 183–185 (1986)
116. Shi, M.M., Chen, H.Z., Sun, J.Z., Ye, J., Wang, M.: Excellent ambipolar photoconductivity of PVK film doped with fluoroperylene diimide. Chem. Phys. Lett. **381**, 666–671 (2003)
117. Horowitz, G., Kouki, F., Spearman, P., Fichou, D., Nogues, C., Pan, X., Garnier, F.: Evidence for n-type conduction in a perylene tetracarboxylic diimide derivative. Adv. Mater. **8**, 242–245 (1996)

118. Malenfant, P.R.L., Dimitrakopoulos, C.D., Gelorme, J.D., Kosbar, L.L., Graham, T.O., Curioni, A., Andreoni, W.: *N*-type organic thin-film transistor with high field-effect mobility based on a N, N′-dialkyl-3,4,9,10-perylene tetracarboxylic diimide derivative. Appl. Phys. Lett. **80**, 2517–2519 (2002)
119. Haas, U., Thalacker, C., Adams, J., Fuhrmann, J., Riethmuller, S., Beginn, U., Ziener, U., Moller, M., Dobrawa, R., Wurthner, F.: Fabrication and fluorescence properties of perylene bisimide dye aggregates bound to gold surfaces and nanopatterns. J. Mater. Chem. **13**, 767–772 (2003)
120. Ranke, P., Bleyl, I., Simmerer, J., Haarer, D., Bacher, A., Schmidt, H.W.: Electroluminescence and electron transport in a perylene dye. Appl. Phys. Lett. **71**, 1332–1334 (1997)
121. Mitchell, R.H., Chaudhary, M., Williams, R.V., Fyles, R., Gibson, J., Smith, M.J.A., Fry, A.J.: Straining strained molecules. 111.′ The spectral and mutagenic properties and an alternate synthesis of diaceperylene and dicyclopenta[l,2,3-cd:11,2′,3′-lmlperylene. Can. J. Chem. **70**, 1015–1021 (1992)
122. Stolarski, R., Fiksinski, K.J.: Fluorescent perylene dyes for liquid crystal displays. Dye. Pigment. **24**, 295–303 (1994)
123. Benning, S., Kitzerow, H.S., Bock, H., Achard, M.F.: Fluorescent columnar liquid crystalline 3,4,9,10-tetra-(n-alkoxycarbonyl)-perylenes. Liq. Cryst. **27**, 901–906 (2000)
124. Archambeau, S., Seguy, I., Jolinat, P., Farenc, J., Destruel, P., Nguyen, T.P., Bock, H., Grelet, E.: Stabilization of discotic liquid organic thin films by ITO surface treatment. Appl. Surf. Sci. **253**, 2078–2086 (2006)
125. Alibert-Fouet, S., Seguy, I., Bobo, J.F., Destruel, P., Bock, H.: Liquid-crystalline and electron-deficient coronene oligocarboxylic esters and imides by twofold benzogenic diels–alder reactions on perylenes. Chem. Eur. J. **13**, 1746–1753 (2007)
126. Takahashi, M., Suzuki, Y., Ichihashi, Y., Yamashita, M., Kawai, H.: 1,3,8,10-Tetrahydro-2,9-diazadibenzo[*cd, lm*]perylenes: synthesis of reduced perylene bisimide analogues. Tetrahedron Lett. **48**, 357–359 (2007)
127. Mo, X., Chen, H.Z., Shi, M.M., Wang, M.: Syntheses and aggregate behaviors of liquid crystalline alkoxycarbonyl substituted perylenes. Chem. Phys. Lett. **417**, 457–460 (2006)
128. Mo, X., Shi, M.M., Huang, J.C., Wang, M., Chen, H.Z.: Synthesis, aggregation and photoconductive properties of alkoxycarbonyl substituted perylenes. Dye. Pigment. **76**, 236–242 (2008)
129. Wurthner, F.: Perylene bisimide dyes as versatile building blocks for functional supramolecular architectures. Chem. Commun. **2004**, 1564–1579 (2004)
130. Struijk, C.W., Sieval, A.B., Dakhorst, J.E.J., van Dijk, M., Kimkes, P., Koehorst, R.B.M., Donker, H., Schaafsma, T.J., Picken, S.J., van de Craats, A.M., Warman, J.M., Zuilhof, H., Sudholter, E.J.R.: Liquid crystalline perylene diimides: architecture and charge carrier mobilities. J. Am. Chem. Soc. **122**, 11057–11066 (2000)
131. Langhals, H., Karolin, J., Johansson, L.B.A.: Spectroscopic properties of new and convenient standards for measuring fluorescence quantum yields. J. Chem. Soc. Faraday Trans. **94**, 2919–2922 (1998)
132. Mende, L.S., Fechtenkotter, A., Mullen, K., Moons, E., Friend, R.H., MacKenzie, J.D.: Self-organized discotic liquid crystals for high-efficiency organic photovoltaics. Science **293**, 1119–1122 (2001)
133. Breeze, A.J., Salomon, A., Ginley, D.S., Gregg, B.A., Tillmann, H., Horhold, H.H.: Polymer—perylene diimide heterojunction solar cells. Appl. Phys. Lett. **81**, 3085–3087 (2002)
134. Kraft, A., Grimsdale, A.C., Holmes, A.B.: Electroluminescent conjugated polymers—seeing polymers in a new light. Angew. Chem. Int. Ed. **37**, 402–428 (1998)
135. Pan, J., Zhu, W., Li, S., Zeng, W., Cao, Y., Tian, H.: Dendron-functionalized perylene diimides with carrier-transporting ability for red luminescent materials. Polymer **46**, 7658–7669 (2005)
136. Karapire, C., Zafer, C., Icli, S.: Studies on photophysical and electrochemical properties of synthesized hydroxy perylenediimides in nanostructured titania thin films. Syn. Met. **145**, 51–60 (2004)

137. Wurthner, F., Schmidt, R.: Electronic and crystal engineering of acenes for solution-processible self-assembling organic semiconductors. ChemPhysChem **7**, 793–797 (2006)
138. Dimitrakopoulos, C.D., Malenfant, P.R.L.: Organic thin film transistors for large area electronics. Adv. Mater. **14**, 99–117 (2002)
139. Jones, B.A., Ahrens, M.J., Yoon, M.H., Facchetti, A., Marks, T.J., Wasielewski, M.R.: High-mobility air-stable n-type semiconductors with processing versatility: dicyanoperylene-3,4:9,10-bis(dicarboximides). Angew. Chem. Int. Ed. **43**, 6363–6366 (2004)
140. Cormier, R.A., Gregg, B.A.: Self-organization in thin films of liquid crystalline perylene diimides. J. Phys. Chem. B **101**, 11004–11006 (1997)
141. Cormier, R.A., Gregg, B.A.: Synthesis and characterization of liquid crystalline perylene diimides. Chem. Mater. **10**, 1309–1319 (1998)
142. Wurthner, F., Thalacker, C., Diele, S., Tschierske, C.: Fluorescent *J*-type aggregates and Thermotropic columnar mesophases of perylene bisimide dyes. Chem. Eur. J. **7**, 2245–2253 (2001)
143. van Herrikhuyzen, J., Syamakumari, A., Schenning, A.P.H.J., Meijer, E.W.: Synthesis of n-type perylene bisimide derivatives and their orthogonal self-assembly with p-type oligo(*p*-phenylene vinylene)s. J. Am. Chem. Soc. **126**, 10021–10027 (2004)
144. Debije, M.G., Chen, Z., Piris, J., Neder, R.B., Watson, M.M., Mullen, K., Wurthner, F.: Dramatic increase in charge carrier lifetime in a liquid crystalline perylene bisimide derivative upon bay substitution with chlorine. J. Mater. Chem. **15**, 1270–1276 (2005)
145. An, Z., Yu, J., Jones, S.C., Barlow, S., Yoo, S., Domercq, B., Prins, P., Siebbeles, L.D.A., Kippelen, B., Marder, S.R.: High electron mobility in room-temperature discotic liquid-crystalline perylene diimides. Adv. Mater. **17**, 2580–2583 (2005)
146. Zucchi, G., Donnio, B., Geerts, Y.H.: Remarkable miscibility between disk- and lathlike mesogens. Chem. Mater. **17**, 4273–4277 (2005)
147. Nolde, F., Pisula, W., Muller, S., Kohl, C., Mullen, K.: Synthesis and self-organization of core-extended perylene tetracarboxdiimides with branched alkyl substituents. Chem. Mater. **18**, 3715–3725 (2006)
148. Li, X.Q., Stepanenko, V., Chen, Z., Prins, P., Siebbeles, L.D.A., Würthner, F.: Functional organogels from highly efficient organogelator based on perylene bisimide semiconductor. Chem. Commun. **2006**, 3871–3873 (2006)
149. Wurthner, F., Chen, Z., Dehm, V., Stepanenko, V.: One-dimensional luminescent nanoaggregates of perylene bisimides. Chem. Commun. **2006**, 1188–1190 (2006)
150. Percec, V., Aqad, E., Peterca, M., Imam, M.R., Glodde, M., Bera, T.K., Miura, Y., Balagurusamy, V.S.K., Ewbank, P.C., Wurthner, F., Heiney, P.A.: Self-assembly of semifluorinated minidendrons attached to electron-acceptor groups into pyramidal columns. Chem. Eur. J. **13**, 3330–3345 (2007)
151. Chen, Z., Baumeister, U., Tschierske, C., Wurthner, F.: Effect of core twisting on self-assembly and optical properties of perylene bisimide dyes in solution and columnar liquid crystalline phases. Chem. Eur. J. **13**, 450–465 (2007)
152. Chen, Z., Stepanenko, V., Dehm, V., Prins, P., Siebbeles, L.D.A., Seibt, J., Marquetand, P., Engel, V., Wurthner, F.: Photoluminescence and conductivity of self-assembled π–π stacks of perylene bisimide dye. Chem. Eur. J. **13**, 436–449 (2007)
153. Dehm, V., Chen, Z., Baumeister, U., Prins, P., Siebbeles, L.D.A., Wurthner, F.: Helical growth of semiconducting columnar dye assemblies based on chiral perylene bisimides. Org. Lett. **9**, 1085–1088 (2007)
154. Seybold, G., Wagenblast, G.: New perylene and violanthrone dyestuffs for fluorescent collectors. Dye. Pigment. **11**, 303–317 (1989)
155. Göltner, C., Pressner, D., Müllen, K., Spiess, H.W.: Liquid-crystalline perylene derivatives as "discotic pigments". Angew. Chem. Int. Ed. **32**, 1660–1662 (1993)
156. Pressner, D., Göltner, C., Spiess, H.W., Müllen, K.: Liquid-crystalline perylene derivatives – orientation and phase variation of discotic dyes. Ber. Buns. Phys. Chem. **97**, 1362–1365 (1993)

157. Müller, G.R.J., Meiners, C., Enkelmann, V., Geerts, Y., Müllen, K.: Liquid crystalline perylene-3,4-dicarboximide derivatives with high thermal and photochemical stability. J. Mater. Chem. **8**, 61–64 (1998)
158. Avlasevich, Y., Li, C., Müllen, K.: Synthesis and applications of core-enlarged perylene dyes. J. Mater. Chem. **20**, 3814–3826 (2010)
159. Cooper, F.C., Partridge, M.W.: Cyclic amidines. Part I. Derivatives of phenhomazine (dibenzo[b, f]-1: 5-diazocine). J. Chem. Soc. **1954**, 3429–3435 (1954)
160. Ponomarev, I.I., Sinichkin, M.K.: Polytricycloquinazolines – a new class of thermostable and heat-resistant cross-linked polymers. Poly. Sci. Ser. A **38**, 951–953 (1996)
161. Yoneda, F., Mera, K.: A novel one-step synthesis of tricycloquinazolines. Chem. Pharm. Bull. **21**, 1610–1611 (1973)
162. Keinan, E., Kumar, S., Singh, S.P., Ghirlando, R., Wachtel, E.J.: New discotic liquid crystals having a tricycloquinazoline core. Liq. Cryst. **11**, 157–173 (1992)
163. Boden, N., Borner, R.C., Bushby, R.J., Clements, J.: First observation of a n-doped quasi-one-dimensional electronically-conducting discotic liquid crystal. J. Am. Chem. Soc. **116**, 10807–10808 (1994)
164. Schonherr, H., Kremer, F.J.B., Kumar, S., Rego, J.A., Wolf, H., Ringsdorf, H., Jaschke, M., Butt, H.J., Bamberg, E.: Self-assembled monolayers of diacotic liquid crystalline thioethers, discoid disulfides, and Thiols on gold: molecular engineering of ordered surfaces. J. Am. Chem. Soc. **118**, 13051–13057 (1996)
165. Uznanski, P., Kryszewski, M.: Photophysical properties of discotic hexa(heptylthio) tricycloquinazoline in crystalline and liquid crystalline phases. Proc. SPIE Int. Soc. Optic. Eng. **3318**, 398–401 (1997)
166. Hiesgen, R., Schonherr, H., Kumar, S., Ringsdorf, H., Meissner, D.: Scanning tunneling microscopy investigation of tricycloquinazoline liquid crystals on gold. Thin Solid Films **358**, 241–249 (2000)
167. Kumar, S., Wachtel, E.J., Keinan, E.: Hexaalkoxytricycloquinazoline: new discotic liquid crystals. J. Org. Chem. **58**, 3821–3827 (1993)
168. Kumar, S.: A simple, rapid, one-step synthesis of aryl poly ethers from aryl acetates: improved synthesis of Hexaalkoxytricycloquinazoline derivatives. Mol. Cryst. Liq. Cryst. **289**, 247–253 (1996)
169. Kumar, S., Rao, D.S.S., Prasad, S.K.: New branched chain tricycloquinazoline derivatives: a room temperature electron deficient discotic system. J. Mater. Chem. **9**, 2751–2754 (1999)
170. Boden, N., Bushby, R.J., Donovan, K., Liu, Q., Lu, Z., Kreouzis, T., Wood, A.: 2,3,7,8,12,13-Hexakis[2-(2-methoxyethoxy)ethoxy]tricycloquinazoline: a discogen which allows enhanced levels of n-doping. Liq. Cryst. **28**, 1739–1748 (2001)
171. Bushby, R.J., Lozman, O.R., Mason, L.A., Taylor, N., Kumar, S.: Cyclic voltammetry studies of discotic liquid crystals. Mol. Cryst. Liq. Cryst. **410**, 171–181 (2004)
172. Battersby, A.R., Fookes, C.J.R., Matcham, G.W.J., McDonald, E.: Biosynthesis of the pigments of life: formation of the macrocycle. Nature **285**, 17–21 (1980)
173. Falk, J.E.: Porphyrins and Metalloporphyrins. Elsevier, Amsterdam (1964)
174. Wrobel, D., Dudkowiak, A.: Porphyrins and phthalocyanines – functional molecular materials for optoelectronics and medicine. Mol. Cryst. Liq. Cryst. **448**, 15–38 (2006)
175. Drain, C.M., Varotto, A., Radivojevic, I.: Self-organized Porphyrinic materials. Chem. Rev. **109**, 1630–1658 (2009)
176. Macdonald, I.J., Dougherty, T.J.: Basic principles of photodynamic therapy. J. Porphyr. Phthalocyanins. **5**, 105–129 (2001)
177. Donnio, B.: Lyotropic metallomesogens. Curr. Opin. Coll. Interface Sci. **7**, 371–394 (2002)
178. Gregg, B.A., Fox, M.A., Bard, A.J.: 2,3,7,8,12,13,17,18-Octakis(.beta.-hydroxyethyl) porphyrin (octaethanolporphyrin) and its liquid crystalline derivatives: synthesis and characterization. J. Am. Chem. Soc. **111**, 3024–3029 (1989)
179. Gregg, B.A., Fox, M.A., Bard, A.J.: Porphyrin octaesters: new discotic liquid crystals. J. Chem. Soc. Chem. Commun. **1987**, 1134–1135 (1987)
180. Dorphin, D. (ed.): The Porphyrins, vol. 1, pp. 1–7. Academic, New York (1978)

181. Rothemund, P., Menotti, A.R.: Porphyrin studies. IV. 1 the synthesis of α,β,γ,δ-tetraphenylporphine. J. Am. Chem. Soc. **63**, 267–270 (1941)
182. Adler, A.D., Longo, F.R., Finarelli, J.D., Goldmacher, J., Assour, J., Korsakoff, L.: A simplified synthesis for meso-tetraphenylporphine. J. Org. Chem. **32**, 476 (1967)
183. Lindsey, J.S., Schreiman, I.C., Hsu, H.C., Kearney, P.C., Marguerettaz, A.M.: Rothemund and Adler-Longo reactions revisited: synthesis of tetraphenylporphyrins under equilibrium conditions. J. Org. Chem. **52**, 827–836 (1987)
184. Arsenault, G.P., Bullock, E., Macdonald, S.F.: Pyrromethanes and porphyrins therefrom. J. Am. Chem. Soc. **82**, 4384–8389 (1960)
185. Serrano, J.L. (ed.): Metallomesogens: Synthesis, Properties and Applications. VCH, Weinheim (1996)
186. Sakurai, T., Shi, K., Sato, H., Tashiro, K., Osuka, A., Saeki, A., Seki, S., Tagawa, S., Sasaki, S., Masunaga, H., Osaka, K., Takata, M., Aida, T.: Prominent electron transport property observed for triply fused metalloporphyrin dimer: directed columnar liquid crystalline assembly by amphiphilic molecular design. J. Am. Chem. Soc. **130**, 13812–13813 (2008)
187. Piechocki, C., Simon, J., Skoulios, A., Guillon, D., Weber, P.: Annelides. 7. Discotic mesophases obtained from substituted metallophthalocyanines. Toward liquid crystalline one-dimensional conductors. J. Am. Chem. Soc. **104**, 5245–5247 (1982)
188. Clar, E., Ironside, C.T.: Hexabenzocoronene. Proc. Chem. Soc. **1958**, 150–151 (1958)
189. Halleux, A., Martin, R.H., King, G.S.D.: Synthèses dans la série des dérivés polycycliques aromatiques hautement condensés. L'hexabenzo-1,12; 2,3; 4,5; 6,7; 8,9; 10,11-Coronène, le tétrabenzo-4,5; 6,7; 11,12; 13,14-péropyrène et le tétrabenzo-1,2; 3,4; 8,9; 10,11-bisanthène. Helv. Chim. Acta. **129**, 1177–1183 (1958)
190. Hendel, W., Khan, Z.H., Schmidt, W.: Hexa-peri-benzocoronene, a candidate for the origin of the diffuse interstellar visible absorption bands. Tetrahedron **42**, 1127–1134 (1986)
191. Simpson, C.D., Wu, J., Watson, M.D., Mullen, K.: From graphite molecules to columnar superstructures – an exercise in nanoscience. J. Mater. Chem. **14**, 494–504 (2004)
192. Berresheim, A.J., Muller, M., Mullen, K.: Polyphenylene nanostructures. Chem. Rev. **99**, 1747–1785 (1999)
193. Watson, M.D., Fechtenkotter, A., Mullen, K.: Big is beautiful–"Aromaticity" revisited from the viewpoint of macromolecular and supramolecular benzene chemistry. Chem. Rev. **101**, 1267–1300 (2001)
194. Grimsdale, A.C., Wu, J., Mullen, K.: New carbon-rich materials for electronics, lithium battery, and hydrogen storage applications. Chem. Commun. **2005**, 2197–2204 (2005)
195. Mullen, K., Rabe, J.P.: Nanographenes as active components of single-molecule electronics and how a scanning tunneling microscope puts them to work. Acc. Chem. Res. **41**, 511–520 (2008)
196. Herwig, P., Kayser, C.W., Mullen, K., Spiess, H.W.: Columnar mesophases of alkylated hexa-*peri*-hexabenzocoronenes with remarkably large phase widths. Adv. Mater. **8**, 510–513 (1996)
197. Fechtenkotter, A., Saalwachter, K., Harbison, M.A., Mullen, K., Spiess, H.W.: Highly ordered columnar structures from hexa-*peri*-hexabenzocoronenes—synthesis, X-ray diffraction, and solid-state heteronuclear multiple-quantum NMR investigations. Angew. Chem. Int. Ed. **38**, 3039–3042 (1999)
198. Weiss, K., Beernink, G., Dotz, F., Birkner, A., Mullen, K., Woll, C.H.: Template-mediated synthesis of polycyclic aromatic hydrocarbons: cyclodehydrogenation and planarization of a hexaphenylbenzene derivative at a copper surface. Angew. Chem. Int. Ed. **38**, 3748–3752 (1999)
199. Wadumethrige, S.H., Rathore, R.: A facile synthesis of elusive alkoxy-substituted hexa-*peri*-hexabenzocoronene. Org. Lett. **10**, 5139–5142 (2008)
200. Fechtenkotter, A., Tchebotareva, N., Watson, M., Mullen, K.: Discotic liquid crystalline hexabenzocoronenes carrying chiral and racemic branched alkyl chains: supramolecular engineering and improved synthetic methods. Tetrahedron **57**, 3769–3783 (2001)

201. Grimsdale, A.C., Bauer, R., Weil, T., Tchebotareva, N., Wu, J., Watson, M., Mullen, K.: The chemical desymmetrisation of two- and three-dimensional polyphenylenes as a key step to functional nanoparticles. Synthesis **2002**, 1229–1238 (2002)
202. Watson, M.D., Debije, M.G., Warman, J.M., Mullen, K.: Peralkylated coronenes via regiospecific hydrogenation of hexa-*peri*-hexabenzocoronenes. J. Am. Chem. Soc. **126**, 766–771 (2004)
203. Pisula, W., Tomovic, Z., Simpson, C., Kastler, M., Pakula, T., Mullen, K.: Relationship between core size, side chain length, and the supramolecular organization of polycyclic aromatic hydrocarbons. Chem. Mater. **17**, 4296–4303 (2005)
204. Feng, X., Pisula, W., Mullen, K.: From helical to staggered stacking of zigzag nanographenes. J. Am. Chem. Soc. **129**, 14116–14117 (2007)
205. Kastler, M., Schmidt, J., Pisula, W., Sebastiani, D., Mullen, K.: From armchair to zigzag peripheries in nanographenes. J. Am. Chem. Soc. **128**, 9526–9534 (2006)
206. Tomovic, Z., Watson, M.D., Mullen, K.: Superphenalene-based columnar liquid crystals. Angew. Chem. Int. Ed. **43**, 755–758 (2004)
207. Feng, X., Liu, M., Pisula, W., Takase, M., Li, J., Mullen, K.: Supramolecular organization and photovoltaics of triangle-shaped discotic graphenes with swallow-tailed alkyl substituents. Adv. Mater. **20**, 2684–2689 (2008)
208. Wasserfallen, D., Kastler, M., Pisula, W., Hofer, W.A., Fogel, Y., Wang, Z., Mullen, K.: Suppressing aggregation in a large polycyclic aromatic hydrocarbon. J. Am. Chem. Soc. **128**, 1334–1339 (2006)
209. Iyer, V.S., Yoshimura, K., Enkelmann, V., Epsch, R., Rabe, J.P., Mullen, K.: A soluble C60 graphite segment. Angew. Chem. Int. Ed. **37**, 2696–2699 (1998)
210. Debije, M.G., Piris, J., de Hass, M.P., Warman, J.M., Tomovic, Z., Simpson, C.D., Watson, M.D., Mullen, K.: The optical and charge transport properties of discotic materials with large aromatic hydrocarbon cores. J. Am. Chem. Soc. **126**, 4641–4645 (2004)
211. Zhang, Y.-D., Jespersen, K.G., Kempe, M., Kornfield, J.A., Barlow, S., Kippelen, B., Marder, S.R.: Columnar discotic liquid-crystalline oxadiazoles as electron-transport materials. Langmuir **19**, 6534–6536 (2003)
212. Kim, B.G., Kim, S., Park, S.Y.: Synthesis of novel discotic mesogen containing electron-transportable oxadiazole moiety. Mol. Cryst. Liq. Cryst. **370**, 391–394 (2001)
213. Manickam, M., Belloni, M., Kumar, S., Varshney, S.K., Rao, D.S.S., Ashton, P.R., Preece, J.A., Spencer, N.: The first hexagonal columnar discotic liquid crystalline carbazole derivatives induced by noncovalent π-π interactions. J. Mater. Chem. **11**, 2790–2800 (2001)

Chapter 5
Charge Transport in Reactive Mesogens and Liquid Crystal Polymer Networks

T. Kreouzis and K.S. Whitehead

5.1 Introduction

At a fundamental level the current flowing through an organic semiconductor device can be linked to the concentration and drift velocity (v_d) of the charge carriers. The drift velocity itself is determined by the charge carrier mobility (μ) and the electric field (E) established within the device. There are several factors that affect the charge transport in such systems, including molecular structure and morphology. Given the many types of device possible, the function of a specific part of a structure will determine its optimum properties. For example, organic light emitting devices (OLEDs) may contain distinct hole and electron transport layers and emission layers all of which need appropriate charge transport characteristics.

Many liquid crystalline materials exhibit highly desirable transport characteristics such as ambipolar long range carrier transport and high carrier mobility [1–5]. There is a very pronounced mobility-phase relationship, where the mobility is enhanced in higher order liquid crystalline phases compared to lower order phases, i.e., observed carrier mobilities in the smectic mesophase of a given liquid crystal are usually higher than those seen in the nematic phase [4]. Given that the liquid crystalline mesophases of thermotropic materials commonly occur at elevated temperatures and given that these mesophases do not form stable solid films by their very nature, one cannot take advantage of the desirable mesophase properties for use as semiconducting layers for operation at room temperature. In the vast

T. Kreouzis (✉)
School of Physics and Astronomy, Queen Mary University of London, C14 NS, London, UK
e-mail: t.kreouzis@qmul.ac.uk

K.S. Whitehead
School of Physics and Astronomy, Queen Mary University of London, C14 NS, London, UK

New York University in London, 6 Bedford Square, WC1B 3RA, London, UK
e-mail: k.s.whitehead@qmul.ac.uk

R.J. Bushby et al. (eds.), *Liquid Crystalline Semiconductors*, Springer Series in Materials Science 169, DOI 10.1007/978-90-481-2873-0_5,

majority of cases crystallisation leads to the formation of grain boundaries and other defects, which trap charge carriers [6, 7]. This is also true of the optical anisotropic properties of liquid crystals.

Some semiconducting polymers also have the ability to form an anisotropic mesophase as well as charge-transporting solid films at room temperature [8]. Problems are encountered using liquid crystalline polymers as they generally possess lower mobilities compared to low-molar-mass (small molecule) liquid crystalline systems. They also tend to have high-temperature mesophases and consequently require high annealing temperatures, which can lead to chemical damage [9]. Difficulties also arise when constructing multilayer polymer devices where orthogonal solvents and polymers must be carefully chosen to allow subsequent layers to be coated.

The key to exploiting the above-mentioned conditions is the preservation of the charge transport properties of the liquid crystalline mesophase in a solid film. There are several approaches. One option would be to form a perfect, defect-free molecular crystal. However, this is a lengthy and non trivial process. Another approach would be to suppress crystallisation on cooling by forming a metastable glassy phase. Unfortunately, problems may arise using this approach, since this state is often unstable [10, 11]. The third approach would be to somehow polymerise the liquid crystal into a network, thereby preserving the mesophase structure. Herein lies the motivation behind RM and liquid crystalline polymer network concepts.

5.1.1 Reactive Mesogens

Figure 5.1 shows a schematic of a calamitic RM molecule. It is comprised of a conjugated core and flexible alkane spacers and, in this respect, it is identical to a calamitic liquid crystalline molecule, and one or two reactive (crosslinkable) end groups. The concept of an RM is to allow the material to form the mesophase required and then induce the crosslinking reaction, thus permanently preserving the mesophase microstructure [12, 13]. This approach has also been applied to discotic liquid crystals [14, 15]. It is worth noting that RMs are widely used for the formation of solid, optically anisotropic functional layers; in the literature we find examples including polarising filters, optical coatings, sensors etc. [16]. The versatility of RMs lies in how the resulting layer can be tailored to suit a particular function. The desirable properties from small liquid crystalline molecules, e.g., low temperature processing, formation of several mesophases, ease of alignment into a monodomain, are combined with those of conjugated polymers, i.e., semiconducting, robust films, light emission, etc. In terms of charge transport, crosslinked RMs maintain and, in some case enhance [17], the long charge carrier range and high mobilities found in liquid crystalline mesophases across a wide temperature range including room temperature, by forming an insoluble, intractable network and preventing crystallisation.

There are several advantages offered by the RM approach. Their flexibility in terms of processing, structure and material characteristics deserve a mention. Liquid

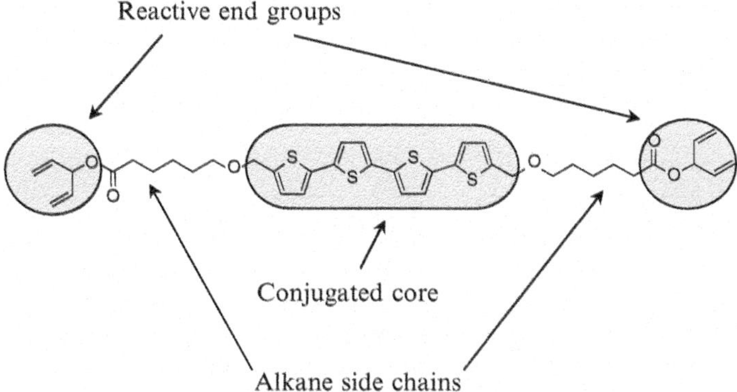

Fig. 5.1 Schematic of a calamitic reactive mesogen for charge transport

crystalline phases at low temperatures, as we have seen, are useful for increasing transport due to molecular ordering. As RMs can contain an emissive chromophore, they can also be used to build polarised light emitting devices [13], see Chap. 6. The choice of reactive end group offers different methods of crosslinking, thermally or photo initiated being the most common [18, 19]. Ultimately this kind of polymerisation brings about the inherent stability of the mesophase and also allows the 'freezing in' of any chosen phase preserving anisotropy. Given that these layers become insoluble following crosslinking, multilayer devices can easily be processed. It may be seen as a disadvantage of the processing that the presence of different end groups and the crosslinking procedure often affect the charge carrier transport but by making careful choice this could be beneficial when tailoring RMs for specific charge transport layers. Photo-crosslinking of the end groups also makes it possible to micropattern RMs so they are also useful for active displays [20]. Side groups may be used to tailor the mesophases and it is also possible to crosslink blends with other RMs and non-reactive liquid crystals. The potential parameter space afforded by this approach is therefore vast.

5.1.2 Liquid Crystalline Polymer Networks

Liquid crystalline polymer networks can also be formed without attaching the reactive end group directly to the mesogen, but by mixing the liquid crystal with a reactive oligomer. The oligomers are crosslinked around the liquid crystalline molecules leaving the latter free to form the relevant mesophases [21–23]. We note that the latter describes the formation of a chemical network around the liquid crystalline molecules, but that physical networks (gels) are also possible using both calamitic [24] and discotic [25, 26] systems.

5.2 Charge Transport Theory

The theoretical modelling of charge transport in RMs and liquid crystalline networks is closely related to that used in amorphous polymers and other organic, e.g., small molecule, semiconductors. For low mobility materials with $\mu \ll 1\text{cm}^2\,\text{V}^{-1}\,\text{s}^{-1}$, carriers are assumed to undergo hopping conduction from one localised site to another, usually within an energetically and positionally disordered site distribution and with the common assumption of the Miller-Abrahams (MA) [27] hopping rate between sites. The individual sites carriers are localised on, and can be identified as, the conjugated molecular cores in liquid crystals and other small molecule systems or one or more monomer units in the case of polymers.

One famous formalism is the Gaussian Disorder Model (GDM) proposed by Bässler and co-workers [27], where the distribution of the site energies is assumed to be Gaussian, width σ, which is referred to as the energetic disorder parameter (units of energy). The sites themselves are positionally displaced from a (cubic) lattice. The displacements from the lattice positions also follow a Gaussian distribution of normalised deviation Σ, referred to as the positional disorder parameter (dimensionless).

By running Monte Carlo (MC) simulations of carriers within such a landscape and by assuming that the probability of a carrier hopping from one site to the other follows the MA expression this model provides us with semi-empirical expressions for the full-temperature and field-dependence of the carrier mobility. Improvements and developments upon this model in common use include the correlated disorder model (CDM) of Novikov and co-workers [28]. Goto and co-workers have also carried out MC simulations using a modified Gay-Berne potential specifically for application to anomalous charge transport in nematic liquid crystals [29].

Charge transport in liquid crystals has also been analysed using a polaronic localisation of the charge in applying the Holstein small polaron theory [30, 31]. This model accounts for the distortion caused by the presence of a charge to its surroundings as an effective lowering of the charge's potential and the formation of a barrier in reaching the next available site. The polaronic features of charges in organic materials have also been combined with the disordered nature of the site energies and positions in work by Parris and co-workers [32]. In contrast to the aforementioned localised charge models, more recently, and in order to be applicable to high mobility liquid crystalline systems, a band transport model has also been proposed by Lever and co-workers [33]. In terms of charge transport in crosslinked reactive mesogens and liquid crystalline systems, the energetic disorder parameters (values of σ) obtained for different systems are particularly useful, when making comparisons between the work of various groups [34–39].

5.2.1 Disordered Hopping Transport

In hopping transport the probability of a carrier moving from site i to site j is given by several expressions, the most commonly used one of which is the Miller-Abraham hopping rate (Eq. 5.1).

$$\begin{aligned} k_{i,j} &= \nu_0 e^{-\beta L} e^{-\frac{(E_i - E_j)}{kT}} && \text{for } E_i < E_j \\ k_{i,j} &= \nu_0 e^{-\beta L} && \text{else} \end{aligned} \tag{5.1}$$

Here, $k_{i,j}$ is the rate of (probability) hopping between sites i and j, the separation between sites is L, and E_i and E_j are their respective energies. The attempt frequency is expressed in ν_0 and β characterises the wave function decay (kT has its usual meaning, with T being the absolute temperature and k Boltzmann's constant). It shows elegantly both the exponential drop in the probability of hopping with distance as well as the Boltzmann drop in probability for hopping to a higher energy site, i.e., it has an intuitive mathematical form. This can be used at the core of MC simulations, where carriers are positioned within an array of sites, in the presence of an electric field, E, which will modify the individual site energies, at a given temperature, and the subsequent carrier motion modelled. From the large body of work in the literature, of particular use for experimentalists are the semi-empirical expressions obtained for the full field and temperature dependence of the mobility. Equation 5.2 shows the expression obtained for the GDM formalism in three spatial dimensions,[1]

$$\mu(T, E) = \mu_0 e^{-\left(\frac{2\sigma}{3kT}\right)^2} e^{C_0 \sqrt{E}\left(\left(\frac{\sigma}{kT}\right)^2 - \Sigma^2\right)} \tag{5.2}$$

Here, C_0 is an empirical constant, which is dependent on the chosen lattice, e.g., cubic, and its spacing, and μ_0 is the prefactor mobility, which can be viewed as a theoretical upper limit to the carrier mobility at zero field and infinite temperature. The energetic and positional disorder parameters are σ and Σ as previously defined. It shows an extremely strong thermal activation of the mobility μ sometimes termed super-Arrhenius. We also note that Eq. 5.2 follows the experimentally obtained general Poole-Frenkel like form for carrier mobility, at a given temperature, found in polymer semiconductors given in Eq. 5.3.

$$\mu(E) = \mu(E - 0)\, e^{\gamma \sqrt{E}} \tag{5.3}$$

---

[1]Depending on the system and especially in liquid crystalline materials the dimensionality of the carrier motion can be suppressed. For instance charge carriers travelling within an individual smectic layer can be viewed as travelling in two dimensions only because of the anisotropy of transport within a layer as opposed to between layers [35]. Similarly one may model carriers moving in columnar mesophases as moving in one dimension only.

Here the two important experimentally determined parameters are the zero field mobility, $\mu(E=0)$, and the field dependence of the mobility, γ. There are two common ways of applying the GDM to experimental data for organic materials. The one approach, e.g., [36, 39], involves a two step process. First plotting the logarithm of the experimentally determined zero field mobility, $\mu(E=0)$, versus $\frac{1}{T^2}$ and extracting μ_0 and σ from the intercept and gradient respectively. Then plotting the experimentally determined field dependence of the mobility $\gamma(T)$ versus $\left(\frac{\sigma}{kT}\right)^2$ and extracting C_0 from the gradient and Σ from the intercept.

The second approach (e.g. [35]), which allows for modifications of the dimensionality of the hopping motion, is to perform computer simulations for fixed assumed values for all but one parameter and vary the parameter of interest, e.g., σ. The resulting families of simulated mobility – electric field results are then compared to the experimentally determined values and the one closest to the experimental result yields the fitted value. In the CDM formalism by Novikov and co-workers correlations are introduced between neighbouring site energies. This was done in order to reproduce the experimentally observed field dependent mobility, at low fields which commonly occurs in semiconducting polymers. They also obtain a semi empirical expression for the full field and temperature dependence of the mobility, see Eq. 5.4, although their parameters differ slightly from the Bässler GDM.

$$\mu\left(T,E\right)=e^{-\left(\frac{3}{5}\hat{\sigma}\right)^2}e^{\left\{C_0\sqrt{\frac{eER}{kT\hat{\sigma}}}\left[\hat{\sigma}^{\frac{3}{2}}-\Gamma\right]\right\}} \tag{5.4}$$

In this expression $\hat{\sigma}$ is the energetic disorder normalised to the temperature, R is the intersite distance and C_0 is a different empirical constant to the one used in the GDM. Γ denotes the positional disorder, although it is defined as a variance rather than a standard deviation, so it approximates Σ^2 in the GDM.

5.2.2 Non-disordered Transport Models

Goto and co-workers use a hopping model, which retains the MA hopping rate, but dispenses with the concept of energetic or positional disorder preferring a fluid approach using a Gay-Berne potential between molecules [29]. The modification is there to account for anomalous temperature dependence in nematic systems, where the fluid motion is significant and yet suppressed in smectic phases. At its most basic form, polaronic transport can be applied without considering any static disorder in the site energies. In liquid crystalline systems it is applicable as a charge will reside upon an individual molecule, with significant polaronic binding energy, because of the distortion it causes on the molecule itself. This approach has been used in analysing transport in discotic systems by Donovan and co-workers [31]. A Holstein small polaron in the non-adiabatic limit, that is where the transfer integral J is small compared to the polaron binding energy E_p the temperature dependence of the mobility is given by Eq. 5.5.

$$\mu = \frac{ea^2}{kT}\frac{1}{\hbar}\left(\frac{\pi}{2E_p kT}\right)^{1/2} J \exp\left(\frac{-E_p}{2kT}\right) \tag{5.5}$$

Where a is the intersite distance, e is the electronic charge and the rest of the symbols have their usual meanings. There is no electric field dependence of the mobility in this model and the polaron binding energy transfer integrals can be extracted easily from a plot of $\ln\left(\mu T^{3/2}\right)$ versus T^{-1} using the field independent mobility measured (the average mobility across a range of fields). Lever and co-workers have applied a band transport model to extremely high mobility discotic liquid crystalline systems ($\mu \sim 1\ \text{cm}^2\ \text{V}^{-1}\ \text{s}^{-1}$) [33]. These systems reside between what is clearly hopping transport ($\mu \ll 1\ \text{cm}^2\ \text{V}^{-1}\ \text{s}^{-1}$) and clearly band transport ($\mu \gg 1\ \text{cm}^2\ \text{V}^{-1}\ \text{s}^{-1}$). In their model the carriers are delocalised across several molecular cores.

5.3 Experimental Determination of Charge Transport

There exist several techniques for determining the charge transport parameters in organic semiconductors. Two techniques that have been widely used to measure carrier mobility in crosslinked systems are the Time of Flight (ToF) technique [40] and Field Effect Transistor (FET) characterisation [41]. These methods will be discussed in detail in the subsections. Other methods used for determining transport in organic systems include Dark Injection Transients [42], Time Resolved Microwave Conductivity Reflectivity [43] and Charge Extraction in a Linearly Increasing Voltage (CELIV) [44]. It is also possible to analyse the *J-V* characteristics of diode devices using Mott-Gurney equation, (Eq. 5.6) to obtain the carrier mobility as a fitting value [45].

$$J = \frac{9\varepsilon\varepsilon_0\mu V^2}{8d^3} \tag{5.6}$$

Here, J is the current density, ε the dielectric constant of the semiconductor, ε_0 the permittivity of free space, μ the mobility of the carriers, V the bias and d the length of the sample. It is also possible to modify this expression to take into account a field dependent mobility, essentially by substituting Eq. 5.3. See Hertel and Bässler [46] for an excellent review of experimental methods applicable to organic systems.

5.3.1 *Time of Flight*

The time of flight technique was originally used on an organic semiconductor in 1960 by Kepler, who was studying charge transport in single crystal anthracene

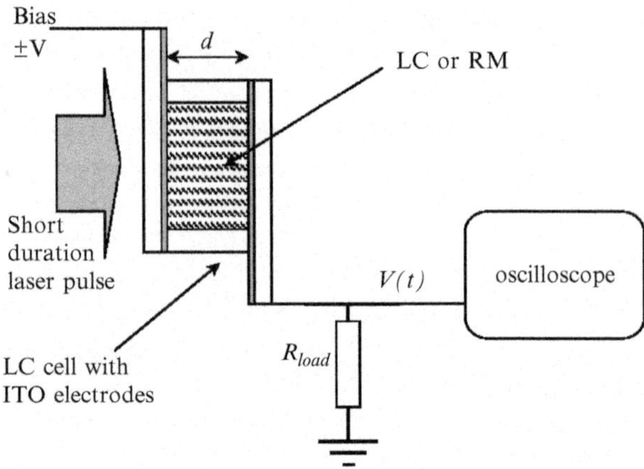

Fig. 5.2 Schematic of the ToF experimental arrangement

[40]. It is a transient photoconductivity technique, where the time evolution of the current induced in an external circuit by photogenerated charges moving through the semiconductor under the action of an electric field is monitored. A schematic of the ToF experimental arrangement is shown in Fig. 5.2.

A semiconductor sample of thickness, *d*, is placed between two electrodes with an externally applied bias, pulsed or DC, of magnitude *V* providing the electric field within the sample. Essentially, the sample assembly forms a capacitor of capacitance *C*. The polarity of the applied bias determines the direction of the electric field and, thus, the polarity of the charge carriers, whose transport is to be studied. A short duration light pulse, of a wavelength chosen to be strongly absorbed by the sample, impinges upon one electrode and generates a sheet of electron–hole pairs in the vicinity of the illuminated electrode. Depending on the direction of the electric field, one sign of carrier will drift through the bulk of the semiconductor, while the other rapidly combines with the charges present on the illuminated electrode. Any drifting charge will induce a voltage across the load resistor in the external circuit and this is monitored using an oscilloscope. We note that this only applies to motion perpendicular to the plane of the electrodes. The current flowing through the sample, $I(t)$, is related to the voltage drop, $V(t)$, across the resistor by Eq. 5.7.

$$I(t) = \frac{1}{R}\left\{V(t) + \tau\frac{dV(t)}{dt}\right\} \tag{5.7}$$

Where τ is the *RC* time constant of the circuit, i.e., $\tau = C\Sigma R$ where ΣR is the sum of all series resistances in the circuit, which is approximately CR_{load}. If one ensures that the time constant is short compared to the timescale of the experiment then the second term in Eq. 5.7 can be neglected and one recovers Ohm's law. This is current-mode time of flight and is the most commonly used. This is because the photocurrent transients it provides are the simplest to interpret. One can also deliberately increase the time constant so that the second term in Eq. 5.7 dominates by simple integration which leads to the charge being related to the voltage. This is

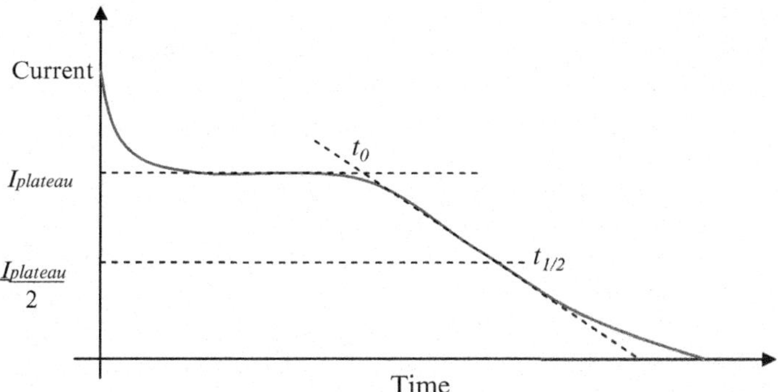

Fig. 5.3 Schematic of a non-dispersive photocurrent transient obtained by ToF

known as integral-mode time of flight. In addition to care being taken with respect to the electronic response time of the circuit, there are other conditions, that must be met for this technique to be successful. One is that the dielectric relaxation time of the sample be much longer than the timescale of the experiment. This condition is easily met by high purity undoped samples. Another is that the sample thickness be considerably greater than the penetration depth δ, of the light ($d \geq 10\ \delta$). The photogenerated charge, q, must also be kept significantly smaller than the charge stored on the capacitor plates (CV) in order to avoid space-charge effects. In current-mode time of flight the resulting photocurrent transient shape will be determined by the nature of the carrier transport. If the carriers drift as part of a coherent (possibly Gaussian) sheet, with a constant drift velocity, through the bulk of the semiconductor the induced current will be constant until recombination at the counter electrode starts taking place. This would be termed a non-dispersive photocurrent and is characterised by a constant current plateau as shown in Fig. 5.3, which clearly shows the constant current plateau due to the holes drifting through the sample.

The initial spike preceding the plateau is due the counter charges being collected at the illuminated electrode. There are two marked times t_0, the arrival time of the fastest carriers which is determined by the inflection point time, and $t_{1/2}$, the average arrival time, which is the time taken for the current to decay to half its plateau value. The spread in arrival times can be quantified by the dimensionless parameter w given by the ratio of the difference between t_0 and $t_{1/2}$ normalised to $t_{1/2}$. If the carriers do not drift as a coherent sheet through the sample they will have a broad distribution of drift velocities and no constant current plateau will be seen in the photocurrent. In such a case, the charge distribution within the sample successively broadens as the fastest carriers move rapidly though the sample leaving the slower carriers behind. It is still possible, however, to determine the transit time of the fastest carriers from these dispersive photocurrents. An apparently featureless photocurrent, when plotted on linear axes, can be re-plotted on a double logarithmic plot as shown in Fig. 5.4.

The inflection point evident on the double logarithmic plot denotes the arrival time of the fastest carriers and is equivalent to the transit time t_0 in the non-dispersive

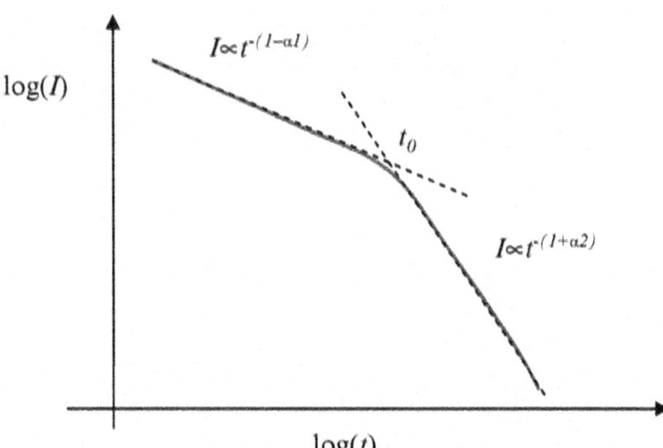

Fig. 5.4 Double logarithmic plot schematic of a dispersive photocurrent transient obtained by ToF

case. In dispersive transport the pre-transit and post-transit data of the double logarithmic plot are governed by Eqs. 5.8a and 5.8b.

$$I \propto t^{-(1-\alpha_1)} \text{ for } t < t_0 \tag{5.8a}$$

$$I \propto t^{-(1+\alpha_2)} \text{ for } t > t_0 \tag{5.8b}$$

Post-transit, the current has to decay faster than t^{-1} as the integral of the photocurrent, when charges cross the whole of the sample, is equal to the photogenerated charge, which is finite. A slower decay than this leads to a nonconvergent integral. In both dispersive and nondispersive transport the transit time can be used to calculate the drift velocity of the carriers, since the sample thickness is known. Whether or not the mobility is field dependent, the transit times at different fields still have to scale correctly with the electric field at least qualitatively, i.e., shorter times at higher fields. Since the mobility is defined as the drift velocity normalised to the electric field, at a given field, it can be calculated using Eq. 5.9.

$$\mu = \frac{d^2}{V t_t} \tag{5.9}$$

Where t_t denotes the transit time, which can be either t_0 or $t_{1/2}$.

This technique allows one to distinguish between hole and electron transport and measure this across a broad range of fields and at different temperatures. As the optical properties of liquid crystals are commonly studied in liquid crystalline cells, which have electrodes, the cells make ideal ToF samples. It is worth noting that application of an electric field across a liquid crystalline cell may disrupt the orientation of the molecules in the mesophase particularly in low-viscosity, high-temperature nematic phases. It is necessary to verify, using polarised microscopy, that such 'switching' is not occurring in samples used for mobility measurements,

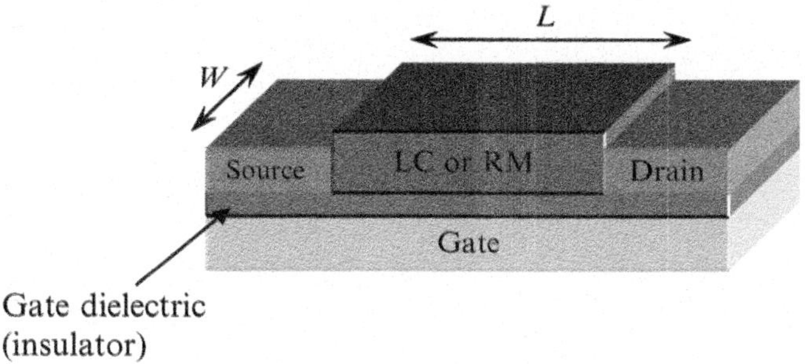

Fig. 5.5 Schematic of a bottom-contact FET

as this will give results that do not correspond to in plane molecular alignment. Samples measured in smectic phases, however, are not prone to this problem.

5.3.2 Field Effect Transistors (FETs)

Field effect transistors are operational semiconductor electronic devices, so any mobility values obtained from them is determined by the performance characteristics of the device. The FET consists of source and drain electrodes connected by an organic semiconductor and a gate electrode separated from the organic semiconductor by a thin dielectric, see Fig. 5.5. A voltage applied to the gate controls the current between the source and drain. The source drain and gate correspond to the emitter, collector and base of a traditional bipolar transistor. There are various possible architectures such as a bottom-gate, top-source drain, but commonly for RMs we use a bottom-gate, bottom-source drain for characterising organic semiconductors, since the channel is accessible and different semiconductors can be allowed to fill it.

Since a FET is a three-electrode device, there are two different ways of characterising its performance. One can keep the gate voltage fixed and sweep the source drain voltage creating an output characteristic, or, keep the source drain voltage fixed and vary the gate voltage producing a transfer characteristic. These are shown in Fig. 5.6.

Figure 5.6a allows us to see the linear and saturation regions of the source drain current I_{sd}, that is the regions where it varies linearly with the source drain bias V_{sd} for a given gate bias V_g, and where it reaches a limiting (saturation) value. The saturation value of the current, although independent of V_{sd}, is determined by the gate bias and increases rapidly with increasing V_g. Figure 5.6b shows the turn-on voltage of the device V_0, this is the gate bias at which the current begins to rapidly rise (at a given V_{sd}) when the transistor begins to conduct. It also shows how the on-off ratio for a transistor can be calculated using the ratio of the value of I_{sd} at high V_g to that below turn on. The linear and saturation regimes can be defined elegantly by considering the relative magnitudes of the source-drain bias and the gate bias, taking into account the turn-on voltage,

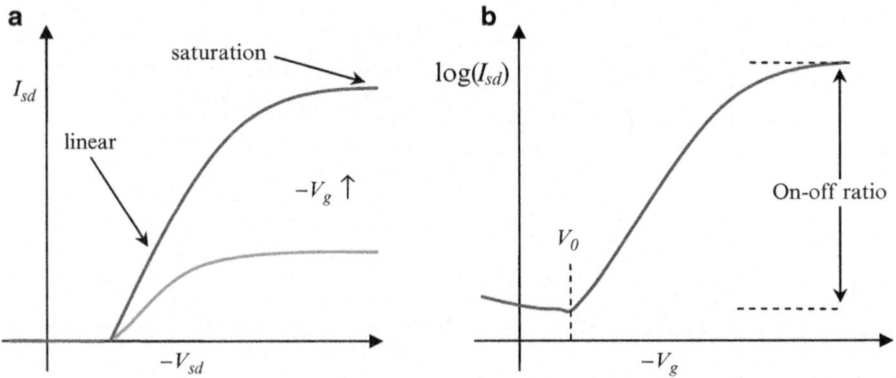

Fig. 5.6 (**a**) Schematic of a FET output characteristic for two different values of gate bias. (**b**) Schematic of a FET transfer characteristic. Note that as the device is p-type both the gate bias and the source-drain voltage are negative, hence the abscissa is plotted as –V

$$V_{sd} < \left(V_g - V_0\right) \quad \text{linear regime} \tag{5.10a}$$

$$V_{sd} > \left(V_g - V_0\right) \quad \text{saturation regime} \tag{5.10b}$$

The ubiquity of inorganic FETs with very well understood characteristics means we can use the existing theoretical expressions for the source-drain current in the linear and saturation regimes in the analysis of organic FETs (Eqs. 5.11 and 5.12).

$$I_{sd}^{lin} = \frac{WC_o}{L}\mu^{lin}(V_g - V_0)V_{sd} \quad \text{linear regime} \tag{5.11}$$

$$I_{sd}^{sat} = \frac{WC_o}{2L}\mu^{sat}(V_g - V_0)^2 \quad \text{saturation regime} \tag{5.12}$$

Where W is the width of the FET, L is the source-drain distance (channel length), C_0 is the capacitance per unit area of the gate dielectric and μ the relevant carrier mobility, see Fig. 5.5. In the linear regime the mobility can be obtained easily by plotting I_{sd} versus either V_g or V_{sd} and taking the slope of the graph (keeping all other quantities fixed). In the saturation regime the mobility is obtained from the slope of a plot of $\sqrt{I_{sd}}$ versus V_g at a fixed (large value) V_{sd}. We note that the device mobility in the linear regime and the saturation regime may differ. Since the linear regime mobility may be affected by parasitic resistances it is common to quote the saturation regime mobility as the FET mobility $\mu_{(FET)}$. When one compares the mobility derived from the ToF method to that derived from a FET device, one would only expect agreement in a purely isotropic semiconductor at best, since the directions in which the carrier transport occurs are orthogonal. In ToF the mobility is measured normal to the plane of the substrate and in FETs it is parallel to the substrate. We also note that in the case of FETs utilising organic semiconductors, see Chap. 8, which are generally undoped (intrinsic),[2] the carriers responsible for the source-drain current are injected from one of the electrodes. Hole injection is generally easier to achieve into the Highest Occupied Molecular Orbital (HOMO), from an anode such as gold, compared to electron injection into the Lowest Unoccupied Molecular Orbital (LUMO), which requires a very low work function cathode. For this reason, hole-based, so called *p*-type, OFETs, are common in the literature. In these devices the gate tends to be negatively biased and $\mu_{(FET)}$ corresponds to an hole mobility.

5.4 Charge Transport Measurements in Real Systems

There are several examples in the literature of crosslinked, highly conjugated liquid crystalline systems being used in successful organic electronic device production, for example in OLEDs [47–55], see Chap. 6, and organic photovoltaics (OPVs) [56–58], see Chap. 7. Although the operation of such devices implies effective charge transport the literature devoted to direct transport measurement is such crosslinked systems is more limited. This section will focus on the work of several key groups who have measured the transport properties of RMs directly. Discotic RMs were studied at the University of Bayreuth [14], see Chap. 3, whereas work on calamitic RMs has been carried out at the University of Hull [34, 59, 60], Queen Mary, University of London (QMUL) [17, 39] and Merck Chemicals UK [61–63], see Chap. 6. In a large number of cases the study of the transport properties has been carried out systematically by measuring the transport properties of a non reactive liquid crystalline model system and comparing these measurements to those carried out on RM substituted compounds both in the liquid crystalline mesophase and post polymerisation (cross-linking). This approach has allowed the effect of all the compound constituents to be analysed in terms of the impact they have on the charge transport. Charge transport has been measured directly by ToF in chemical networks of a calamitic liquid crystal and a cross-linkable acrylate [21, 22, 64]. Physical gels consisting of a discotic liquid crystal and hydrogen bonded fibres have also been studied by ToF [25, 26].

5.4.1 Discotic RMs

In this study the well-known, alkyl-substituted triphenylenes HAT4 and HAT5 were synthesised with a single acrylate reactive end group [14]. Several molecules were prepared having alkyl chains of different lengths between the core and the reactive end group The hole transport was measured by time of flight in the

[2] This is in contrast to inorganic FETs where the type of mobile carrier is determined by the dopant.

Table 5.1 Summary of hole transport parameters obtained for discotic RM systems

Compound	Mesophase mobility ($cm^2\ V^{-1}\ s^{-1}$)	Cross-linked mobility ($cm^2 V^{-1}\ s^{-1}$)
HAT4-monoacrylate 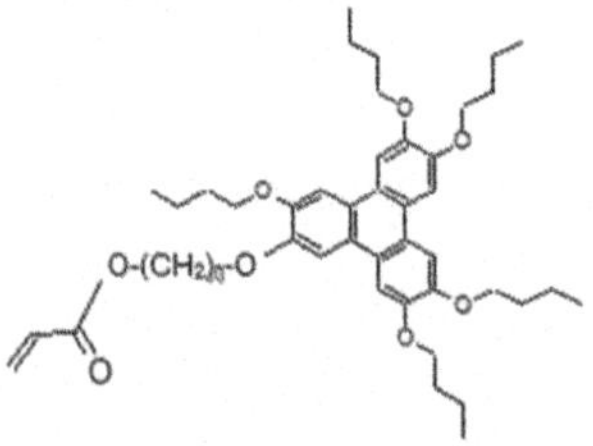	$n = 2$, $\mu_h = 1.5 \times 10^{-3}$ by ToF $n = 3$, $\mu_h = 6 \times 10^{-4}$ by ToF Field independent	$n = 2$, $\mu_h = 1 \times 10^{-5}$ at RT by ToF Thermally activated with 200 meV activation energy by simple Arrhenius $\sigma = 60$ meV using GDM
HAT5-monoacrylate 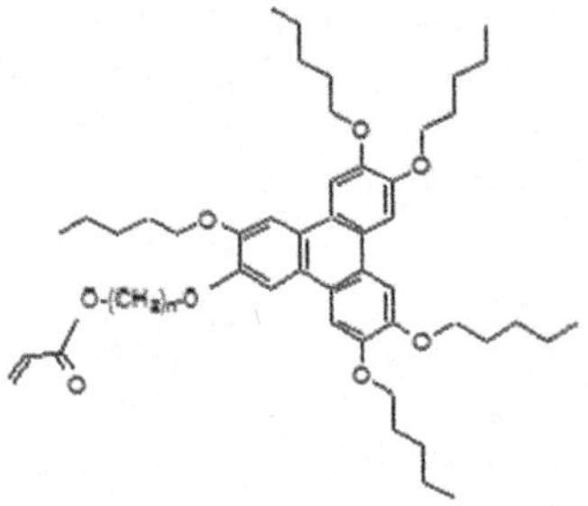	$n = 2$, $\mu_h = 3 \times 10^{-4}$ by ToF $n = 4$, $\mu_h = 2 \times 10^{-4}$ by ToF $n = 6$, $\mu_h = 5 \times 10^{-5}$ by ToF Field independent	

reactive mesogens both pre- and post-polymerisation. They measured the transport in the columnar mesophase for various alkyl chain lengths finding a maximum hole mobility of 1.5×10^{-3} $cm^2\ V^{-1}\ s^{-1}$ in the butyloxy-triphenylene with the shortest alkyl spacer linking the molecular core to the acrylate group. This value of the charge mobility compared very favourably with that of the original liquid crystal ($\sim 1 \times 10^{-3}$ $cm^2\ V^{-1}\ s^{-1}$). In general the RM mobility dropped as the length of the alkyl chain linking the acrylate end group was increased. These measurements are summarised in Table 5.1. The RM mesophase mobility was found to be electric-field independent over the range of field strengths measured, but did display strong temperature dependence. By a simple Arrhenius analysis of the temperature dependence of the mobility, the authors extracted a thermal activation energy of 200 meV as a quantitative measure of the dependence. A GDM parameter, $\sigma = 60$ meV, was also presented in the analysis. The study also looked at the effect of the presence of a photoinitiator and crosslinking on the two fastest systems. The presence of photoinitiator did not significantly reduce the mobility compared to the pure RM, but a one order of magnitude drop was measured upon cross-linking. There is one important thing to note, which is that as an RM concept these are successful results, because they show long-range transport at room temperature in the crosslinked system, which is outside of the mesophase window of the compound. Columnar triphenylene discotic liquid crystals are known to display charge carrier trapping in the polycrystalline solid at room temperature [7]. The authors also show that the transport properties of the crosslinked film are very sensitive to the crosslinking conditions, e.g., nitrogen purge or vacuum, as the presence of oxygen inhibits polymerisation of the acrylate end group.

5.4.2 *Calamitic RMs*

5.4.2.1 Nematic Systems

Researchers at the University of Hull undertook comparisons between model liquid crystalline compounds and related reactive mesogens, which form nematic mesophases using two different conjugated cores containing the fluorene moiety, see Chaps. 7 and 8. In one study, based on a biphenyl-bithiophene–fluorene (compound H1), core time of flight measurements were carried out in the nematic phase of the model compound, the nematic phase of the diene substituted RM and on the crosslinked compound at room temperature [34]. They found a large drop in hole mobility between the model compound and the nematic phase of the RM from 8×10^{-4} to 1.5×10^{-5} $cm^2\ V^{-1}\ s^{-1}$. There was a small recovery in hole mobility upon crosslinking, rising to 3×10^{-5} $cm^2\ V^{-1}\ s^{-1}$, which was attributed to physical shrinkage as a result of crosslinking. The RM mesophase hole mobility and the crosslinked film hole mobility were field independent over the range of fields studied. The mobility in the crosslinked film was less temperature dependent than the nematic RM liquid crystalline phase. The field and temperature dependence of the hole mobility was analysed using the correlated disorder model CDM yielding an energetic order parameter $\sigma = 89$ meV and a positional disorder of $\Gamma = 5.7$ in the cross-linked film.

One can compare the drop in mobility from model compound to RM found in compound H1 to that measured in systems containing a larger conjugated core (compound H2). Hole transport was measured in the latter by time of flight both in the nematic phase of the model liquid crystal and the nematic phase of the diene substituted RM [60]. As expected there was a drop in hole mobility from the model compound to the RM system (from 7×10^{-4} to 3×10^{-4} $cm^2\ V^{-1}\ s^{-1}$ as measured at ~120 °C where both materials are in the nematic phase) but this drop was proportionately smaller (a ~50 % decrease rather than a drop of a factor of 50 as seen in H1). Unfortunately, no transport data is presented for the cross-linked diene RM of H2. The results for compounds H1 and H2 are summarized in Table 5.2.

5.4.2.2 Smectic Systems

Comprehensive studies were carried out at QMUL and Merck Chemicals UK using a variety of conjugated cores and end-groups using both time of flight and FET measurements to quantify the charge transport. The approach taken in these studies followed the pattern of measuring the charge transport in a model liquid crystal

Table 5.2 Summary of hole transport parameters obtained for Nematic phase forming compounds

Compound	Mobility ($cm^2 V^{-1} s^{-1}$)	Cross-linked mobility ($cm^2 V^{-1} s^{-1}$)
Model liquid crystalline for H1 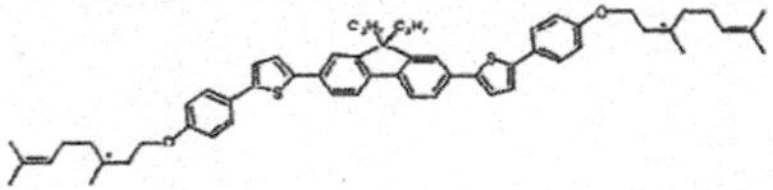	$\mu_h = 1 \times 10^{-4}$ by ToF at RT Field independent	N/A
H1 (diene) 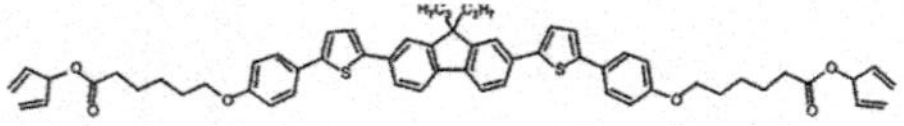	$\mu_h = 1.5 \times 10^{-5}$ by ToF at RT Field independent	$\mu_h = 3 \times 10^{-5}$ by ToF Field independent Weak temperature dependence $\sigma = 89$ meV $\Gamma = 5.7$ using CDM
Model liquid crystalline for H2 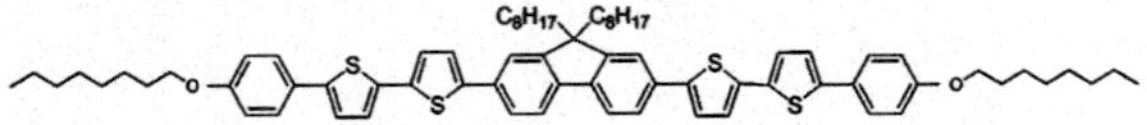	$\mu_h = 7 \times 10^{-4}$ by ToF at ~120 °C	N/A
H2 (diene) 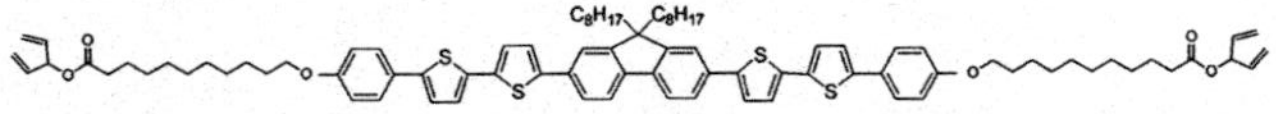	$\mu_h = 3 \times 10^{-4}$ by ToF at ~120 °C Field independent Temperature dependent $\mu_h = 3 \times 10^{-5}$ by ToF at RT	N/A

Table 5.3 Summary of hole transport parameters obtained for phenyl-naphthalene based compounds

Compound	Mesophase mobility ($cm^2\ V^{-1}\ s^{-1}$)	Cross-linked mobility ($cm^2\ V^{-1}\ s^{-1}$)
PNP 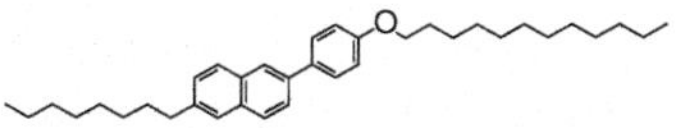	$\mu_h = 2.7 \times 10^{-4}$ by ToF in SmA $\sigma = 60$ meV using 2D-GDM	N/A
PNP-A 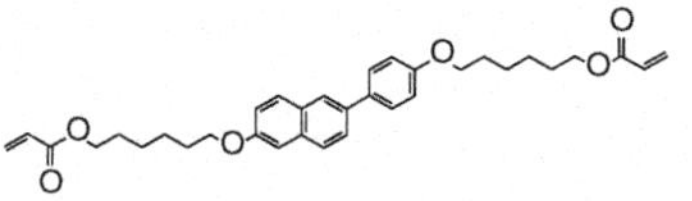	$\mu_h = 4.9 \times 10^{-5}$ by ToF in SmA Field independent	N/A
PNP-D 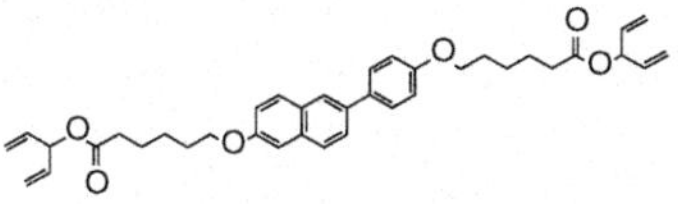	$\mu_h = 7.6 \times 10^{-5}$ by ToF in SmC Field independent	N/A

and reactive end group substituted RMs in the smectic phase as well as in cross linked films where possible. One family of compounds was based on the well known phenylnaphthalene (PNP) liquid crystalline core. Time of flight measurements showed a decrease in hole mobility in both acrylate and diene substituted mesogens compared to the model compound in the smectic phase [39]. Hole mobility dropped from $2.7 \times 10^{-4}\ cm^2\ V^{-1}\ s^{-1}$ in the smectic A phase of the model compound to $7.6 \times 10^{-5}\ cm^2\ V^{-1}\ s^{-1}$ in the smectic C phase of the diene substituted RM and to $4.9 \times 10^{-5}\ cm^2\ V^{-1}\ s^{-1}$ in smectic A acrylate substituted case. All three compounds displayed field-independent hole mobilities in the mesophase. No electron mobility or crosslinked data was presented. Mobility results for PNP based compounds are summarized in Table 5.3.

The second family of compounds based on biphenyl-bithiophene (PTTP) liquid crystalline core was studied in more detail, both by ToF and FET, see Table 5.4 [17, 39, 61, 62]. The model liquid crystals exhibit fast ambipolar transport in the smectic G phase, namely $\mu_h = 4.4 \times 10^{-2}\ cm^2\ V^{-1}\ s^{-1}$ and $\mu_e = 7.1 \times 10^{-2}\ cm^2\ V^{-1}\ s^{-1}$. In the case of the smectic G phase diene substituted RM, the hole mobility suffers a modest reduction compared the model liquid crystalline dropping to $\mu_h = 6.6 \times 10^{-3}\ cm^2\ V^{-1}\ s^{-1}$ whereas the electron mobility is reduced by more than two orders of magnitude to $\mu_e = 1.7 \times 10^{-4}\ cm^2\ V^{-1}\ s^{-1}$. Upon crosslinking the hole mobility suffers a further reduction to $\mu_h = 7.1 \times 10^{-4}\ cm^2\ V^{-1}\ s^{-1}$ in contrast to the electron mobility, which increases compared to the mesophase value to $7.5 \times 10^{-4}\ cm^2\ V^{-1}\ s^{-1}$. We note that the crosslinking process made long range transport measurements possible at room temperature, outside the mesophase range of the mesogens, validating the RM approach. Hole and electron mobilities in the liquid crystal and RM compounds are field independent and the crosslinked material exhibits weakly temperature-dependent transport characteristics. The charge transport for holes in the crosslinked material was analysed using the GDM and

Table 5.4 Summary of charge transport parameters obtained for BiPhenyl-BiThiophene based compounds

Compound	Mesophase mobility ($cm^2\ V^{-1}\ s^{-1}$)	Cross-linked mobility ($cm^2\ V^{-1}\ s^{-1}$)
PTTP 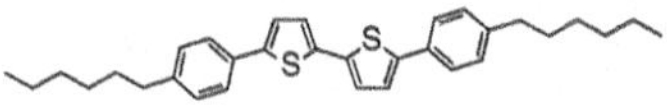	$\mu_h = 4.4 \times 10^{-2}$ by ToF in SmG $\mu_e = 7.1 \times 10^{-2}$ by ToF in SmG Both field independent	N/A
PTTP-D 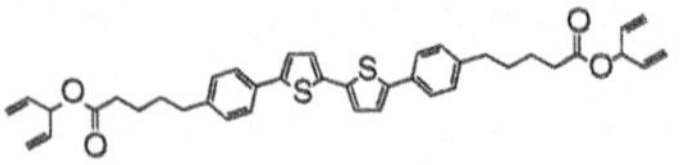	$\mu_h = 6.6 \times 10^{-3}$ by ToF in SmG $\mu_e = 1.7 \times 10^{-4}$ by ToF in SmG Both weakly field dependent $\mu_h = 1 \times 10^{-2}$ by FET in SmG	$\mu_h = 7.1 \times 10^{-4}$ by ToF at RT $\mu_e = 7.5 \times 10^{-4}$ by ToF at RT Weakly temperature dependent $\sigma_h = 26$ meV using GDM $\mu_h = 1 \times 10^{-3}$ by FET at RT
PTTP-Ox 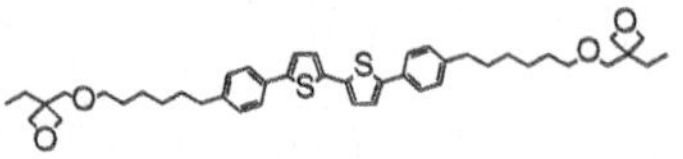	$\mu_h = 9.5 \times 10^{-3}$ by ToF in SmB $\mu_e = 7.5 \times 10^{-3}$ by ToF in SmB Both weakly field dependent $\mu_h = 8 \times 10^{-5}$ by FET in SmB	$\mu_h = 1.6 \times 10^{-2}$ by ToF at RT $\mu_e = 2.8 \times 10^{-2}$ by ToF at RT weakly temperature dependent $\sigma_h = 24$ meV using GDM $\sigma_e = 27$ meV using GDM $\mu_h = 2 \times 10^{-4}$ by FET at RT

Holstein small polaron models. The GDM yields an energetic disorder parameter $\sigma = 26$ meV and a prefactor mobility μ_0 of 4×10^{-4} $cm^2\ V^{-1}\ s^{-1}$. In the case of the small polaron model the hole binding energy was calculated at 102 meV with a 1 meV transfer integral. FET hole mobilities were also extracted for the smectic phase diene substituted RM and for the resulting cross-linked material at the same temperature as the mesophase. These were in broad agreement to those measured by ToF being $1 \times 10^{-2}$and 1×10^{-3} $cm^2\ V^{-1}\ s^{-1}$ for the smectic phase and crosslinked cases, respectively.

For the RM formed using the oxetane end-group both hole and electron mobilities are slightly reduced compared to the model compound in the smectic B mesophase being $\mu_h = 9.5 \times 10^{-3}$ $cm^2\ V^{-1}\ s^{-1}$ and $\mu_e = 7.5 \times 10^{-3}$ $cm^2\ V^{-1}\ s^{-1}$ and are electric-field independent over the range of fields measured. The crosslinked oxetane RM compound charge transport properties approaches those seen in the mesophase of the model compound, being $\mu_h = 1.6 \times 10^{-2}$ $cm^2\ V^{-1}\ s^{-1}$ and $\mu_e = 2.8 \times 10^{-2}$ $cm^2\ V^{-1}\ s^{-1}$ and displaying very weak field dependence and weak temperature dependence. Both hole and electron mobilities were analysed using GDM and the Holstein small polaron models. The energetic disorder parameters are small, just like in the case of the crosslinked diene, being $\sigma_h = 24$ meV and $\sigma_e = 27$ meV although the prefactor mobilities were much higher $\mu_{0h} = 2.9 \times 10^{-2}$ $cm^2\ V^{-1}\ s^{-1}$ and $\mu_{0e} = 2.1 \times 10^{-2}$ $cm^2\ V^{-1}\ s^{-1}$. The small polaron model was also used, yielding a hole binding energy of 194 meV and hole transfer integral of 19 meV. For electrons a smaller binding energy was calculated, 78 meV, with a transfer integral of 18 meV. FET hole mobility values were also measured in the smectic mesophase of the oxetane substituted RM and its resulting crosslinked material. In contrast the diene substituted RM case there appears a large discrepancy (approximately two orders of magnitude) between the FET and

Table 5.5 Summary of charge transport parameters obtained for QuaterThiophene based compounds

Compound	Mesophase mobility ($cm^2\ V^{-1}\ s^{-1}$)	Cross-linked mobility ($cm^2\ V^{-1}\ s^{-1}$)
QT 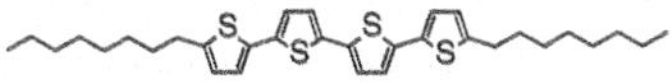	$\mu_h = 6.3 \times 10^{-2}$ by ToF in SmG $\mu_e = 7.7 \times 10^{-2}$ by ToF in SmG Both field independent	N/A
QT-D 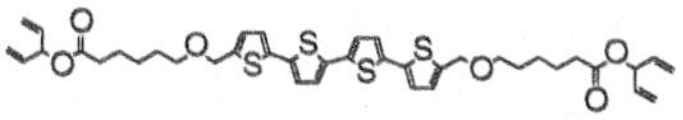	$\mu_h = 4.6 \times 10^{-3}$ by ToF in SmG Field independent	N/A
QT-Ox 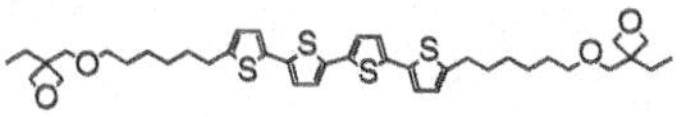	$\mu_h = 5.8 \times 10^{-3}$ by ToF in SmB $\mu_e = 5.4 \times 10^{-3}$ by ToF in SmB Weakly field dependent	$\mu_h = 1.6 \times 10^{-3}$ by ToF at RT $\mu_e = 1.1 \times 10^{-3}$ by ToF at RT Weakly temperature dependent $\sigma_h = 18$ meV using GDM $\sigma_e = 16$ meV using GDM

ToF values in both cases with the FET mobilities being much smaller, namely $\mu_h = 8 \times 10^{-5}\ cm^2\ V^{-1}\ s^{-1}$ in the smectic phase and $\mu_e = 2 \times 10^{-4}\ cm^2\ V^{-1}\ s^{-1}$ for the crosslinked material.

A third family of compounds, based on a quaterthiophene (QT) core, was also studied by the time of flight technique, see Table 5.5 [39, 63]. Once again the model liquid crystalline compound displays fast ambipolar transport properties in the smectic G mesophase, yielding $\mu_h = 6.3 \times 10^{-2}\ cm^2\ V^{-1}\ s^{-1}$ and $\mu_e = 7.7 \times 10^{-2}\ cm^2\ V^{-1}\ s^{-1}$. The hole mobility in the smectic G phase of the diene substituted RM is significantly reduced compared to the model compound, being $\mu_h = 4.6 \times 10^{-3}\ cm^2\ V^{-1}\ s^{-1}$ and the electron transport is undetectable due to trapping. This is in contrast to the oxetane substituted RM, where smectic B hole and electron mobilities were essentially balanced, although reduced compared to that of the model liquid crystal ($\mu_h = 5.8 \times 10^{-3}\ cm^2\ V^{-1}\ s^{-1}$ and $\mu_e = 5.1 \times 10^{-3}\ cm^2\ V^{-1}\ s^{-1}$). Again in the smectic phases of all three quaterthiophenes both the electron and hole mobilities are field independent. Both hole and electron mobilities are reduced upon crosslinking the oxetane substituted quaterthiophene to $\mu_h = 1.7 \times 10^{-3}\ cm^2\ V^{-1}\ s^{-1}$ and $\mu_e = 1.1 \times 10^{-3}\ cm^2\ V^{-1}\ s^{-1}$ (at room temperature) and both display weak temperature dependence. Analysis using GDM produces $\sigma_h = 18$ meV and $\sigma_e = 16$ meV. No cross-linked material results were presented for the diene.

Kelly and coworkers at the University of Hull studied the charge transport properties of a smectic bis-diene substituted RM based on a (4-octyloxyphenyl)pyrimidine core by ToF [59]. Although they do not compare the RM to a model liquid crystalline compound and present no crosslinked results, they report ambipolar transport in the smectic C phase. In contrast to other smectic RMs, the mobilities are quite low, $\mu_h \approx \mu_e = 1.5 \times 10^{-5}\ cm^2\ V^{-1}\ s^{-1}$ at room temperature, but they do report significant temperature dependence of the electron mobility. In fact, close to the upper boundary of the smectic phase the hole mobility rises to $\sim 2 \times 10^{-4}\ cm^2\ V^{-1}\ s^{-1}$. Although no attempt is made to fit the temperature dependence of the electron mobility, we note that it appears exponential [59].

5.4.3 Liquid Crystalline Chemical Networks

An approach to forming solid films incorporating liquid crystalline semiconductors based on the formation of a network consisting of a liquid crystalline semiconductor and an insulating crosslinkable small molecule has been pioneered by researchers at the Tokyo Institute of Technology. It consists of a mixture of the well known calamitic semiconductor 8-PNP-O12 and an acrylate end-capped alkyl molecule HDA. Crosslinking this mixture results in the formation of a solid film incorporating the liquid crystal, which continues to undergo thermotropic phase transitions, even though it is encapsulated in a thin solid film. When the polymerisation is carried out in the isotropic phase, the hole transport behaviour of the liquid crystalline chemical network follows exactly the behaviour of the original liquid crystal up to a polymer content of 10 wt%. At 20 wt% the hole mobility is reduced significantly compared to that of the pure liquid crystal in the smectic B phase, dropping from 2×10^{-3} to $\sim 1 \times 10^{-4}$ cm^2 V^{-1} s^{-1} [21]. We note that in using this approach, high-mobility, long-range transport is measured at elevated temperatures, for example in the smectic B phase between 70 and 100 °C, and that crystallisation still results in trapping behaviour at room temperature. The approach, however, is perfectly valid as it is, in principle, possible to form thin solid films using materials displaying room temperature smectic mesophases. The mesophase, in which the crosslinking has an effect on the amount of HDA, will affect the hole mobility. If the polymerisation is carried out in the smectic A phase, then a 20 wt% content has no effect on the hole transport even in the higher-order, higher-mobility smectic B phase [22]. The electron mobility in the composites, in contrast, is very severely reduced, dropping from 1.6×10^{-3} cm^2 V^{-1} s^{-1} in the smectic B phase of the pure liquid crystal to 10^{-6}–10^{-5} cm^2 V^{-1} s^{-1}, even at 5 wt% HDA [64].

5.4.4 Liquid Crystalline Physical Gels

Yet another approach to forming a solid, which preserves the semiconducting properties of thermotropic liquid crystals, consists of a network of discotic semiconductors, e.g., triphenylenes, and a fibrous aggregate (gelator) resulting in a soft solid. The charge transport behaviour of the resulting solid broadly follows the behaviour of the original liquid crystal, i.e., enhanced transport within a columnar mesophase, although the presence of the gelator results in enhanced hole mobility. In the columnar mesophase the original hole mobility in the liquid crystal HAT6 is measured to be 4.5×10^{-4} cm^2 V^{-1} s^{-1}, whereas, in the HAT6 gel, it increases to 1.3×10^{-3} cm^2 V^{-1} s^{-1} [25]. The degree of mobility-enhancement depends on the length of the alkyl chains attached to the triphenylene liquid crystal [26]. This comes about as all the gels measured yield hole mobilities of the order of 10^{-3} cm^2 V^{-1} s^{-1}. Since triphenylenes with longer alkyl substituents tend to display lower hole mobilities within the columnar mesophase, e.g., 1.6×10^{-3} cm^2 V^{-1} s^{-1} for HAT5 versus 8.7×10^{-5} cm^2 V^{-1} s^{-1} for HAT8, the mobility-enhancement ratio in a gel can be as high as 24.

5.5 Discussion and Conclusions

The most important result is that the RM concept is a valid approach to forming thin solid films, which possess the desirable semiconducting properties found in liquid crystalline mesophases and that these properties are displayed across a broad range of temperature, including room temperature, having overcome the problems associated with crystallisation of liquid crystals. Figures 5.7 and 5.8 demonstrate fast, ambipolar and long range transport in a crosslinked RM and we note that the mobility appears field independent, as is often seen within liquid crystalline mesophases and crosslinked RM networks in the literature. The temperature range over which this desirable transport is seen in crosslinked films far exceeds those of any given mesophase, as demonstrated in Fig. 5.9. Please note that in this particular RM crystallisation leads to trapping at room temperature and that the isotropic carrier mobilities are of the order of 10^{-5} cm^2 V^{-1} s^{-1} before cross-linking. As the resulting solid preserves the microscopic order present at the point of cross-linking it is therefore necessary to choose carefully the phase during which polymerisation occurs. The success of the RM approach in forming effective charge-transport layers

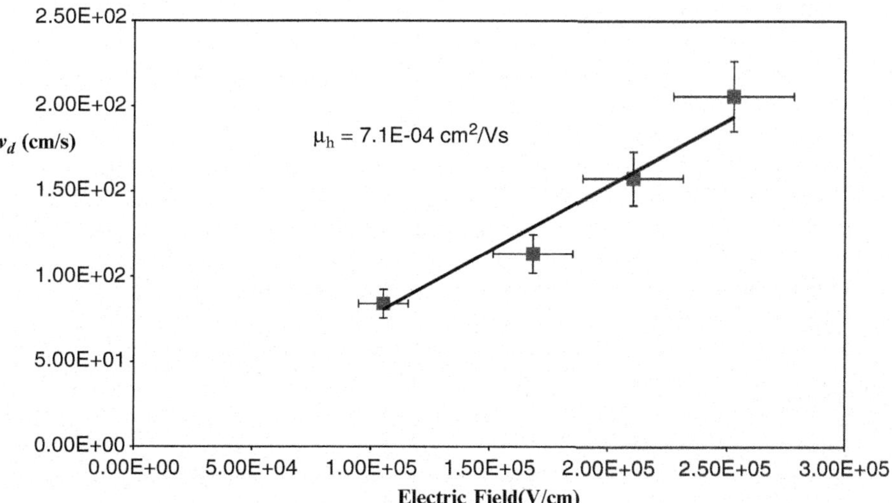

Fig. 5.7 Room temperature hole drift velocity versus electric field as measured by ToF in a 1.9 μm sample of cross-linked PTTP-D. An average mobility is given by the gradient. The cross-linking was previously carried out at 135 °C (Smectic phase)

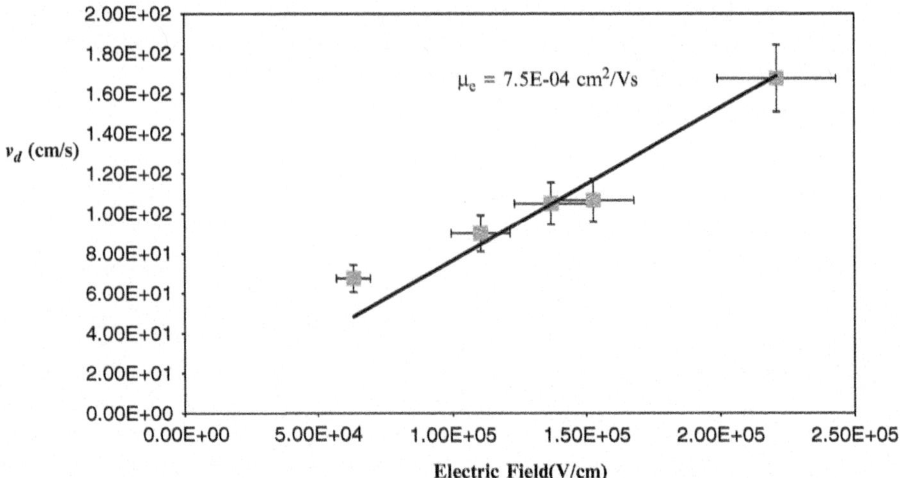

Fig. 5.8 Room temperature electron drift velocity versus electric field as measured by ToF in a 1.9 μm sample of cross-linked PTTP-D. An average mobility is given by the gradient. The cross-linking was previously carried out at 135 °C (Smectic phase)

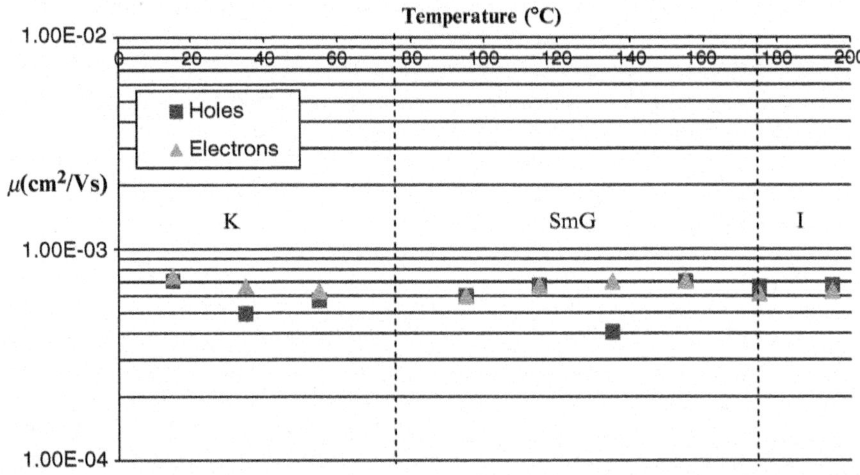

Fig. 5.9 Average hole and electron mobility measured by ToF in a cross-linked sample of PTTP-D (smectic phase polymerisation). The *dotted lines* correspond to the phase transitions of the original (unreacted) mesogen

is quite general and has been demonstrated in a variety of different systems, i.e., columnar [14], nematic [34] and smectic [39]. The charge transport measured within a crosslinked RM does not necessarily correspond exactly to that measured in the original RM mesophase from which the film is formed. There are notable cross-linking effects in most systems.

Crosslinking may have a detrimental effect on hole transport, such as the large drop in mobility measured in crosslinked acrylate containing triphenylene RMs (Table 5.1). We note that the authors can attribute this directly to the act of crosslinking, rather than the presence of impurities, such as photoinitiator. A similar reduction in hole mobility resulting from crosslinking is also seen in the diene containing smectic system PTTP-D, see Table 5.4 and, to a lesser effect in the oxetane substituted smectic system QT-Ox, see Table 5.5. In these cases, however, the RM was measured pure, i.e., without photoinitiator, and so one cannot distinguish between the presence of photoinitiator and the act of crosslinking. Crosslinking does not necessarily reduce the hole mobility and which may remain essentially unchanged, for example in the oxetane substituted smectic PTTP-Ox, see Table 5.4, or even increase as in the nematic diene substituted H1, see Table 5.2. The latter is attributed to conformational changes due to the formation of a rigid backbone by the authors [34].

The effect of crosslinking on electron transport also varies, e.g., it can lead to improved electron mobility, such as in the smectic systems PTTP-D and PTTP-Ox, see Table 5.4 or to minor reduction, such as that seen in QT-Ox, see Table 5.5. In the case of the improvement in electron mobility in PTTP-D on crosslinking, given that diene substituted RMs do not usually exhibit effective electron transport, the authors attribute the mobility rise to the absence of a trapping effect of the diene end group itself. Crosslinking effects notwithstanding, a common approach has been to compare charge transport in different RM systems to a non-reactive liquid crystalline model compound containing the same conjugated core in order to investigate the effect of the reactive end group. It is widely agreed that the presence of reactive end groups invariably detrimentally affects the charge transport in reactive mesogens compared to model compounds. At first glance this may appear surprising, as the charge transport occurs in hops from core to core and the end-groups are physically separated from the cores by the alkane spacers.

The presence of the end groups, however, is believed to be able to affect the core packing and, thus, the charge transport, indeed, molecular conformational simulations of the diene end group show that it possesses a large volume of rotation, which can easily affect the intermolecular packing [60]. This is consistent with the observation that the drop in hole mobility from model compound to RM measured in nematic systems, compounds H1 and H2, is more pronounced in the smaller core material H1, see Table 5.2. The reduction in hole mobility from model liquid crystal to diene-containing RM is a general feature, which also occurs in smectic systems. It occurs as a drop by a factor of $\sim$10 or less in RMs of similar core size, see Tables 5.3, 5.4, and 5.5. There is also evidence that the diene end-group affects electron transport more than hole transport. The electron mobility is over one order of magnitude smaller than the hole mobility in PTTP-D, even though the model compound, PTTP, displays essentially balanced transport, see Table 5.4. For the quaterthiophene based compounds the effect is more pronounced. The QT displays balanced transport, yet electron mobility is not measurable due to electron trapping in smectic QT-D, see Table 5.5. In general, the presence of diene end-group can result in very poor electron transport whereas it doesn't affect hole transport as dramatically. Electron transport can, however, recover post polymerisation, see Table 5.4. The acrylate end-group is found to reduce hole mobility compared to model compounds in both discotic, see Table 5.1 and calamitic, see Table 5.3 systems. In addition to this, it suffers from unwanted thermal crosslinking effects making it somewhat unsuitable in RMs for charge transport. Evidence suggests that, where ambipolar transport is required, oxetane end-groups should be chosen. Both oxetane-substituted smectic systems, PTTP-Ox and QT-Ox, display balanced charge transport, see Tables 5.4 and 5.5. The electron and hole mobilities in these systems are again reduced, compared to that of the model compound, by a factor of $\sim$10 or less. There is strong evidence that the charge transport in crosslinked RM solids maintains the anisotropy found in liquid crystal mesophases and this may explain discrepancies between mobility values obtained using different techniques, such as time of flight and FET. For example, in the case of PTTP-D there is good agreement between ToF and FET mobility values, whereas for PTTP-Ox the disagreement is two orders of magnitude, see Table 5.4.

The electric-field and temperature dependence of the charge carrier mobility in crosslinked RMs has been theoretically analysed and one common approach across the literature is to use some form of disordered hopping transport model. Although the details of any particular model vary, one can in general make meaningful comparisons between, for example, the energetic disorder parameters calculated for different systems. The energetic disorder values measured for cross-linked RM systems span a large range, from 16 meV (PTTP-Ox, electrons) to 89 meV (H1, holes). These values are qualitatively consistent with the models as compounds with lower, room-temperature mobilities correspond to higher-energetic disorder parameters and vice versa. For example, cross-linked H1 has a room-temperature hole mobility of 3×10^{-5} cm^2 V^{-1} s^{-1} and an energetic disorder of 89 meV, whereas crosslinked QT-Ox has a room temperature hole mobility of 1.6×10^{-3} cm^2 V^{-1} s^{-1} and an energetic disorder parameter of 18 meV. The models rely on the temperature dependence of the mobility in order to calculate the energetic disorder, therefore, the occurrence of smaller disorder values for less temperature-dependent transport is at one level tautological. The commonly occurring, field-independent transport in crosslinked RMs is in contrast to polymer semiconductors and is thought to come about as a result of the combination of both energetic and positional disorder in these systems. Small energetic disorder values measured in crosslinked RM systems are also consistent with a reduced distribution of delocalisation lengths, compared to polymers, with carriers residing on a single, conjugated core in RMs. These low-energetic disorder parameters go some way to explaining the susceptibility to impurities that liquid crystalline semiconductors display, see, for example [65]. We note that in the susceptibility of charge transport to impurities the relative energetic positioning of the impurities and the charge transport levels of the semiconductor are of paramount importance. Thus, compound HAT4-monoacrylate ($n = 2$) has a measured energetic disorder of 60 meV, which is the same magnitude as that reported in ref [65], yet the charge transport is not susceptible to the presence of photoinitiator or inhibitor. Further internal consistency of these models comes from the energetic disorder parameters being comparable to photoluminescence peak widths. The disordered model approach, however, is not entirely satisfactory. Often fits to these models do not return all the parameters one would expect. For example, positional disorder parameters are rarely reported and prefactor mobilities can be unphysically small.

Solid semiconducting films are successfully formed using liquid crystalline chemical networks and, in these cases, the hole transport follows the phase behaviour of the original liquid crystal. The liquid crystalline chemical network overcomes the problem of hole mobility reduction due to the presence of the end groups by incorporating a separate cross-linkable compound, which physically distances them from the charge transporting cores. As in the RM approach, the crosslinking conditions, e.g., the nature of the mesophase, play an important role on the charge transport properties of the resulting film.

Where phase separation allows physical distancing of the crosslinkable component, as is the case in smectic phase polymerisation, the hole transport suffers no reduction compared to the model compound. This occurs at the same concentration of crosslinkable compound, where hole mobility reduction is observed in the higher order smectic phase, if one carries out isotropic polymerisation. Unfortunately, electron transport *is* severely affected in these networks due to trapping at chemical impurity sites occurring as a result of the polymerisation reactions. As in the case of crosslinked RMs, the hole mobility appears field independent in these systems. In liquid crystalline physical gels hole transport appears enhanced and this is attributed to the gelator affecting the molecular dynamics of the liquid crystal. The enhancement results in films displaying hole mobilities of the order of 10^{-4} $cm^2\ V^{-1}\ s^{-1}$, irrespective of the length of alkyl chain on the original liquid crystal.

References

1. Adam, D., et al.: Fast photoconduction in the highly ordered columnar phase of a discotic liquid crystal. Nature **371**, 141–143 (1994)
2. Funahashi, M., Hanna, J.: High ambipolar carrier mobility in self-organizing terthiophene derivative. Appl. Phys. Lett. **76**, 2574 (2000)
3. Bushby, R.J., Lozman, O.R.: Photoconducting liquid crystals. Curr. Opin. Solid State Mater. Sci. **6**(6), 569–578 (2002)
4. Takayashiki, Y., et al.: Ambipolar carrier transport in terphenyl derivative. Mol. Cryst. Liq. Cryst. **480**, 295–301 (2008)
5. Funahashi, M., Hanna, J.I.: High carrier mobility up to 0.1 cm(2)V(−1)s(−1) at ambient temperatures in thiophene-based smectic liquid crystals. Adv. Mater. **17**(5), 594 (2005)
6. Maeda, H., Funahashi, M., Hanna, J.I.: Electrical properties of domain boundary in photoconductive smectic mesophases and their crystal phases. Mol. Cryst. Liq. Cryst. **366**, 2221–2228 (2001)
7. Donovan, K.J., Kreouzis, T., Boden, N., Clements, J.: One dimensional carrier trapping in the crystalline phase of a columnar liquid crystal. J. Chem. Phys. **109**(23), 10400–10408 (1998)
8. Grell, M., Redecker, M., Whitehead, K.S., Bradley, D.D.C., Inbasekaran, M., Woo, E.P., Wu, W.: Monodomain alignment of thermotropic fluorene copolymers. Liq. Cryst. **26**(9), 1403–1407 (1999)

9. Cho, S., Seo, J.H., Park, S.H., Beaupre´, S., Leclerc, M., Heeger, A.J.: A thermally stable semiconducting polymer. Adv. Mater. **22**, 1253–1257 (2010)
10. Kreouzis, T., Scott, K., Donovan, K.J., Boden, N., Bushby, R.J., Lozman, O.R., Liu, Q.: Enhanced electronic transport properties in complementary binary discotic liquid crystal systems. Chem. Phys. **262**(2–3), 489–497 (2000)
11. Vlachos, P., et al.: Charge-transport in crystalline organic semiconductors with liquid crystalline order. Chem. Commun. **23**, 2921–2923 (2005)
12. Bacher, A., et al.: Synthesis and characterisation of a conjugated reactive mesogen. J. Mater. Chem. **9**(12), 2985–2989 (1999)
13. Bacher, A., et al.: Conjugated reactive mesogens. Synth. Met. **111**, 413–415 (2000)
14. Bleyl, I., et al.: Photopolymerization and transport properties of liquid crystalline triphenylenes. Mol. Cryst. Liq. Cryst. Sci. Technol. Section A Mol. Cryst. Liq. Cryst. **299**, 149–155 (1997)
15. Kastler, M., et al.: Nanostructuring with a crosslinkable discotic material. Small **3**(8), 1438–1444 (2007)
16. Broer, D.J., Lub, J., Mol, G.N.: Wide-band reflective polarisers from cholesteric polymer networks with a pitch gradient. Nature **378**, 467–469 (1995)
17. Kreouzis, T., et al.: High mobility ambipolar charge transport in a cross-linked reactive mesogen at room temperature. Appl. Phys. Lett. **87**(17), 172110 (2005)
18. Wilderbeek, H.T.A., et al.: Photoinitiated bulk polymerization of liquid crystalline thiolene monomers. Macromolecules **35**(24), 8962–8968 (2002)
19. Thiem, H., et al.: Photopolymerization of reactive mesogens. Macromol. Chem. Phys. **206**(21), 2153–2159 (2005)
20. O'Neill, M., Kelly, S.M.: Liquid crystals for charge transport, luminescence, and photonics. Adv. Mater. **15**(14), 1135–1146 (2003)
21. Yoshimoto, N., Hanna, J.: A novel charge transport material fabricated using a liquid crystalline semiconductor and crosslinked polymer. Adv. Mater. **14**(13–14), 988–991 (2002)
22. Yoshimoto, N., Funahashi, M., Hanna, J.: Charge transport in liquid crystalline semiconductor and crosslinked polymer composite. Mol. Cryst. Liq. Cryst. **409**, 493–504 (2004)
23. Prasad, S.K., et al.: Polymer network as a template for control of photoconductivity of a liquid crystal semiconductor. Liq. Cryst. **31**(9), 1265–1270 (2004)
24. Mizoshita, N., et al.: Smectic liquid-crystalline physical gels. Anisotropic self-aggregation of hydrogen-bonded molecules in layered structures. Chem. Commun. **9**, 781–782 (1999)
25. Mizoshita, N., et al.: The positive effect on hole transport behaviour in anisotropic gels consisting of discotic liquid crystals and hydrogen-bonded fibres. Chem. Commun. **5**, 428–429 (2002)
26. Hirai, Y., et al.: Enhanced hole-transporting behavior of discotic liquid-crystalline physical gels. Adv. Funct. Mater. **18**(11), 1668–1675 (2008)
27. Bässler, H.: Charge transport in disordered organic photoconductors, a Monte Carlo simulation study. Phys. Status Solidi (b) **175**(1), 15–56 (1993)
28. Novikov, S., Dunlap, D., Kenkre, V., Parris, P., Vannikov, A.: Essential role of correlations in governing charge transport in disordered organic materials. Phys. Rev. Lett. **81**, 4472–4475 (1998)
29. Goto, M., Takezoe, H., Ishikawa, K.: Carrier transport simulation of anomalous temperature dependence in nematic liquid crystals. Phys. Rev. E **76**(4), 200710 (2007)
30. Holstein, T.: Studies of polaron motion, parts I, II, III. Mol. Cryst. Model Ann. Phys. **8**(3), 325–342 (November 1959)
31. Kreouzis, T., et al.: Temperature-independent hole mobility in discotic liquid crystals. J. Chem. Phys. **114**(4), 1797–1802 (2001)
32. Parris, P.E., Kenkre, V.M., Dunlap, D.H.: Nature of charge carriers in disordered molecular solids: are polarons compatible with observations? Phys. Rev. Lett. **87**(12), 126601 (2001)
33. Lever, L.J., Kelsall, R.W., Bushby, R.J.: Band transport model for discotic liquid crystals. Phys. Rev. B **72**(3), 35130 (2005)

34. Farrar, S.R., et al.: Nondispersive hole transport of liquid crystalline glasses and a cross-linked network for organic electroluminescence. Phys. Rev. B **66**(12), 5 (2002)
35. Ohno, A., Hanna, J.: Simulated carrier transport in smectic mesophase and its comparison with experimental result. Appl. Phys. Lett. **82**(5), 751–753 (2003)
36. Kawamoto, M., et al.: Charge carrier transport properties in polymer liquid crystals containing oxadiazole and amine moieties in the same side chain. J. Phys. Chem. B **109**(19), 9226–9230 (2005)
37. Iino, H., et al.: Hopping conduction in the columnar liquid crystal phase of a dipolar discogen. J. Appl. Phys. **100**(4), 043716 (2006)
38. Ohno, A., et al.: Charge-carrier transport in smectic mesophases of biphenyls. J. Appl. Phys. **102**(8), 83711 (2007)
39. Baldwin, R.J., et al.: A comprehensive study of the effect of reactive end groups on the charge carrier transport within polymerized and nonpolymerized liquid crystals. J. Appl. Phys. **101**(2), 023713 (2007)
40. Kepler, R.G.: Phys. Rev. **119**(4), 1226 (1960)
41. Sze, S.M.: Physics of Semiconductor Devices, 2nd edn. Wiley, New York (1981). ISBN ISBN-10, 0471056618
42. Many, A., Rakavi, G.: Theory of transient space-charge-limited currents in solids in the presence of trapping. Phys. Rev. **126**, 1980–1988 (1962)
43. Dicker, G., et al.: Electrodeless time-resolved microwave conductivity study of charge-carrier photogeneration in regioregular poly(3-hexylthiophene) thin films. Phys. Rev. B **70**, 045203 (2004)
44. Juška, G., et al.: Charge transport in π-conjugated polymers from extraction current transients. Phys. Rev. B **62**, 016235 (2000)
45. Lampert, M.A., Mark, P.: Current Injection in Solids. Academic, New York (1970)
46. Hertel, D., Bässler, H.: ChemPhysChem **9**, 666–688 (2008)
47. Bayerl, M.S., et al.: Crosslinkable hole-transport materials for preparation of multilayer organic light emitting devices by spin-coating. Macromol. Rapid Commun. **20**(4), 224–228 (1999)
48. Contoret, A.E.A., et al.: Photopolymerisable nematic liquid crystals for electroluminescent devices. Synth. Met. **14**(4), 1629–1630 (2001)
49. Contoret, A.E.A., et al.: The photopolymerization and cross-linking of electroluminescent liquid crystals containing methacrylate and diene photopolymerizable end groups for multilayer organic light-emitting diodes. Chem. Mater. **14**(4), 1477–1487 (2002)
50. Aldred, M.P., et al.: Light-emitting fluorene photoreactive liquid crystals for organic electroluminescence. Chem. Mater. **16**(24), 4928–4936 (2004)
51. Aldred, M.P., et al.: Linearly polarised organic light-emitting diodes (OLEDs), synthesis and characterisation of a novel hole-transporting photoalignment copolymer. J. Mater. Chem. **15**(31), 3208–3213 (2005)
52. Aldred, M.P., et al.: Organic electroluminescence using polymer networks from smectic liquid crystals. Liq. Cryst. **33**(4), 459–467 (2006)
53. Zacharias, P., et al.: New crosslinkable hole conductors for blue-phosphorescent organic light-emitting diodes. Angew. Chem. Int. Ed. **46**(23), 4388–4392 (2007)
54. Rehmann, N., et al.: Advanced device architecture for highly efficient organic light-emitting diodes with an orange-emitting crosslinkable iridium(III) complex. Adv. Mater. **20**(1), 129 (2008)
55. Liedtke, A., et al.: White-light OLEDs using liquid crystal polymer networks. Chem. Mater. **20**(11), 3579–3586 (2008)
56. Carrasco-Orozco, M., et al.: New photovoltaic concept, liquid-crystal solar cells using a nematic gel template. Adv. Mater. **18**(13), 1754–1758 (2006)
57. Carrasco-Orozco, M.A., et al.: Superlattices of organic/inorganic semiconductor nanostructures from liquid-crystal templates. Phys. Rev. B **75**(3), 5 (2007)
58. Tsoi, W.C., et al.: Distributed bilayer photovoltaics based on nematic liquid crystal polymer networks. Chem. Mater. **19**(23), 5475–5484 (2007)

59. Vlachos, P., et al.: Electron-transporting and photopolymerisable liquid crystals. Chem. Commun. **8**, 874–875 (2002)
60. Woon, K.L., et al.: Electronic charge transport in extended nematic liquid crystals. Chem. Mater. **18**(9), 2311–2317 (2006)
61. McCulloch, I., et al.: Designing solution-processable air-stable liquid crystalline crosslinkable semiconductors. Philos. Trans. R. Soc. Math. Phys. Eng. Sci. **364**(1847), 2779–2787 (2006)
62. McCulloch, I., et al.: Electrical properties of reactive liquid crystal semiconductors. Jpn. J. Appl. Phys. **47**(1), 488–491 (2008)
63. McCulloch, I., et al.: Polymerisable liquid crystalline organic semiconductors and their fabrication in organic field effect transistors. J. Mater. Chem. **13**(10), 2436–2444 (2003)
64. Yoshimoto, N., Hanna, J.I.: Preparation of a novel organic semiconductor composite consisting of a liquid crystalline semiconductor and crosslinked polymer and characterization of its charge carrier transport properties. J. Mater. Chem. **13**(5), 1004–1010 (2003)
65. Ahn, H., Ohno, A., Hanna, J.I.: Impurity effects on charge transport in various mesophases of smectic liquid crystals. J. Appl. Phys. **102**, 093718 (2007)

Chapter 6
Optical Properties of Light-Emitting Liquid Crystals

Mary O'Neill and Stephen M. Kelly

6.1 Introduction

Over many years the liquid crystal display (LCD) industry has stimulated an enormous research effort in the development of new liquid crystals, mostly required to be transparent, highly insulating as well as easily reoriented in an electric field. A wide range of chemically, photochemically and electrochemically stable liquid crystals and additives were developed to meet these criteria over the last four decades. However, the continued dominance of LCDs as a flat panel display technology is now being challenged by rapid improvements in Organic Light-Emitting Displays (OLEDs), which represent an emissive and, potentially, more efficient display device technology [1]. Unsurprisingly, OLEDs have a different spectrum of material requirements, e.g., some must be highly conjugated for light-emission and their molecular energies must be engineered for electronic injection and transport. The organic materials should not move in electric fields, so that an extremely high viscosity is required for stable OLEDs leading to the formation of glassy liquid crystalline phases. These commercially important developments in OLEDs are occurring concurrently with rapid progress in organic photovoltaics [2–4], and solid-state lasers [5], leading to a renaissance in the study of light-emitting organic materials. Although many of these new materials are not liquid crystalline, the self-assembling properties of liquid crystals can be used to enhance the performance of light-emitting organic devices. This chapter discusses the optical properties of light-emitting liquid crystalline materials and systems using them [6].

M. O'Neill (✉)
Department of Physics and Mathematics, University of Hull, Hull HU6 7RX, UK
e-mail: m.oneill@hull.ac.uk

S.M. Kelly
Department of Chemistry, University of Hull, Hull HU6 7RX, UK
e-mail: s.m.kelly@hull.ac.uk

R.J. Bushby et al. (eds.), *Liquid Crystalline Semiconductors*, Springer Series in Materials Science 169, DOI 10.1007/978-90-481-2873-0_6,

In Sect. 6.2 we briefly describe the mechanism of electroluminescence for OLEDs, in order to identify the main parameters which influence performance. In Sects. 6.3 and 6.4, we discuss the luminescence of liquid crystalline semiconductors and the out-coupling of light from thin films with particular reference to the parameters identified in Sect. 6.2. The extended length of the aromatic core of liquid crystalline semiconductors results in a very high birefringence, as outlined in Sect. 6.5, which also discusses possible applications of birefringence for thin film lasers and photonic band-gap structures. Progress in chiral liquid crystal lasers is reviewed in Sect. 6.6 followed by a brief discussion and conclusion in Sect. 6.7.

6.2 OLEDs

OLEDs based on conjugated molecular or polymeric thin film devices were first pioneered by groups in Rochester and Cambridge [7, 8]. They are increasingly used in small area displays for mobile phones, organizers, etc., and are also considered future competitors of LCD TVs particularly for high-end markets. OLEDs offer the advantages of wider viewing angles, higher efficiency, very high contrast and darker gray levels as well as faster response times compared to LCDs.

The simplest OLED consists of a light-emitting conjugated organic layer sandwiched between two dissimilar electrodes. Light-emission occurs on application of a voltage in forward-bias. An energy level diagram of the device is given in Fig. 6.1, which shows the built-in field across the organic layer. This originates from the difference in the work-functions of the two electrodes [1]. Electrons are injected from the cathode into the lowest unoccupied molecular orbital (LUMO) of the molecules at the interface. Similarly holes are injected from the anode into the highest occupied molecular orbital (HOMO), i.e., electrons are removed from the HOMO. The charged carriers drift in the applied field by a thermally-activated hopping mechanism from molecule to molecule, at a rate depending on the strength of the electronic coupling between the HOMO (LUMO) of adjacent molecules. An exciton is formed by the columbic interaction between an electron and hole, normally located on the same molecule or region of a single polymer chain. Excitons are neutral, bound electron-hole pairs with energy given as

$$E_{ex} = E_{HOMO} - E_{LUMO} - E_{BE}, \tag{6.1}$$

where E_{BE} is the exciton binding energy. Organic materials have large binding energies, $\approx$ 0.5 eV, because they are highly localized. The exciton recombines with the emission of a photon of energy $\approx E_{ex}$. The color of the emitted light can be tuned from blue to red by modification of E_{HOMO} and E_{LUMO} using chemical design [9]. The emitted light exits the device through one of the electrodes. Indium Tin Oxide (ITO) is often used as a transparent anode. It has a low work function of about 4.7 eV, which presents a large barrier for hole injection for most organic materials. The conductive polymer poly (3-4-ethylenedioxythiophene)/poly (styrene

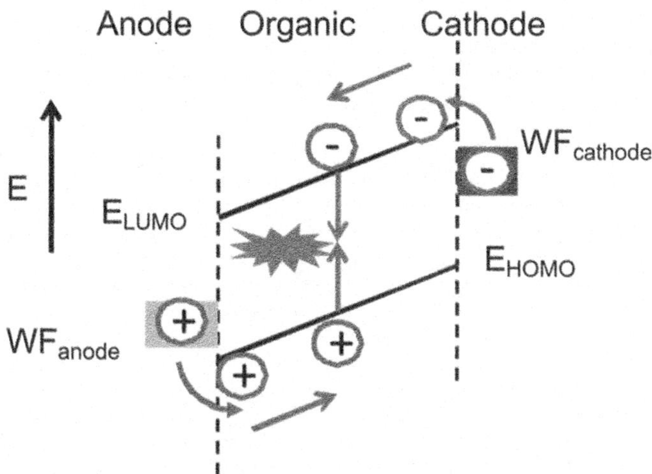

Fig. 6.1 Energy level diagram of an OLED showing the HOMO and LUMO energy levels, E_{HOMO} and E_{LUMO}, of an organic thin film sandwiched between an anode and a cathode of work functions WF_{anode} and $WF_{cathode}$ respectively. The levels are tilted because of the built-in field. A photon of energy $\approx E_{ex}$ is emitted from the OLED

sulfonate) (PEDOT:PSS), which has a work function tunable over a limited range about 5 eV, is often inserted between the anode and the organic light-emitting film to improve the injection of holes [10]. Low-work-function metals, e.g., calcium (Ca) or magnesium (Mg), are often used as cathodes to minimize the barrier for electron injection. However, these are unstable and must be capped with an inert metal, such as aluminium (Al) or silver (Ag). Alternatively high work-function metals such as aluminium are used in conjunction with a thin insulating buffer layer, e.g., lithium or cesium fluorides (LiF or CsF, respectively), which improve electron injection by band bending [11]. Multiple organic layers are often used to stagger the injection barriers, which are particularly large for blue emission. Hole/electron blocking layers at the cathode/anode may be introduced to confine carriers in the emission layer. Exciton blocking layers are used to prevent excitons from reaching the electrodes, where they can be quenched. Thermal evaporation under high vacuum is the standard method to deposit thin films of small conjugated molecules, which are mostly insoluble. In this case multi -layer devices can be made using multiple evaporation sources. Polymeric films are normally deposited by solution processing, e.g., spin or drop casting, so that orthogonal solvents are needed to avoid layer mixing in multi-layer devices.

The external efficiency of the OLED is given by

$$\eta_{\mathrm{EL}} = \gamma \chi_{S(T)} \eta_{\mathrm{PL}} \eta_{\mathrm{C}} \tag{6.2}$$

where γ is the ratio of the number of excitons within the device to the number of electrons in the external circuit and $\chi_{S(T)}$ is the fraction of singlet or triplet

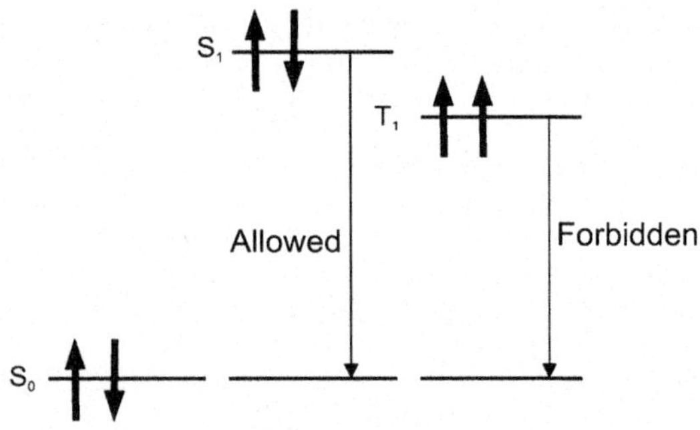

Fig. 6.2 Schematic of ground singlet state (S_0), first excited singlet (S_1) and triplet (T_1), states. The $S_1 \rightarrow S_0$ transition is allowed and the $T_1 \rightarrow S_0$ transition is forbidden

excitons for OLEDs based on fluorescence or phosphorescence, respectively, η_{PL} is the photoluminescence quantum efficiency of the emitter and η_C is the fraction of photons coupled out of the device.

6.3 Luminescence in Liquid Crystalline Semiconductors

6.3.1 Fluorescence

γ in Eq. (6.2) depends on the injection and charge transport properties of the OLED. An equal number of electrons and holes are required for efficient exciton formation so that it is important to balance the injection of electron and holes. The mobility for both types of charge carriers should be high, so that space charge does not significantly impede charge injection. Chapters 2, 3 and 5 discuss the mechanism of charge transport in liquid crystal semiconductors. In an operational OLED the exciton is formed from two independent charges of spin $\pm 1/2$ giving either an excited singlet state with a total spin quantum number $S = 0$ or a triplet state with $S = 1$, as illustrated in Fig. 6.2. The ground state is a singlet with the HOMO containing two electrons of opposite spins. The formation of exciton triplet states by electron-hole capture is three times more likely than singlet formation, if the process can be assumed to be spin independent. Therefore $\chi_S = 0.25$ and $\chi_T = 0.75$ by statistics, although some studies suggest that χ_S may be higher in extended molecules or polymers [12]. Spin-allowed radiative emission (fluorescence) is from the singlet only. Triplet excitons can produce light by phosphorescence, but this process is forbidden and, therefore, inefficient in most organic materials, with exceptions as discussed in more detail below. Therefore, the formation of triplets represents a significant energy loss, especially since the energy difference between the excited singlet and triplet states, the so-called exchange energy, is large, inhibiting cross-over from the red-shifted triplet to the singlet.

Figure 6.3 shows a typical absorption and fluorescence spectrum of a light-emitting liquid crystal semiconductor, compound **1**, whose chemical structure is

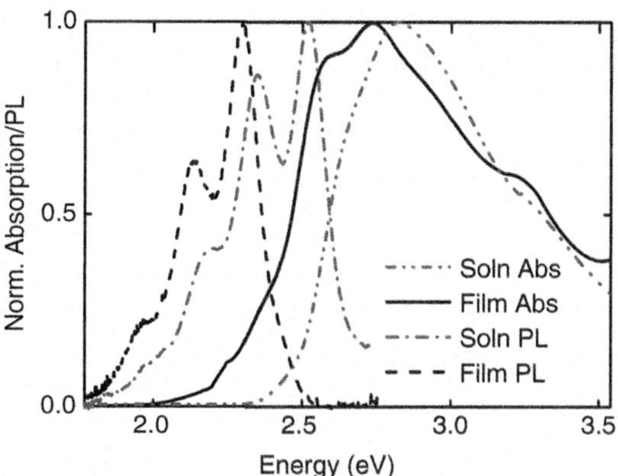

Fig. 6.3 Absorption and PL spectra of a solution and thin film of **1** (Reprinted with permission from [13]. Copyright (2010) American Chemical Society)

given in Table 6.1 [13]. A thin film of **1** forms a nematic glassy phase at room temperature. In solution the isolated molecules adopt a twisted conformation in the ground state, which explains its featureless absorption spectrum, broadened by conformational disorder, primarily associated with a distribution of torsion angles between the rings along the chains, since the low-frequency torsional modes are easily populated by thermal excitations at room temperature [14]. This disorder is reduced in the stiffer excited-state potential, where the molecules tend to adopt a more planar conformation with a relaxed semiquinoid geometry along the chain. This rationalizes the appearance of a well-resolved vibronic structure seen as multiple peaks in the emission spectra. The prominent red shift of both the absorption and emission bands in thin films results from the planarisation of the chains induced by solid state packing. The red shift of 0.12 eV for absorption is fully consistent with the calculated value of 0.15 eV going from a twisted to planar geometry [13]. Compound **1** shows green emission. The colour of the emitted light can be tuned from blue to red by modification of E_{HOMO} and E_{LUMO}, see Eq. (6.1), using chemical design, e.g., an electron donating group can be added to increase E_{HOMO}, whereas electron-withdrawing groups decrease E_{LUMO}. Compound **2** is an oligofluorene with blue emission. The emission is shifted to green light by inserting a 2,1,3-benzothiadiazolyl group between two oligofluorene units to give rise to compound **3**. Compound **4** has a thiophene group at each end of the benzothiadiazolyl moiety, which further shifts the light-emission to the red part of the electromagnetic spectrum. The polarized electroluminescence spectra from devices incorporating compounds **2–4** are shown in Fig. 6.4.

6.3.2 Aggregation

η_{PL} depends on the relative size of the radiative and non-radiative decay rates of the exciton. Interestingly, measurements of η_{PL} for a number of nematic compounds

Table 6.1 Chemical structures and phase transition temperatures of some liquid crystalline semiconductors **1–6**

1	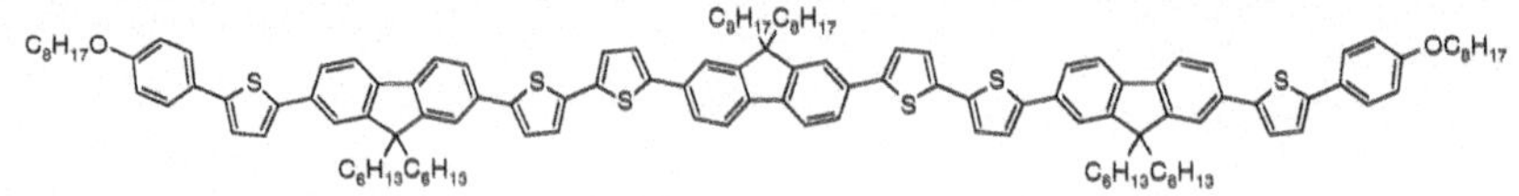 $C_8H_{17}O$; $C_8H_{17}C_8H_{17}$; OC_8H_{17}; $C_6H_{13}C_6H_{13}$; $C_6H_{13}C_6H_{13}$ $t_g \approx 55°C$; Cr-I $\approx 235°C$
2	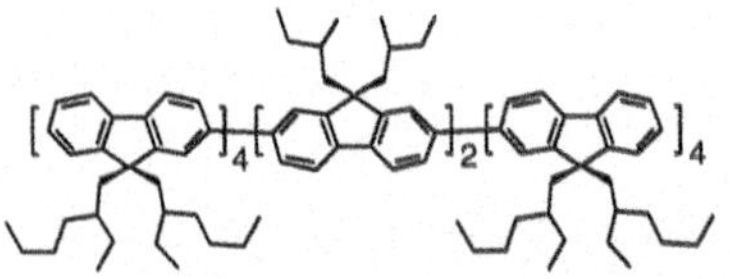 F(MB)10F(EH)2: t_g-N 123°C, N-I >375°C
3	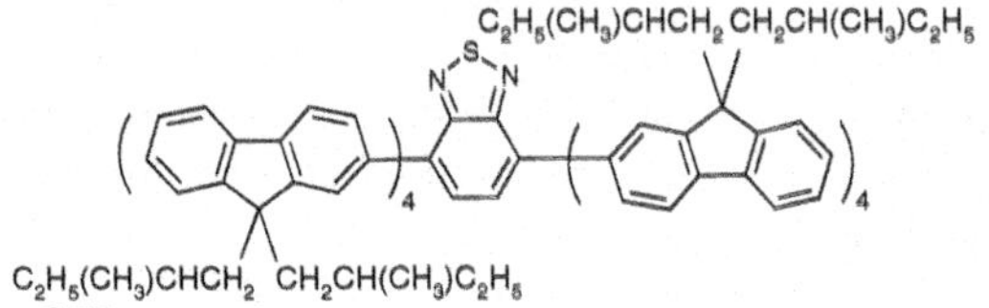 $C_2H_5(CH_3)CHCH_2\ CH_2CH(CH_3)C_2H_5$; $C_2H_5(CH_3)CHCH_2\ CH_2CH(CH_3)C_2H_5$ OF1:
4	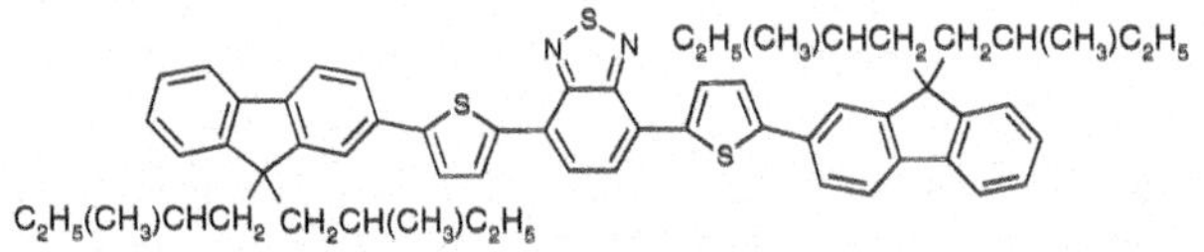 $C_2H_5(CH_3)CHCH_2\ CH_2CH(CH_3)C_2H_5$; $C_2H_5(CH_3)CHCH_2\ CH_2CH(CH_3)C_2H_5$ OF2:
5	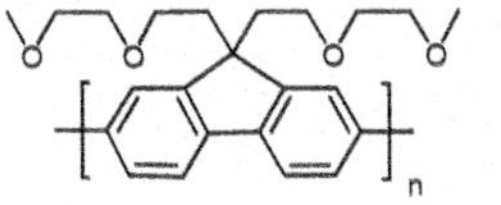 BDOH-PF
6	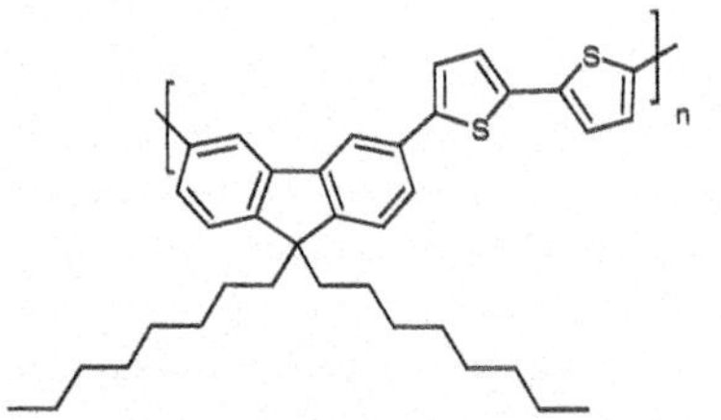 F8T2: Cr-N ≈ 265 °C; N-I > 300 °C

show no correlation with the theoretically calculated oscillator strength, which is directly related to the radiative decay rate [13]. Non-radiative decay can be induced by rotation of the aliphatic end-chains of the liquid crystals, by exciton diffusion to traps, etc. Intermolecular interactions can also quench luminescence and are particularly important in liquid crystals, which have orientational order with or without positional order. The impact of intermolecular interactions between

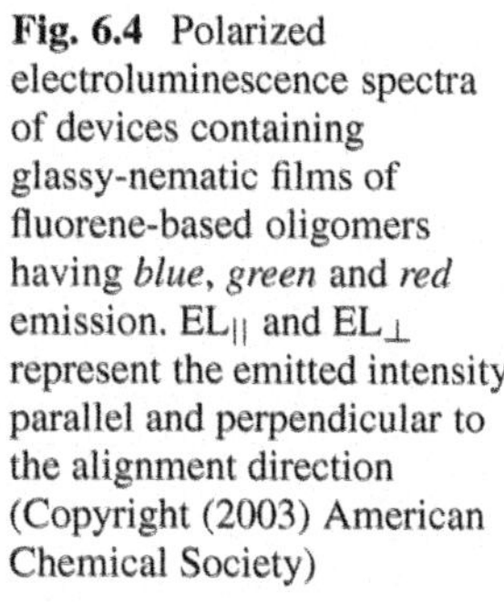

Fig. 6.4 Polarized electroluminescence spectra of devices containing glassy-nematic films of fluorene-based oligomers having *blue*, *green* and *red* emission. $EL_{||}$ and $EL_{\perp}$ represent the emitted intensity parallel and perpendicular to the alignment direction (Copyright (2003) American Chemical Society)

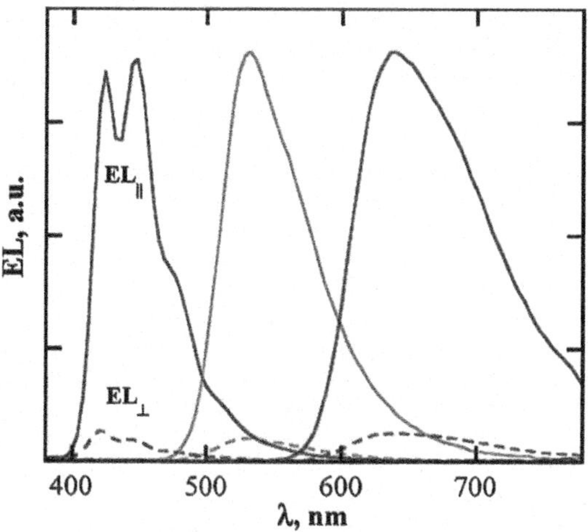

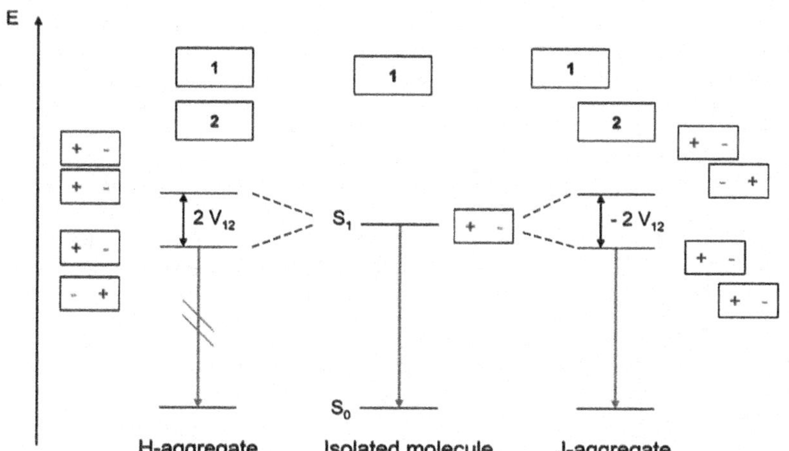

Fig. 6.5 Illustration of the splitting of the lowest excited state of an isolated molecule in H and J aggregates. The combination of the transition dipoles makes the emission symmetry-allowed in the J-aggregate and forbidden in the H-aggregate (Reprinted with permission from [13]. Copyright (2010) American Chemical Society)

adjacent pairs of rod-shaped molecules has been quantified by calculating the excitonic coupling energy, V_{12}, between the lowest excited state in the framework of an excitonic model [13, 15]. Molecules stack in a ladder configuration in an H-aggregate whereas a J-aggregate has neighbouring molecules displaced in a brick-wall configuration, as illustrated in Fig. 6.5. A positive value of the excitonic coupling implies the formation of an H-aggregate, i.e., the splitting of the lowest excited state in the dimer leads to (i) an optically forbidden lower-lying excited state resulting from the antisymmetric combination of the two transition dipoles and (ii) an optically allowed higher-lying excited state built from a symmetric combination

Fig. 6.6 Evolution of the exciton coupling as a function of the degree of translation for a dimer of molecule 1. When $V_{12} > 0$, a H aggregate is formed, whilst $V_{12} < 0$ indicates a J-aggregate

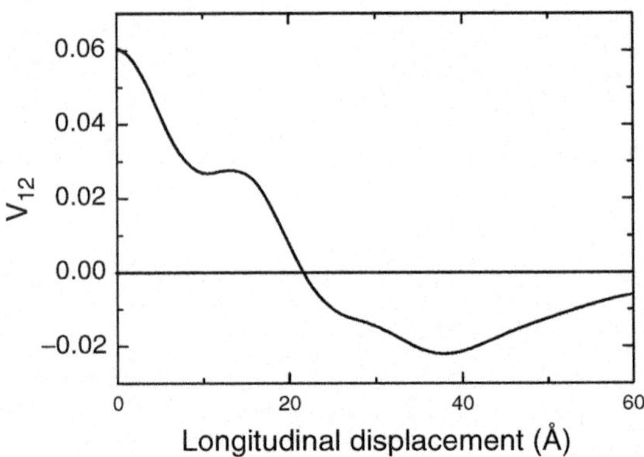

of the dipoles. H-aggregates are, therefore, only weakly emissive with radiative decay partly allowed as a result of localisation of the excitation over one unit due to lattice relaxations in the excited state and/or energetic/positional disorder [16]. In contrast, a negative value of the exciton coupling corresponds to the formation of J-aggregates with an optically allowed low-energy state resulting from the symmetric combination of the dipoles and an optically forbidden high-energy state as the corresponding antisymmetric combination.

The formation of J-aggregates is highly desirable to optimise the emission properties and might be triggered by shifting longitudinally one molecule of the dimer with respect to the other. Figure 6.6 shows V_{12} for a pair of molecules of **1**, each of length about 56 Å, as a function of longitudinal displacement from a head-to-head orientation. The molecules are separated by 4.5 Å. The transition from an H- to a J-aggregate occurs at a displacement of about 22 Å. Nematic materials have orientational order and no positional order and so will form both H- and J-aggregates depending on the displacement of the nearest neighbours.

Many conjugated polymers or oligomers have nematic or other liquid crystalline phases [17–20], so that some H-aggregation is expected to reduce their photoluminescence quantum efficiency. The molecules of smectic liquid crystals are organized in parallel layers, which represent the ideal configuration for the formation of H-aggregates. This may explain the relatively low efficiency of smectic light-emitters in OLEDs [21, 22]. A strategy to improve η_{PL} is to decrease V_{12} for example by increasing the intermolecular separation. It is no coincidence that the most efficient light-emitting nematics in thin films contain 2,7-disubstituted-9,9-dialkylfluorene groups, since the two alkyl chains at the bridging benzylic position of the moiety gives rise to a relatively large intermolecular distance, of at least 4.5 Å [23]. Hence, the high PLQE of 60% for a thin film of a nematic oligofluorene having branched sidechains and a similar chemical structure to **2** [24]. Photoluminescence quantum efficiencies greater than 70% have also been obtained from the fluorene polymer **5** with dioxyheptyl (PEG) side-chains, illustrated in Table 6.1 [25]. It is clear from the discussion in Chaps. 3 and 5 that good charge transport requires small intermolecular separations and so may be incompatible with high photoluminescence quantum efficiency because of aggregation. The liquid crystalline polymer **6**, has recently been identified as one that possesses a desirable combination of charge transport and light emission properties for OLEDs [26]. The bi-thiophene moiety in **6** affords good hole-transporting properties, with hole mobility in unaligned films within organic FET structures of up to 5×10^{-3} cm^2 V^{-1} s^{-1} [27]. Hole mobility can be further increased to 0.01–0.02 ^{3} cm^2 V^{-1} s^{-1} by aligning the polymer chains in the direction of transport [28]. The solid-state PLQE of **6** is 21% and green OLEDs with efficacies of 1.2 cd A^{-1} at 1,000 cd m^{-2} were demonstrated. Efficient OLEDs have also been fabricated using extended molecules and oligomers with nematic phases. Many of these materials form long-lived nematic glasses at room temperature on annealing to remove the solvent from a solution-cast thin film [29–39]. A green OLED incorporating a light-emitting molecule with a chemical structure similar to compound **1** has a maximum efficacy of 11.1 cd A^{-1} [38]. Polarised OLEDs with a yield of 6.4 cd A^{-1} were obtained from the nematic glassy molecule **4** [30].

There have been a number of attempts to make OLEDs based on discotic liquid crystals although performance is disappointing probably because of aggregation. Wendorff and co-workers demonstrated monoesters of triphenylene **7**, see Table 6.2, in their Col$_p$ phase and mainchain polymers of triphenylene in the Col$_h$ phase in electroluminescent devices. High electric fields were required, $\approx 10^5$ V cm^{-1}, and the lifetimes of the devices are probably not very long [40, 41]. The bilayer devices made by Bock and co-workers, e.g., ITO/triphenylene **8** (hole transporter)/perylene **9** (electron transporter)/aluminium [43, 44] exploit materials that have liquid crystal phases above room temperature [42, 43]. Simple variants on these structures were synthesised, which had green, blue and sometimes almost white light emission [44].

6.3.3 Phosphorescence

As discussed above, χ_T is 0.75, so that phosphorescent OLEDs are potentially more efficient than fluorescent devices. Metallo-organic complexes are often used as phosphorescent dopants in OLEDs [45]. The spin-orbit coupling produced by the heavy atom increases the rate of the forbidden transition from the triplet to the singlet ground state. Although the phosphorescent decay rate is low, light-emission is efficient when the non-radiative decay rate is even lower. Indeed, white light electrophosphorescence is now considered a practical and efficient method for solid-state lighting with major advances in the high performance, blue, green and red emitting phosphors [46, 47]. There has been very little effort to combine these efficient emitters with liquid crystallinity [48–50]. Pt^{II} complexes have a planar core and complexes with a 2,6-di(4-alkoxyphenyl)pyridine **l** ligand and diketonate co-ligands show high temperature smectic phases, e.g., the transition temperatures for compound **10**, also shown in Table 6.2, are Cr 171 °C SmA 231 °C I. The complexes exhibit photoluminescence quantum efficiencies exceeding 0.5 in solution, the highest values yet reported for materials of this type [51]. Aggregation

Table 6.2 Chemical structures and phase transition temperatures of some liquid crystalline semiconductors **7–9** and **12** and phosphorescent emitters **10** and **11**

7	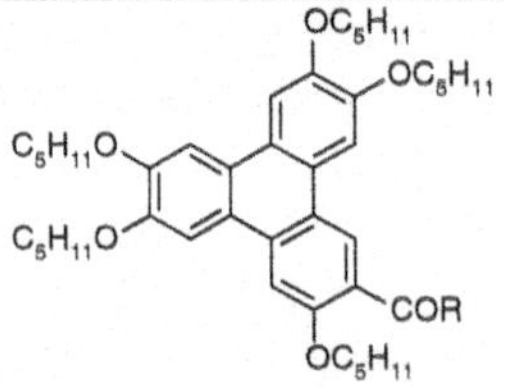 Triphenylene: R = C_6H_4CN, C_6H_{11}, $C(CH_3)_3$ or adamantyl
8	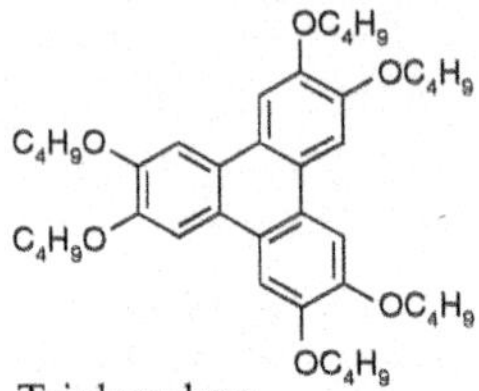 Triphenylene
9	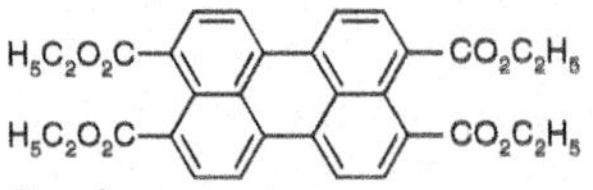 Perylene
10	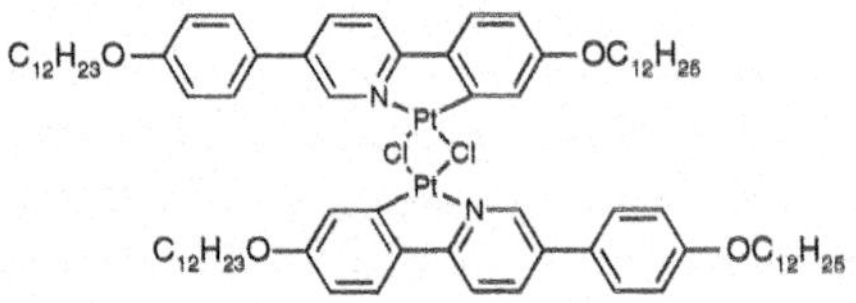 Platinum chromophore
11	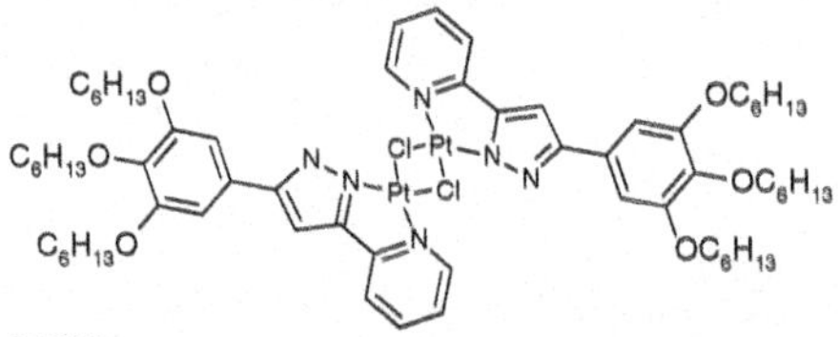 N200
12	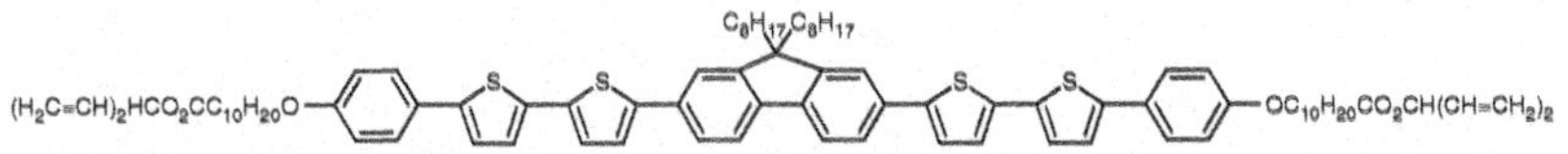 t_g = 0 °C; Cr-N = 52 °C; N-I = 143 °C

or excimer formation in thin films at high concentrations results in red-shifted phosphorescence spectra. A polarised phosphorescence OLED was demonstrated using a blend of the Pt^{II} complex **10** dispersed in an oligofluorene host [52]. **11** has a Col_h phase between 98 and 340 °C. In the host–guest film, the mesogenic Pt(II) complex tends to aggregate and self-assemble into the columnar stacking arrangement, exhibiting metal-metal-to-ligand charge transfer emission of the Pt(II) complex. A rubbed conducting polymer was used as the alignment layer. The emission polarised perpendicular to the alignment direction was approximately twice as intense as that parallel polarized component.

6.4 Out Coupling of Light

η_C of the OLED depends critically on the device configuration. The rear electrode acts as a mirror setting up standing waves so the optical mode pattern and recombination zone must be matched for efficient out-coupling of the light. η_C is also limited by total internal reflection, which traps photons emitted at large angles to the normal of the device within the organic thin films and substrate [53]. The critical angle for total internal reflection, θ_c, depends on the film refractive index, n, according to $\sin\theta_c = 1/n$ for a film-air interface. Light-emitting conjugated materials have a refractive index >2 in the visible, which give rise to low critical angles, <30°. η_c is affected by the distribution of the emitting dipoles, since oscillating dipoles only emit light transverse to their orientation. Numerical calculations showed that the total intensity coupled out of the device varies as $0.75\ n^{-2}$ for the isotropic case, and as $1.2\ n^{-2}$ for the in-plane case. Hence, the out coupling efficiency is a factor of 1.6 higher, when the emitters are aligned in the plane compared with an isotropic orientation [53]. This is a significant improvement in terms of device efficiency. Many low-molar-mass, calamitic liquid crystals spontaneously adopt homogeneous (in-plane) alignment in the nematic or smectic state and so can promote this enhanced output coupling. Main-chain liquid crystalline polymers also tend to lie in the plane especially for high molecular weight materials [54, 55]. The total internally reflected light propagates in film or substrate waveguide modes and out-couples from the edges of the device. The waveguide modes are lossy because of the long propagation pathways and materials absorption. Light scattering by liquid-crystal multi-domains was used to suppress the transverse electric waveguide mode [56]. The multi-domains were created by thermal annealing of thin liquid crystalline light-emitting polymer films. The orientational order of calamitic liquid crystals provides another advantage in guest-host systems based on Förster energy transfer. This involves a dipole-dipole interaction between an excited donor and ground-state acceptor so that excitation is spatially transferred to the latter. The rate of energy transfer is proportional to the orientational factor for dipole-dipole interaction. This is six times larger for perfect parallel alignment of the donor and acceptor compared with a random alignment [57]. Calamitic liquid crystals adopt a roughly parallel configuration and should exhibit an enhanced Förster transfer rate. Efficient energy transfer in a liquid crystalline polymer guest-host system results in an extremely low threshold fluence of 3 nJ cm^{-2} per pulse for optically pumped lasing [58]. A microcavity has been used to provide feedback in the device.

6.5 Optical Anisotropy

6.5.1 Extremely High Birefringence

Visible light-emitting, calamitic, liquid crystals, see examples in Table 6.1, have extended aromatic cores and so are highly anisotropic. Polarised luminescence measurement have been used to investigate the orientational order of polyfluorene [59]. According to the Onsager theory the order parameter of nematics increases with molecular length [60]. *R* gives the ratio of the measured luminescence of polarized light parallel and perpendicular to the alignment direction. *R* values >30 have been found for aligned films of compound **1** and an oligofluorene with 24 connected aromatic rings [32, 38]. One of the major applications is polarised electroluminescence, discussed in Chap. 7, which is obtained by uniform planar alignment of molecules. Another consequence of the high conjugation is an extremely high birefringence. Figure 6.7 shows the birefringence Δn and refractive index of the extraordinary and ordinary rays, n_e and n_o, respectively, of **1**, plotted as a function of wavelength [38, 61]. **1** has 14 rings in its aromatic core. The highly dispersive nature of $n_e(\lambda)$ results from the $\pi - \pi^*$ absorption resonance, which peaks at 452 nm. The dipole moment of the transitions lies along the long axes of the molecules so that n_e rather than n_o is affected by absorption. A maximum Δn value of 1.1 is obtained. A similar molecule with six aromatic rings has a maximum Δn value of 0.7, confirming the expected increase of birefringence with molecular length [61]. Spectroscopic ellipsometry shows that materials similar to **2** and **3** also have very high values of n_e and the birefringence [20, 24].

6.5.2 Benefits of Anisotropy for Thin Film Lasers

Although not widely investigated, the anisotropic properties of light-emitters have the potential to reduce the laser threshold in thin film lasers. In such an organic

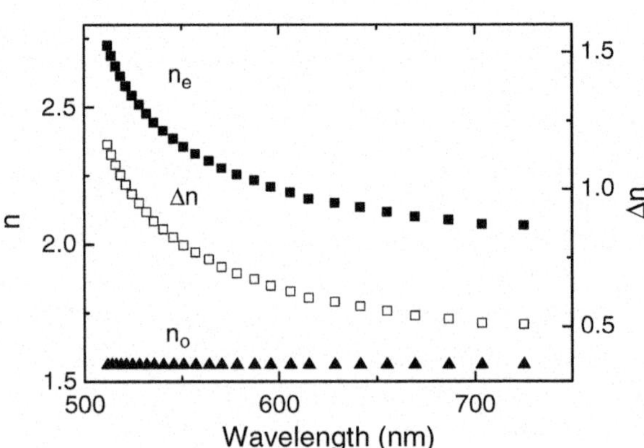

Fig. 6.7 Birefringence, Δn, and refractive index of the extraordinary and ordinary rays, n_e and n_o, of 1

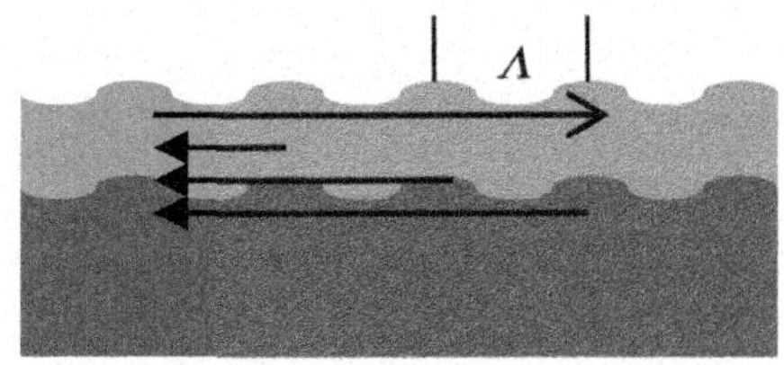

Fig. 6.8 Schematic structure of an organic thin film laser with distributed feedback provided by corrugations of period Λ etched in the underlying silica substrate [6]

laser, gain is provided by optical pumping of a light-emitting polymer or small molecule, which forms all or just a component of the active thin film. Many different feedback approaches have been demonstrated including Fabry-Perot resonators, micro-cavities, distributed feedback (DFB) [5]. An example of the latter is illustrated in Fig. 6.8, where a 1- or 2- dimensional surface relief grating of period Λ is imprinted onto the light-emitting thin film. This grating induces a periodic variation in refractive index, which provides feedback by backward Bragg scattering at wavelengths λ, so that

$$m\lambda = 2n_{eff}\Lambda \tag{6.3}$$

where m is an integer and n_{eff} is the effective refractive index of the film.

The laser output from first-order gratings is coupled from the side facet, whilst second-order gratings can provide a surface-emitted output coupling of the laser light, while providing in-plane feedback *via* second-order diffraction. Recent studies have found that the optical gain of *poly*(9,9-dioctylfluorene-co-benzothiadiazole) is higher, when prepared as a highly oriented nematic glass rather than as a standard, spin-cast, thin film [62, 63]. The material gain coefficient is approximated by

$$g_{mat} = N\sigma_{em}$$

where N is the exciton density and σ_{em} the stimulated emission cross-section. The latter parameter is higher in an oriented thin film, because of the greater degree of overlap between the polarization direction of the stimulated emission and the transition dipole moments of the ensemble of chromophores [64]. Unlike randomly oriented, rod-like emitters, all excited emitters are able to contribute to lasing. The oriented film shows lower loss because of the higher optical quality of a monodomain sample. DFB lasers have also been constructed using oriented light-emitting polyfluorene and the lasing threshold was minimized when the incident excitation beam was polarized parallel to the chain alignment axis and grating lines [64].

6.5.3 Photo-embossing

The nanoscale patterning of the surfaces of organic semiconductors is a relatively young research area. Current applications include the provision of distributed

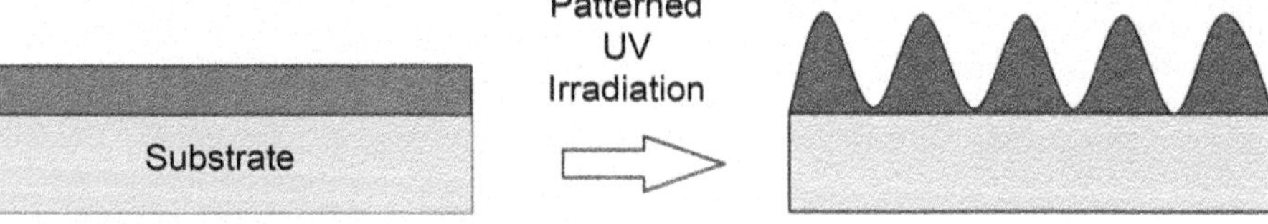

Fig. 6.9 A nanostructured surface is spontaneously formed by ultraviolet irradiation through a phase mask of a thin film of the reactive mesogen 11. The amplitude of the grating produced is often larger than the original thickness of the film. The material is not washed or etched

feedback in organic lasers [5, 65], enhanced efficiency from OLEDs with corrugated structures [66], and the formation of channel waveguides [67]. Embossing and nanoimprinting techniques are often used, whereby patterns are created by mechanical deformation of the organic thin film using a mold with topographic features [68–70]. We have recently demonstrated the spontaneous and single-step formation of surface relief gratings by photoinduced mass-transfer in thin films of the reactive mesogen **12**, whose chemical structure is shown in Table 6.2 [71]. Reactive mesogens are low mass liquid crystals with polymerisable groups separated from the core of the liquid crystal by spacer units [72–76]. Reactive mesogens have low viscosity and can be designed with mesophases just above room temperature, so they can be easily ordered or macroscopically aligned in thin films at low temperatures. The order is then permanently fixed by photochemical or thermal cross-linking of adjacent molecules either by irradiation with ultraviolet light or thermal polymerization. The resultant highly crosslinked polymer network is insoluble and intractable.

The photoembossing process occurs spontaneously, as illustrated in Fig. 6.9, for materials with nematic phases near room temperature, such as compound **12** (Cr–N = 52 °C; N–I = 143 °C) with non-conjugated, photopolymerisable, diene end-groups is such an appropriate reactive mesogen. Light from an ultraviolet HeCd laser was spatially modulated on transmission through a phase mask of period Λ_{PM} and incident to a thin film of **12**. The polymerization-induced change of chemical potential drives mass transfer from regions, where there are troughs in irradiance and less photopolymerisation towards the regions containing peaks. A surface grating is formed spontaneously of period Λ_{PM} or $\Lambda_{PM}/2$, as illustrated in Fig. 6.10, depending on the distance of the phase mask from the film [77]. Deeper gratings are formed, when the polarisation direction of the incident light is perpendicular, rather than parallel, to the grating grooves. This observation is related to the anisotropic shear viscosity of nematic liquid crystals [78]. Irradiation with plane polarised light gives a photoinduced reorientation of the molecules along the polarisation direction during the photochemical crosslinking process. Monomers lying perpendicular to the grating grooves can flow more easily into the more highly crosslinked regions of the film resulting in co-operative reorientation and deeper grooves. Gratings deeper than the original film thickness are created with a period

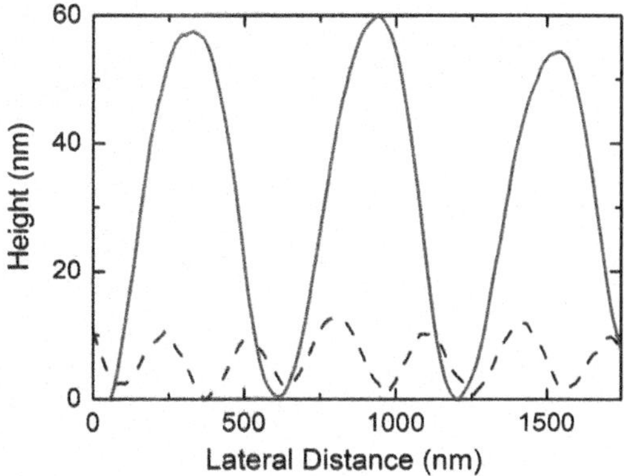

Fig. 6.10 Surface profile of the surface morphology of two thin film samples of **12** irradiated with ultraviolet light through a phase mask of period 0.53 μm (Reprinted with permission from [71]. Copyright (2010) American Chemical Society)

as small as 265 nm. This period approaches that required, ≈ 250 nm [61], to make a second-order, distributed-feedback structure for lasing near the emission peak of the material at 510 nm, so that the photo-embossing process has very significant potential for applications in photonics.

6.6 Optical Properties of Light-Emitting Chiral Liquid Crystals

Organic solid-state lasers offer the prospect of compact, low-cost (even disposable) coherent, visible-light sources suitable for applications from point-of-care diagnostics to sensing. Chiral liquid crystals can provide mirrorless lasing, since distributed feedback is provided by the helix. This research field has sparked enormous interest because of the compact size of the lasers, their ease of manufacture and wide tunability. Gain originates from the homogeneous doping of a dye through the helix [79] or from the light-emitting liquid crystal itself [80].

Figure 6.11 shows the periodic helical structure of a dye-doped cholesteric liquid crystal, which forms a one-dimensional photonic crystal. Spontaneous emission from the dye is suppressed in the photonic stop-band, when it has the same handedness of circular polarization as the helix. However, it is enhanced *via* the Purcell effect at resonance frequencies about the edges of the stop-band [81], as illustrated in Fig. 6.12 for a light-emitting, chiral nematic liquid crystal, whose emission spectrum overlaps the stopband [82, 83]. The density of photon states diverges at the stopband edge with a corresponding increase in the photon dwell time. The chiral laser lases at the frequency of one of the resonant modes at stop-band edge. In an alternative configuration, a thin defect layer can be inserted between two chiral reflectors. In the latter case, laser emission occurs within the band gap at a frequency associated with the allowed state due to the defect [84, 85]. The output wavelength of the laser is easily tuned by altering the concentration

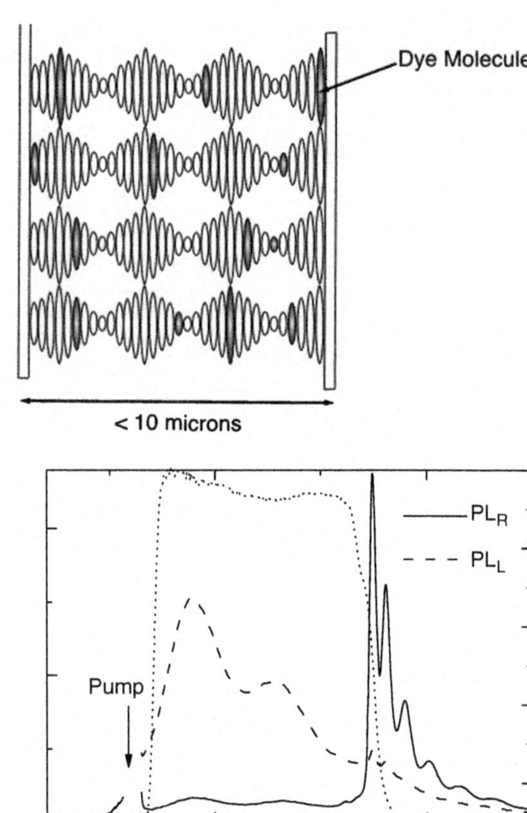

Fig. 6.11 Schematics of distributed feedback system for a chiral liquid crystal laser. The uniformly aligned helical structure of a chiral nematic liquid crystal provides distributed feedback for 1-dimensional lasing. Gain is provided by the optical pumping of dye molecules doped in the liquid crystal helix, shown by *shading*

Fig. 6.12 Right (PL_R) and left (PL_L) circularly polarized photoluminescence from a light-emitting chiral nematic liquid crystal, whose photoluminescence spectrum overlap its stop-band. The *dotted line* shows the stop-band, where transmission of right-hand circularly polarized light is forbidden

of chiral material and the periodicity of the helix. Chiral liquid crystal lasers were first demonstrated in 1998 [79] and have been recently reviewed [86–89]. Most of the research to date is based on chiral nematic phases, but lasing from smectic and blue phases, the latter of which form 3D photonic crystals, has also been observed [90, 91]. Non-reciprocal lasing has also been demonstrated, where the spectrum of the circularly polarised laser emission depends on the direction of observation [92]. There have been many advances in tuning the laser output using electric fields [93, 94], temperature [95], pH [96], light [97–101] and even mechanical deformation in the case of a dye-doped elastomer [102, 103]. The photoinduced *cis-trans* isomerisation was used to alter the pitch length reversibly resulting in tunable laser emission over a 70 nm range [100]. A spatially-graded helical period was obtained by altering the concentration of the chiral dopant to spatially tune the wavelength [97, 104]. Spatially varying the temperature of ultraviolet curing of an elastomer gives a similar tuning [96]. Position-dependent, tunable lasing over most of the visible spectrum was obtained by spatially grading both the optical pitch and dye composition in a two-dye polymer-stabilised system [105, 106]. Two research groups have simultaneously obtained three-colour lasing, in the red, green and blue. In the first approach, thin layers of polyvinyl alcohol formed defects in a polymeric chiral liquid crystal. These defects split the broad stop-band and so provided separate reflection bands for red, green and blue light [107, 108]. A dye-doped nematic liquid crystal was sandwiched between two such composite layers to give simultaneous red, green and blue lasing [109]. The second approach combined gradient-pitch, liquid crystalline samples with lens-array-based pumping techniques to achieve a multi-coloured laser array, consisting of separate red, green, yellow and blue beams in a two-dimensional array. The device was photo-pumped at a single fixed pump wavelength [110, 111]. Display, medical and lab-on-a-chip applications are anticipated. A compact chiral laser system requires a small pump laser. Hence, a frequency tripled Nd:YAG laser was investigated as a pump source instead of a laser-pumped optical parametric oscillator [105]. A cascaded system, where a chiral laser pumps a second longer wavelength chiral laser, has also been demonstrated [112]. A paintable, band-edge chiral laser was recently demonstrated [113]. The lasing medium consists of dye-doped, chiral nematic droplets dispersed within a polymer matrix that spontaneously aligned as the film dried. Such lasers can be easily formed on single substrates with no alignment layers.

There is a large ongoing research effort to obtain continuous-wave (cw) lasing from chiral liquid crystal systems. This requires that both the dye and chiral host must be stable to pump-induced temperature increases. Glassy, rather than fluid, films have been used to improve temporal stability by avoiding temperature-dependent pitch dilation and material flow [114, 115]. Improved temperature-stability was obtained using a photopolymerized chiral nematic polymer network [116], which forms a solid film removable from the substrate [117]. Blue phases normally occur over very limited temperature ranges. However, polymer stabilisation has been used to extend the temperature range of the blue phase with lasing observed over a range of more than 35 °C [118]. Threshold pump energies as low as 20 nJ per pulse have been obtained and the slope efficiencies can be as high as 32% [119]. The transition dipole moment of dopant molecules can be oriented along the director of the chiral nematic phase to maximise coupling between the emitted laser light and the distributed resonator and so reduce the threshold for lasing [120, 121]. A significant challenge is to lower the laser threshold to allow cw lasing, especially since photobleaching of the dye due to triplet formation results in additional loss. New perylene and anthracene dyes have been recently introduced, which give very low thresholds [122]. The electrical pumping of chiral lasers has not yet been achieved. Organic materials have low values of charge carrier mobility giving rise to space-charge limited injection, so that it is very difficult to inject and transport high densities of charge carriers over the typical cavity lengths of some microns. Hence, much lower threshold voltages and careful cavity design are necessary for electrical pumping.

6.7 Conclusions

The optical properties of light-emitting liquid crystals is a relatively immature research field with some very promising areas of activity. Anisotropic gain and feedback have not been widely investigated, but have been identified as tools to improve the performance of organic thin film lasers and may help overcome the challenges to cw and electrically pumped lasing. Chiral lasers have been demonstrated using feedback intrinsic to chiral nematic, smectic and blue phases with tuning ranges over most of the visible spectrum. A multicoloured laser array and a paintable laser have been shown. Many liquid crystalline nematic semiconductors have been used in OLEDs with reasonably high efficiencies. There have also been many demonstrations of polarised OLEDs as well as photolithographic patterned multi-coloured OLEDs, using nematic light-emitters, as discussed in more detail in Chap. 7. A key issue for efficient light-emission is the avoidance of H-aggregation, so that nematic materials, which have no positional order are better light-emitter than smectics. Phosphorescence from metallo-organic complexes with liquid crystalline hosts is a relatively unexplored area of study.

References

1. Friend, R.H., Gymer, R.W., Holmes, A.B., Burroughes, J.H., Marks, R.N., Taliani, C., Bradley, D.D.C., Dos Santos, D.A., Bredas, J.L., Logdlund, M., Salaneck, W.R.: Electroluminescence in conjugated polymers. Nature **397**(6715), 121–128 (1999)
2. Park, S.H., Roy, A., Beaupré, S., Cho, S., Coates, N., Moon, J.S., Moses, D., Leclerc, M., Lee, K., Heeger, A.J.: Bulk heterojunction solar cells with internal quantum efficiency approaching 100%. Nat. Photon. **3**(5), 297–303 (2009)
3. Chen, H.Y., Hou, J., Zhang, S., Liang, Y., Yang, G., Yang, Y., Yu, L., Wu, Y., Li, G.: Polymer solar cells with enhanced open-circuit voltage and efficiency. Nat. Photon. **3**(11), 649–653 (2009)
4. Liang, Y., Xu, Z., Xia, J., Tsai, S.T., Wu, Y., Li, G., Ray, C., Yu, L.: For the bright future-bulk heterojunction polymer solar cells with power conversion efficiency of 7.4%. Adv. Mater. **22**(20), E135–E138 (2010)
5. Samuel, I.D.W., Turnbull, G.A.: Organic semiconductor lasers. Chem. Rev. **107**(4), 1272–1295 (2007)
6. O'Neill, M., Kelly, S.M.: Ordered materials for organic electronics and photonics. Adv. Mater. **23**(5), 566–584 (2011)
7. Burroughes, J.H., Bradley, D.D.C., Brown, A.R., Marks, R.N., Mackay, K., Friend, R.H., Burns, P.L., Holmes, A.B.: Light-emitting diodes based on conjugated polymers. Nature **347**(6293), 539–541 (1990)
8. Tang, C.W., Vanslyke, S.A.: Organic electroluminescent diodes. Appl. Phys. Lett. **51**(12), 913–915 (1987)
9. Grimsdale, A.C., Chan, K.L., Martin, R.E., Jokisz, P.G., Holmes, A.B.: Synthesis of light-emitting conjugated polymers for applications in electroluminescent devices. Chem. Rev. **109**(3), 897–1091 (2009)
10. Nardes, A.M., Kemerink, M., de Kok, M.M., Vinken, E., Maturova, K., Janssen, R.A.J.: Conductivity, work function, and environmental stability of PEDOT:PSS thin films treated with sorbitol. Org. Electron. **9**(5), 727–734 (2008)

11. Hung, L.S., Tang, C.W., Mason, M.G.: Enhanced electron injection in organic electroluminescence devices using an Al/LiF electrode. Appl. Phys. Lett. **70**(2), 152–154 (1997)
12. Wilson, J.S., Dhoot, A.S., Seeley, A.J.A.B., Khan, M.S., Kohler, A., Friend, R.H.: Spin-dependent exciton formation in Ï€-conjugated compounds. Nature **413**(6858), 828–831 (2001)
13. Liedtke, A., O'Neill, M., Kelly, S.M., Kitney, S.P., Van Averbeke, B., Boudard, P., Beljonne, D., Cornil, J.: Optical properties of light-emitting nematic liquid crystals: a joint experimental and theoretical study. J. Phys. Chem. B **114**, 11975–11982 (2010)
14. Gierschner, J., Cornil, J., Egelhaaf, H.J.: Optical bandgaps of Ï€-conjugated organic materials at the polymer limit: experiment and theory. Adv. Mater. **19**(2), 173–191 (2007)
15. Kasha, M.: Energy transfer mechanisms and the molecular exciton model for molecular aggregates. Radiat. Res. **20**, 55–70 (1963)
16. Cornil, J.: Influence of interchain interactions in the absorption and luminescence of conjugated oligomers and polymers: a quantum-chemical characterization. J. Am. Chem. Soc. **120**(6), 1289–1299 (1998)
17. Grell, M., Bradley, D.D.C., Ungar, G., Hill, J., Whitehead, K.S.: Interplay of physical structure and photophysics for a liquid crystalline polyfluorene. Macromolecules **32**(18), 5810–5817 (1999)
18. Gather, M.C., Heeney, M., Zhang, W., Whitehead, K.S., Bradley, D.D.C., McCulloch, I., Campbell, A.J.: An alignable fluorene thienothiophene copolymer with deep-blue electroluminescent emission at 410 nm. Chem. Commun. **9**, 1079–1081 (2008)
19. Yang, S.H., Hsu, C.S.: Liquid crystalline conjugated polymers and their applications in organic electronics. J. Polymer Sci., Part A: Polymer Chem. **47**(11), 2713–2733 (2009)
20. Geng, Y., Chen, A.C.A., Ou, J.J., Chen, S.H., Klubek, K., Vaeth, K.M., Tang, C.W.: Monodisperse glassy-nematic conjugated oligomers with chemically tunable polarized light emission. Chem. Mater. **15**(23), 4352–4360 (2003)
21. Aldred, M.P., Carrasco-Orozco, M., Contoret, A.E.A., Dong, D.W., Farrar, S.R., Kelly, S.M., Kitney, S.P., Mathieson, D., O'Neill, M., Tsoi, W.C., Vlachos, P.: Organic electroluminescence using polymer networks from smectic liquid crystals. Liq. Cryst. **33**(4), 459–467 (2006). doi:10.1080/02678290500487073
22. Tokuhisa, H., Era, M., Tsutsui, T.: Polarized electroluminescence from smectic mesophase. Appl. Phys. Lett. **72**(21), 2639–2641 (1998)
23. Droege, S., Khalifah, M.S.A., O'Neill, M., Thomas, H.E., Simmonds, H.S., Macdonald, J.E., Aldred, M.P., Vlachos, P., Kitney, S.P., Loebbert, A., Kelly, S.M.: Grazing incidence X-ray diffraction of a photoaligned nematic semiconductor. J. Phys. Chem. B **113**(1), 49–53 (2009)
24. Geng, Y., Culligan, S.W., Trajkovska, A., Wallace, J.U., Chen, S.H.: Monodisperse oligofluorenes forming glassy-nematic films for polarized blue emission. Chem. Mater. **15**(2), 542–549 (2003)
25. Yang, Y., Pei, Q.: Efficient blue-green and white light-emitting electrochemical cells based on poly[9,9-bis(3,6-dioxaheptyl)-fluorene-2,7-diyl]. J. Appl. Phys. **81**(7), 3294–3296 (1997)
26. Levermore, P.A., Jin, R., Wang, X., De Mello, J.C., Bradley, D.D.C.: Organic light-emitting diodes based on poly(9,9-dioctylfluorene-co- bithiophene) (F8T2). Adv. Funct. Mater. **19**(6), 950–957 (2009)
27. Chua, L.L., Zaumseil, J., Chang, J.F., Ou, E.C.W., Ho, P.K.H., Sirringhaus, H., Friend, R.H.: General observation of n-type field-effect behaviour in organic semiconductors. Nature **434**(7030), 194–199 (2005)
28. Sirringhaus, H., Wilson, R.J., Friend, R.H., Inbasekaran, M., Wu, W., Woo, E.P., Grell, M., Bradley, D.D.C.: Mobility enhancement in conjugated polymer field-effect transistors through chain alignment in a liquid-crystalline phase. Appl. Phys. Lett. **77**(3), 406–408 (2000)
29. Zeng, L., Yan, F., Wei, S.K.H., Culligan, S.W., Chen, S.: Synthesis and processing of monodisperse oligo(fluorene-co-bithiophene)s into oriented films by thermal and solvent annealing. Adv. Funct. Mater. **19**(12), 1978–1986 (2009)
30. Chen, A.C.A., Culligan, S.W., Geng, Y., Chen, S.H., Klubek, K.P., Vaeth, K.M., Tang, C.W.: Organic polarized light-emitting diodes via foÌrster energy transfer using monodisperse conjugated oligomers. Adv. Mater. **16**(9–10), 783–788 (2004)

31. Chen, A.C.A., Wallace, J.U., Wei, S.K.H., Zeng, L., Chen, S.H., Blanton, T.N.: Light-emitting organic materials with variable charge injection and transport properties. Chem. Mater. **18**(1), 204–213 (2006)
32. Culligan, S.W., Geng, Y., Chen, S.H., Klubek, K., Vaeth, K.M., Tang, C.W.: Strongly polarized and efficient blue organic light-emitting diodes using monodisperse glassy nematic oligo(fluorene)s. Adv. Mater. **15**(14), 1176–1180 (2003)
33. Liedtke A.: Liquid crystals for light emitting diodes. University of Hull, UK (2009)
34. Wallace, J.U., Chen, S.H.: Fluorene-based conjugated oligomers for organic photonics and electronics. Adv. Polym. Sci. **212**, 145–186 (2008)
35. Aldred, M.P., Eastwood, A.J., Kelly, S.M., Vlachos, P., Contoret, A.E.A., Farrar, S.R., Mansoor, B., O'Neill, M., Tsoi, W.C.: Light-emitting fluorene photoreactive liquid crystals for organic electroluminescence. Chem. Mater. **16**(24), 4928–4936 (2004). doi:10.1021/cm0351893
36. Contoret, A.E.A., Farrar, S.R., Jackson, P.O., Khan, S.M., May, L., O'Neill, M., Nicholls, J.E., Kelly, S.M., Richards, G.J.: Polarized electroluminescence from an anisotropic nematic network on a non-contact photoalignment layer. Adv. Mater. **12**(13), 971–974 (2000)
37. Contoret, A.E.A., Farrar, S.R., O'Neill, M., Nicholls, J.E.: The photopolymerization and cross-linking of electroluminescent liquid crystals containing methacrylate and diene photopolymerizable end groups for multilayer organic light-emitting diodes. Chem. Mater. **14**(4), 1477–1487 (2002). doi:10.1021/cm011111f
38. Woon, K.L., Contoret, A.E.A., Farrar, S.R., Liedtke, A., O'Neill, M., Vlachos, P., Aldred, M.P., Kelly, S.M.: Material and device properties of highly birefringent nematic glasses and polymer networks for organic electroluminescence. J. Soc. Inf. Disp. **14**(6), 557–563 (2006)
39. Woon, K.L., Liedtke, A., O'Neill, M., Aldred, M.P., Kitney, S.P., Vlachos, P., Bruneau, A., Kelly, S.M.: Photopolymerization studies of a light-emitting liquid crystal with methacrylate reactive groups for electroluminescence. In: Proceedings of SPIE – The International Society for Optical Engineering, Bellingham, USA, p. 70500E (2008)
40. Lüssem, G., Wendorff, J.H.: Liquid crystalline materials for light-emitting diodes. Polym. Adv. Technol. **9**(7), 443–460 (1998)
41. Stapff, I.H., Stumpflen, V., Wendorff, J.H., Spohn, D.B., Mobius, D.: Preliminary communication multilayer light emitting diodes based on columnar discotics. Liq. Cryst. **23**(4), 613–617 (1997)
42. Seguy, I., Destruel, P., Bock, H.: All-columnar bilayer light-emitting diode. Synth. Met. **111**, 15–18 (2000)
43. Seguy, I., Jolinat, P., Destruel, P., Farenc, J., Mamy, R., Bock, H., Ip, J., Nguyen, T.P.: Red organic light emitting device made from triphenylene hexaester and perylene tetraester. J. Appl. Phys. **89**(10), 5442–5448 (2001)
44. Hassheider, T., Benning, S.A., Kitzerow, H.S., Achard, M.F., Bock, H.: Color-tuned electroluminescence from columnar liquid crystalline alkyl arenecarboxylates. Angew. Chem. Int. Ed. **40**(11), 2060–2063 (2001)
45. Baldo, M.A., O'Brien, D.F., You, Y., Shoustikov, A., Sibley, S., Thompson, M.E., Forrest, S.R.: Highly efficient phosphorescent emission from organic electroluminescent devices. Nature **395**(6698), 151–154 (1998)
46. D'Andrade, B.W., Forrest, S.R.: White organic light-emitting devices for solid-state lighting. Adv. Mater. **16**(18), 1585–1595 (2004)
47. Xiao, L., Chen, Z., Qu, B., Luo, J., Kong, S., Gong, Q., Kido, J.: Recent progresses on materials for electrophosphorescent organic light-emitting devices. Adv. Mater. **23**(8), 926–952 (2011)
48. Kozhevnikov, V.N., Donnio, B., Bruce, D.W.: Phosphorescent, terdentate, liquid-crystalline complexes of platinum(II): stimulus-dependent emission. Angew. Chem. Int. Ed. **47**(33), 6286–6289 (2008)
49. Thomas Iii, S.W., Yagi, S., Swager, T.M.: Towards chemosensing phosphorescent conjugated polymers: cyclometalated platinum(II) poly(phenylene)s. J. Mater. Chem. **15**(27–28), 2829–2835 (2005)

50. Venkatesan, K., Kouwer, P.H.J., Yagi, S., Meuller, P., Swager, T.M.: Columnar mesophases from half-discoid platinum cyclometalated metallomesogens. J. Mater. Chem. **18**(4), 400–407 (2008)
51. Santoro, A., Whitwood, A.C., Williams, J.A.G., Kozhevnikov, V.N., Bruce, D.W.: Synthesis, mesomorphism, and luminescent properties of calamitic 2-phenylpyridines and their complexes with platinum(II). Chem. Mater. **21**(16), 3871–3882 (2009)
52. Liu, S.H., Lin, M.S., Chen, L.Y., Hong, Y.H., Tsai, C.H., Wu, C.C., Poloek, A., Chi, Y., Chen, C.A., Chen, S.H., Hsu, H.F.: Polarized phosphorescent organic light-emitting devices adopting mesogenic host-guest systems. Org. Electron. Phys. Mater. Appl. **12**(1), 15–21 (2011)
53. Kim, J.S., Ho, P.K.H., Greenham, N.C., Friend, R.H.: Electroluminescence emission pattern of organic light-emitting diodes: implications for device efficiency calculations. J. Appl. Phys. **88**(2), 1073–1081 (2000)
54. Ramsdale, C.M., Greenham, N.C.: Ellipsometric determination of anisotropic optical constants in electroluminescent conjugated polymers. Adv. Mater. **14**(3), 212–215 (2002). doi:10.1002/1521-4095(20020205)14:3<212::aid-adma212>3.0.co;2-v
55. Winfield, J.M., Donley, C.L., Kim, J.S.: Anisotropic optical constants of electroluminescent conjugated polymer thin films determined by variable-angle spectroscopic ellipsometry. J. Appl. Phys. **102**(6), 06305 (2007)
56. Lee, T.W., Park, O.O., Kim, Y.C.: Control of emission outcoupling in liquid-crystalline fluorescent polymer films. Org. Electron. Phys. Mater. Appl. **8**(4), 317–324 (2007)
57. Rabek, J.F.: Mechanisms of Photophysical Process and Photochemical Reactions in Polymers: Theory and Applications. Wiley, Chichester (1987)
58. Lee, T.W., Park, O.O., Cho, H.N., Kim, D.Y., Kim, Y.C.: Low-threshold lasing in a microcavity of fluorene-based liquid-crystalline polymer blends. J. Appl. Phys. **93**(3), 1367–1370 (2003)
59. Schartel, B., Wachtendorf, V., Grell, M., Bradley, D.D.C., Hennecke, M.: Polarized fluorescence and orientational order parameters of a liquid-crystalline conjugated polymer. Phys. Rev. B. Condens. Matter Mater. Phys. **60**(1), 277–283 (1999)
60. Knaapila, M., Stepanyan, R., Lyons, B.P., Torkkeli, M., Hase, T.P.A., Serimaa, R., Güntner, R., Seeck, O.H., Scherf, U., Monkman, A.P.: The influence of the molecular weight on the thermotropic alignment and self-organized structure formation of branched side chain hairy-rod polyfluorene in thin films. Macromolecules **38**(7), 2744–2753 (2005)
61. Woon, K.L., O'Neill, M., Vlachos, P., Aldred, M.P., Kelly, S.M.: Highly birefringent nematic and chiral nematic liquid crystals. Liq. Cryst. **32**(9), 1191–1194 (2005). doi:10.1080/02678290500286863
62. Heliotis, G., Xia, R., Whitehead, K.S., Turnbull, G.A., Samuel, I.D.W., Bradley, D.D.C.: Investigation of amplified spontaneous emission in oriented films of a liquid crystalline conjugated polymer. Synth. Met. **139**(3), 727–730 (2003)
63. Xia, R., Campoy-Quiles, M., Heliotis, G., Stavrinou, P., Whitehead, K.S., Bradley, D.D.C.: Significant improvements in the optical gain properties of oriented liquid crystalline conjugated polymer films. Synth. Met. **155**(2), 274–278 (2005)
64. Song, M.H., Wenger, B., Friend, R.H.: Tuning the wavelength of lasing emission in organic semiconducting laser by the orientation of liquid crystalline conjugated polymer. J. Appl. Phys. **104**(3), 033107 (2008)
65. McGehee, M.D., Diaz-Garcia, M.A., Hide, F., Gupta, R., Miller, E.K., Moses, D., Heeger, A.J.: Semiconducting polymer distributed feedback lasers. Appl. Phys. Lett. **72**(13), 1536–1538 (1998)
66. Ziebarth, J.M., Saafir, A.K., Fan, S., McGehee, M.D.: Extracting light from polymer light-emitting diodes using stamped Bragg gratings. Adv. Funct. Mater. **14**(5), 451–456 (2004)
67. Gather, M.C., Ventsch, F., Meerholz, K.: Embedding organic light-emitting diodes into channel waveguide structures. Adv. Mater. **20**(10), 1966–1971 (2008). doi:10.1002/adma.200702837

68. Chou, S.Y., Krauss, P.R., Renstrom, P.J.: Imprint lithography with 25-nanometer resolution. Science **272**(5258), 85–87 (1996)
69. Meier, M., Dodabalapur, A., Rogers, J.A., Slusher, R.E., Mekis, A., Timko, A., Murray, C.A., Ruel, R., Nalamasu, O.: Emission characteristics of two-dimensional organic photonic crystal lasers fabricated by replica molding. J. Appl. Phys. **86**(7), 3502–3507 (1999)
70. Lawrence, J.R., Turnbull, G.A., Samuel, I.D.W.: Polymer laser fabricated by a simple micromolding process. Appl. Phys. Lett. **82**(23), 4023–4025 (2003)
71. Liedtke, A., Chunhong, L., O'Neill, M., Dyer, P.E., Kitney, S.P., Kelly, S.M.: One-step photoembossing for submicrometer surface relief structures in liquid crystal semiconductors. ACS Nano **4**(6), 3248–3253 (2010). doi:10.1021/nn100012g
72. Hikmet, R.A.M., Lub, J.: Anisotropic networks and gels obtained by photopolymerisation in the liquid crystalline state: synthesis and applications. Prog. Polym. Sci. (Oxford) **21**(6), 1165–1209 (1996)
73. Kelly, S.M.: Anisotropic networks, elastomers and gels. Liq. Cryst. **24**(1), 71–82 (1998)
74. Hikmet, R.A.M., Lub, J., Broer, D.J.: Anisotropic networks formed by photopolymerization of liquid-crystalline molecules. Adv. Mater. **3**(7–8), 392–394 (1991)
75. Broer, D.J., Boven, J., Mol, G.N., Challa, G.: In-situ photopolymerization of oriented liquid-crystalline acrylates. Makromol. Chem. **190**, 2255–2268 (1989)
76. Kelly, S.M.: Anisotropic networks. J. Mater. Chem. **5**(12), 2047–2061 (1995)
77. Trout, T.J., Schmieg, J.J., Gambogi, W.J., Weber, A.M.: Optical photopolymers: design and applications. Adv. Mater. **10**(15), 1219–1224 (1998)
78. De Gennes, P.G., Prost, J.: The Physics of Liquid Crystals. Clarendon, Oxford (1993)
79. Kopp, V.I., Fan, B., Vithana, H.K.M., Genack, A.Z.: Low-threshold lasing at the edge of a photonic stop band in cholesteric liquid crystals. Opt. Lett. **23**(21), 1707–1709 (1998)
80. Muňoz, F.A., Palffy-Muhoray, P., Taheri, B.: Ultraviolet lasing in cholesteric liquid crystals. Opt. Lett. **26**(11), 804–806 (2001)
81. Schmidtke, J., Stille, W.: Fluorescence of a dye-doped cholesteric liquid crystal film in the region of the stop band: theory and experiment. Eur. Phys. J. B **31**(2), 179–194 (2003)
82. Woon, K.L., O'Neill, M., Richards, G.J., Aldred, M.P., Kelly, S.M.: Stokes parameter studies of spontaneous emission from chiral nematic liquid crystals as a one-dimensional photonic stopband crystal: Experiment and theory. Phys. Rev. E **71**(4), doi:04170610.1103/PhysRevE.71.041706 (2005)
83. Woon, K.L., O'Neill, M., Richards, G.J., Aldred, M.P., Kelly, S.M., Fox, A.M.: Highly circularly polarized photoluminescence over a broad spectral range from a calamitic, hole-transporting, chiral nematic glass and from an indirectly excited dye. Adv. Mater. **15**(18), 1555–1558 (2003). doi:10.1002/adma.200304960
84. Schmidtke, J., Stille, W., Finkelmann, H.: Defect mode emission of a dye doped cholesteric polymer network. Phys. Rev. Lett. **90**(8), 083902 (2003)
85. Jeong, S.M., Ha, N.Y., Takanishi, Y., Ishikawa, K., Takezoe, H., Nishimura, S., Suzaki, G.: Defect mode lasing from a double-layered dye-doped polymeric cholesteric liquid crystal films with a thin rubbed defect layer. Appl. Phys. Lett. **90**(26), 211106 (2007)
86. Kopp, V.I., Zhang, Z.Q., Genack, A.Z.: Lasing in chiral photonic structures. Prog. Quantum Electron. **27**(6), 369–416 (2003)
87. Palffy-Muhoray, P., Cao, W., Moreira, M., Taheri, B., Munoz, A., Lacey, D., Sambles, J.R., Gleeson, H.F., Pivnenko, M.N.: Photonics and lasing in liquid crystal materials. Philos. Trans. R. Soc. A: Math, Phys. Eng. Sci. **364**(1847), 2747–2761 (2006)
88. Ford, A.D., Morris, S.M., Coles, H.J.: Photonics and lasing in liquid crystals. Mater. Today **9**(7–8), 36–42 (2006)
89. Coles, H., Morris, S.: Liquid-crystal lasers. Nat. Photon. **4**(10), 676–685 (2010)
90. Ozaki, M., Kasano, M., Ganzke, D., Haase, W., Yoshino, K.: Mirrorless lasing in a dye-doped ferroelectric liquid crystal. Adv. Mater. **14**(4), 306–309 (2002)
91. Cao, W., Muňoz, A., Palffy-Muhoray, P., Taheri, B.: Lasing in a three-dimensional photonic crystal of the liquid crystal blue phase II. Nat. Mater. **1**(2), 111–113 (2002)

92. Hwang, J., Song, M.H., Park, B., Nishimura, S., Toyooka, T., Wu, J.W., Takanishi, Y., Ishikawa, K., Takezoe, H.: Electro-tunable optical diode based on photonic bandgap liquid-crystal heterojunctions. Nat. Mater. **4**(5), 383–387 (2005)
93. Kasano, M., Ozaki, M., Yoshino, K., Ganzke, D., Haase, W.: Electrically tunable waveguide laser based on ferroelectric liquid crystal. Appl. Phys. Lett. **82**(23), 4026–4028 (2003)
94. Ozaki, M., Kasano, M., Kitasho, T., Ganzke, D., Haase, W., Yoshino, K.: Electro-tunable liquid-crystal laser. Adv. Mater. **15**(12), 974–977 (2003)
95. Funamoto, K., Ozaki, M., Yoshino, K.: Discontinuous shift of lasing wavelength with temperature in cholesteric liquid crystal. Jpn. J. Appl. Phys, Part 2: Lett. **42**(12 B), L1523–L1525 (2003)
96. Shibaev, P.V., Madsen, J., Genack, A.Z.: Lasing and narrowing of spontaneous emission from responsive cholesteric films. Chem. Mater. **16**(8), 1397–1399 (2004)
97. Chanishvili, A., Chilaya, G., Petriashvili, G., Barberi, R., Bartolino, R., Cipparrone, G., Mazzulla, A., Oriol, L.: Lasing in dye-doped cholesteric liquid crystals: two new tuning strategies. Adv. Mater. **16**(9–10), 791–795 (2004)
98. Ilchishin, I.P., Yaroshchuk, O.V., Gryshchenko, S.V., Shaydiuk, E.A.: Influence of the light induced molecular transformations on the helix pitch and lasing spectra of cholesteric liquid crystals. In: Proceedings of SPIE – The International Society for Optical Engineering, Bellingham, USA, pp. 229–234 (2004)
99. Shibaev, P.V., Sanford, R.L., Chiappetta, D., Milner, V., Genack, A., Bobrovsky, A.: Light controllable tuning and switching of lasing in chiral liquid crystals. Opt. Express **13**(7), 2358–2363 (2005)
100. Chilaya, G., Chanishvili, A., Petriashvili, G., Barberi, R., Bartolino, R., Cipparrone, G., Mazzulla, A., Shibaev, P.V.: Reversible tuning of lasing in cholesteric liquid crystals controlled by light-emitting diodes. Adv. Mater. **19**(4), 565–568 (2007)
101. Fuh, A.Y.G., Lin, T.H., Liu, J.H., Wu, F.C.: Lasing in chiral photonic liquid crystals and associated frequency tuning. Opt. Express **12**(9), 1857–1863 (2004)
102. Finkelmann, H., Kim, S.T., Muñoz, A., Palffy-Muhoray, P., Taheri, B.: Tunable mirrorless lasing in cholesteric liquid crystalline elastomers. Adv. Mater. **13**(14), 1069–1072 (2001)
103. Hirota, Y., Ji, Y., Serra, F., Tajbakhsh, A.R., Terentjev, E.M.: Effect of crosslinking on the photonic bandgap in deformable cholesteric elastomers. Opt. Express **16**(8), 5320–5331 (2008)
104. Huang, Y., Chen, L.P., Doyle, C., Zhou, Y., Wu, S.T.: Spatially tunable laser emission in dye-doped cholesteric polymer films. Appl. Phys. Lett. **89**(11) (2006). 111106
105. Manabe, T., Sonoyama, K., Takanishi, Y., Ishikawa, K., Takezoe, H.: Toward practical application of cholesteric liquid crystals to tunable lasers. J. Mater. Chem. **18**(25), 3040–3043 (2008)
106. Chanishvili, A., Chilaya, G., Petriashvili, G., Barberi, R., Bartolino, R., Cipparrone, G., Mazzulla, A., Gimenez, R., Oriol, L., Pinol, M.: Widely tunable ultraviolet-visible liquid crystal laser. Appl. Phys. Lett. **86**(5), 1–3 (2005)
107. Ha, N.Y., Ohtsuka, Y., Jeong, S.M., Nishimura, S., Suzaki, G., Takanishi, Y., Ishikawa, K., Takezoe, H.: Fabrication of a simultaneous red-green-blue reflector using single-pitched cholesteric liquid crystals. Nat. Mater. **7**(1), 43–47 (2008)
108. Ha, N.Y., Takanishi, Y., Ishikawa, K., Takezoe, H.: Simultaneous RGB reflections from single-pitched cholesteric liquid crystal films with Fibonaccian defects. Opt. Express **15**(3), 1024–1029 (2007)
109. Ha, N.Y., Jeong, S.M., Nishimura, S., Suzaki, G., Ishikawa, K., Takezoe, H.: Simultaneous red, green, and blue lasing emissions in a single-pitched cholesteric liquid-crystal system. Adv. Mater. **20**(13), 2503–2507 (2008)
110. Hands, P.J.W., Morris, S.M., Wilkinson, T.D., Coles, H.J.: Two-dimensional liquid crystal laser array. Opt. Lett. **33**(5), 515–517 (2008)
111. Morris, S.M., Hands, P.J.W., Findeisen-Tandel, S., Cole, R.H., Wilkinson, T.D., Coles, H.J.: Polychromatic liquid crystal laser arrays towards display applications. Opt. Express **16**(23), 18827–18837 (2008)

112. Chanishvili, A., Chilaya, G., Petriashvili, G., Barberi, R., Bartolino, R., Cipparrone, G., Mazzulla, A.: Laser emission from a dye-doped cholesteric liquid crystal pumped by another cholesteric liquid crystal laser. Appl. Phys. Lett. **85**(16), 3378–3380 (2004)
113. Gardiner, D.J., Morris, S.M., Hands, P.J.W., Mowatt, C., Rutledge, R., Wilkinson, T.D., Coles, H.J.: Paintable band-edge liquid crystal lasers. Opt. Express **19**(3), 2432–2439 (2011)
114. Wei, S.K.H., Chen, S.H., Dolgaleva, K., Lukishova, S.G., Boyd, R.W.: Robust organic lasers comprising glassy-cholesteric pentafluorene doped with a red-emitting oligofluorene. Appl. Phys. Lett. **94**(4) (2009)
115. Shibaev, P.V., Kopp, V., Genack, A., Hanelt, E.: Lasing from chiral photonic band gap materials based on cholesteric glasses. Liq. Cryst. **30**(12), 1391–1400 (2003)
116. Schmidtke, J., Stille, W., Finkelmann, H., Kim, S.T.: Laser emission in a dye doped cholesteric polymer network. Adv. Mater. **14**(10), 746–749 (2002). 693
117. Matsui, T., Ozaki, R., Funamoto, K., Ozaki, M., Yoshino, K.: Flexible mirrorless laser based on a free-standing film of photopolymerized cholesteric liquid crystal. Appl. Phys. Lett. **81**(20), 3741–3743 (2002)
118. Yokoyama, S., Mashiko, S., Kikuchi, H., Uchida, K., Nagamura, T.: Laser emission from a polymer-stabilized liquid-crystalline blue phase. Adv. Mater. **18**(1), 48–51 (2006)
119. Mowatt, C., Morris, S.M., Song, M.H., Wilkinson, T.D., Friend, R.H., Coles, H.J.: Comparison of the performance of photonic band-edge liquid crystal lasers using different dyes as the gain medium. J. Appl. Phys. **107**(4), 043101 (2010)
120. Amemiya, K., Shin, K.C., Takanishi, Y., Ishikawa, K., Azumi, R., Takezoe, H.: Lasing in cholesteric liquid crystals doped with oligothiophene derivatives. Jpn. J. Appl. Phys, Part 1: Regul. Pap. Short Notes Rev. Pap. **43**(9 A), 6084–6087 (2004)
121. Shin, K.C., Araoka, F., Park, B., Takanishi, Y., Ishikawa, K., Zhu, Z., Swager, T.M., Takezoe, H.: Advantages of highly ordered polymer-dyes for lasing in chiral nematic liquid crystals. Jpn. J. Appl. Phys, Part 1: Regul. Pap. Short Notes Rev. Pap. **43**(2), 631–636 (2004)
122. Uchimura, M., Watanabe, Y., Araoka, F., Watanabe, J., Takezoe, H., Konishi, G.I.: Development of laser dyes to realize low threshold in dye-doped cholesteric liquid crystal lasers. Adv. Mater. **22**(40), 4473–4478 (2010)

Chapter 7
Organic Light-Emitting Diodes (OLEDs) with Polarised Emission

E. Scheler and P. Strohriegl

In this chapter the use of reactive mesogens in polarized organic light emitting diodes (OLEDs) is described. The first part also serves as an introduction to the basic principles of OLEDs, which have attracted enormous interest since 1986 when Tang and van Slyke described the first thin layer OLED with high brightness and a low operating voltage.

7.1 Organic Light Emitting Diodes: Basic Principles

The materials used in Tang's OLED [1] and its basic structure are shown in Fig. 7.1. Two different organic materials have been used, the electron conductor and green emitter Alq3 and the hole conducting aromatic amine TAPC. Both materials are deposited on top of an indium tin oxide (ITO) glass by vacuum evaporation and covered with a magnesium/silver cathode. The OLED emits green light with a maximum brightness of more than 1,000 cd m^{-2} and an efficiency of 1.5 lm W^{-1}.

Meanwhile OLEDs have reached a high level of brightness and exhibit high efficiencies. This goal has been reached by optimizing each single layer and by adding additional layers to form multi-layer OLEDs. State-of-the-art OLEDs possess up to seven functional layers, which are responsible for example, for hole injection, hole transport, light emission, electron transport and electron injection, respectively. Additional hole- and electron-blocking layers are also frequently used [2–7].

The development of phosphorescent iridium organometallic complexes was another major breakthrough [8]. Such iridium complexes are triplet emitters and enhance the theoretical quantum efficiency of OLEDs from 25% for fluorescent

E. Scheler • P. Strohriegl (✉)
Lehrstuhl fuer Makromolekular Chemie I, Bayreuther Institut fuer Makromolekulforschung BIMF, University of Bayreuth, Bayreuth D-95440, Germany
e-mail: Peter.Strohriegl@uni-bayreuth.de

R.J. Bushby et al. (eds.), *Liquid Crystalline Semiconductors*, Springer Series in Materials Science 169, DOI 10.1007/978-90-481-2873-0_7,

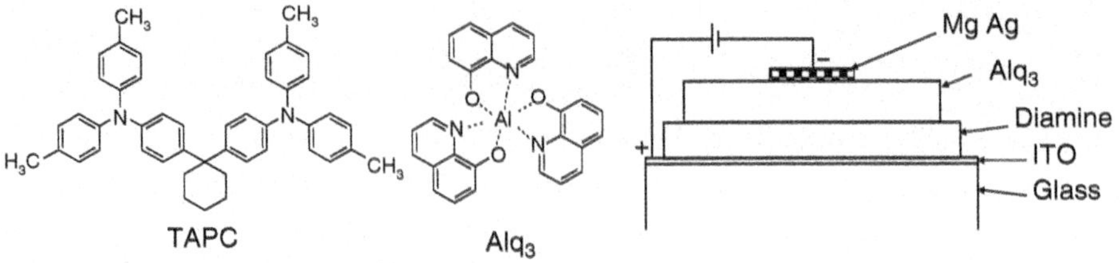

Fig. 7.1 Basic structure and materials used in the first efficient thin layer OLED [1]

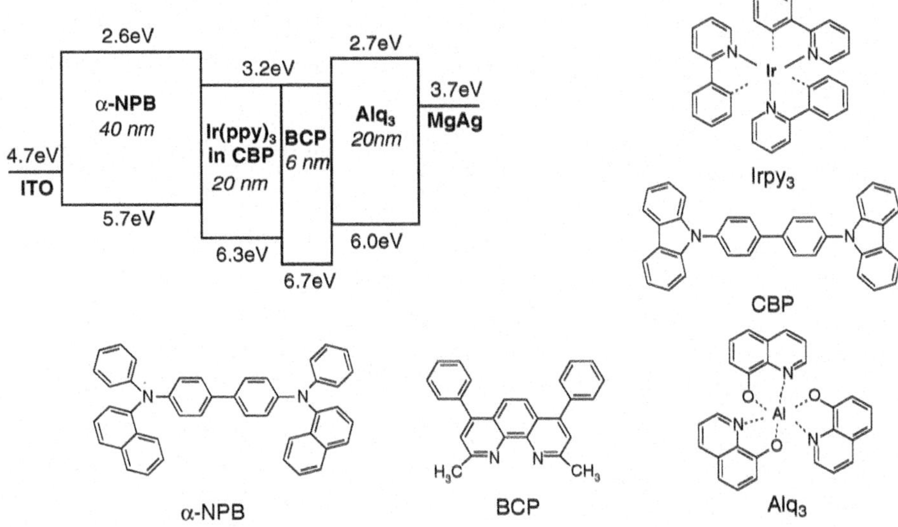

Fig. 7.2 Energy level diagram and materials used in an efficient green phosphorescent OLED [8]

singlet emitters to 100%. The energy level diagram and the materials for a green phosphorescent OLED are shown in Fig. 7.2. Holes are injected from the ITO anode into the aromatic amine α-NPB, a typical hole conductor. From the magnesium/silver cathode, electrons are injected into Alq3, which acts as an electron conductor in this setup. Carrier recombination takes place in the emission layer, which consists of 6% of the green phosphorescent emitter $Irpy_3$ in a CBP matrix. The carbazole containing matrix material CBP is chosen since it has a high triplet level which ensures efficient energy transfer from the matrix to the emitter. The thin layer of BCP prevents holes from escaping to the Alq3 which would reduce the efficiency of the phosphorescent OLED.

Today, the major applications of OLEDs are small displays for MP3 players, mobile telephones and digital cameras, although some OLED TVs are available in the market. Full-colour OLED displays consist of a very large number of single pixels in the three primary colours of red, green and blue (RGB). In most commercial OLEDs the pixels are defined by vacuum evaporation of low-molar-mass materials, such as Alq3 and TAPC, through shadow masks. If polymers are used to fabricate OLED displays, the pixels are usually made by various printing techniques, but with these techniques it is still more difficult to fabricate OLED displays. White light-emitting phosphorescent OLEDs have emerged as a potential highly commercially important application in lighting. Using sophisticated multilayer stacks, white OLEDs with efficiencies up to 64 lm W^{-1} have been recently described [9].

7.2 Introduction to OLED Materials

There are two main classes of materials from which OLED devices can be made, low-molar-mass compounds, often referred to as small molecules, which are deposited by vacuum evaporation [1] and mainchain, highly conjugated polymers, which are processed from solution by ink-jet printing, for example [10]. A hole injection layer is often used as the first layer on top of the conductive indium tin oxide (ITO) coated substrate. This layer can be either an evaporated small molecule, such as copper phthalocyanine, or a conductive polymer like poly(3,4-ethylenedioxythiophene) (PEDT) shown in Fig. 7.3 [11]. The role of the hole injection layer is to match the HOMO levels of the ITO and the hole transport materials, such as aromatic amines, and, especially in the case of PEDT, to planarise the rough ITO surface.

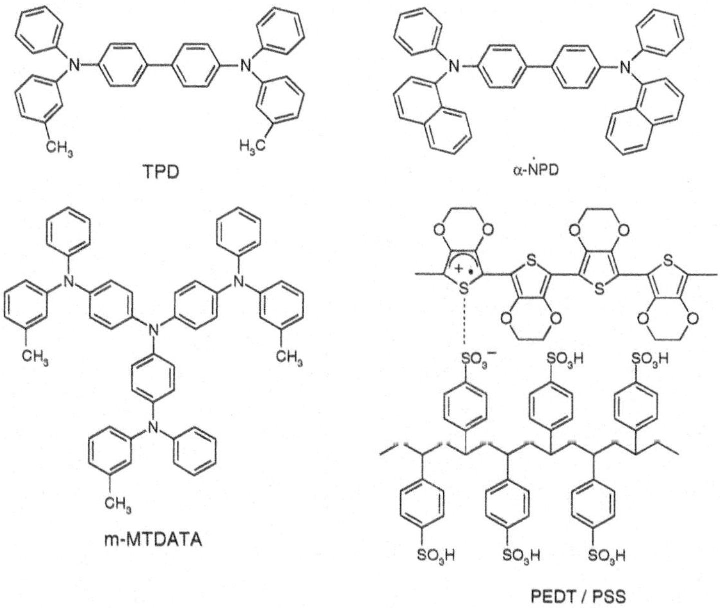

Fig. 7.3 Examples of typical hole transport materials used in OLEDs and molecular structure of the conductive hole injecting polymer PEDT/PSS

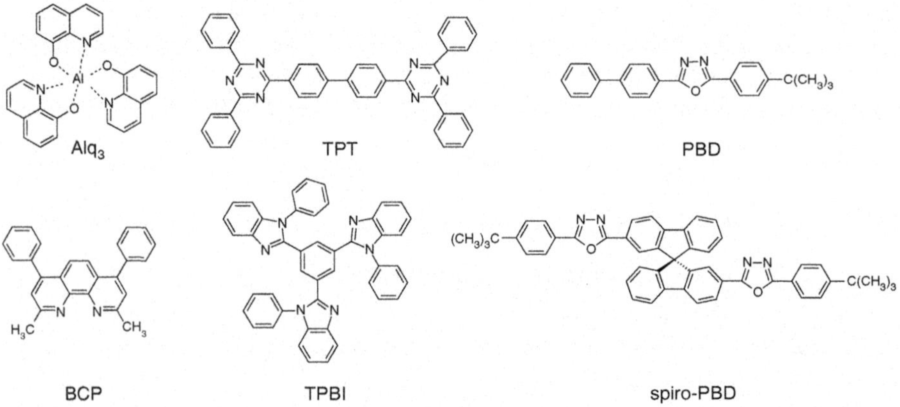

Fig. 7.4 Chemical structures of typical electron conducting and hole blocking materials

Aromatic amines are the most common hole transport materials in OLEDs. In Fig. 7.3 a number of aromatic amines are shown. TPD is a well known organic semiconductor and was used in photoconductor drums long before it became a popular hole conducting material in OLEDs [12]. Since TPD has a relatively low glass transition temperature (Tg = 60°C) it has been replaced by NPD with a higher glass transition temperature ($T_g = 98°C$). An important aspect that affects the performance of an OLED is the morphological stability of the organic thin film layers. The star-shaped aromatic amine *m*-MTDATA forms a stable amorphous phase and functions as excellent hole transport material. *m*-MTDATA is an example of a group of materials often called 'molecular glasses' introduced by Shirota [4, 13].

Figure 7.4 shows the structures of typical electron transport organic semiconductors. Alq3, which can act both as an electron conductor and green emitter, was used by Tang in his first OLEDs and is still in use today. Besides metal complexes electron deficient heterocycles are frequently used as electron transport materials. The triazine derivative TPT [14, 15], the oxadiazole PBD and its spiro analogue spiro-PBD [16, 17] and the benzimidazole TPBI serve as typical examples. 2,9-Dimethyl-4,7-diphenyl-1,10-phenanthroline (BCP) has a very low HOMO level and is often used as a hole blocking layer.

Some common emitter materials are shown in Fig. 7.5. Dimethylquinacridone and DCM are fluorescent green and red emitters, which are doped into suitable matrix materials like Alq3. BCzVBI is an efficient fluorescent blue emitter and is used in a matrix of the distyrylarylene DPVBI [18]. Spiro-sexiphenyl [16, 17] also emits in the blue region of the electromagnetic spectrum.

The internal quantum efficiency is limited to 25% in fluorescent OLEDs due to spin statistics. With phosphorescent emitters both singlet- and triplet excitons contribute to the electroluminescence and hence the theoretical limit of the quantum efficiency rises to 100%. In 1999 Forrest and Thompson introduced the green

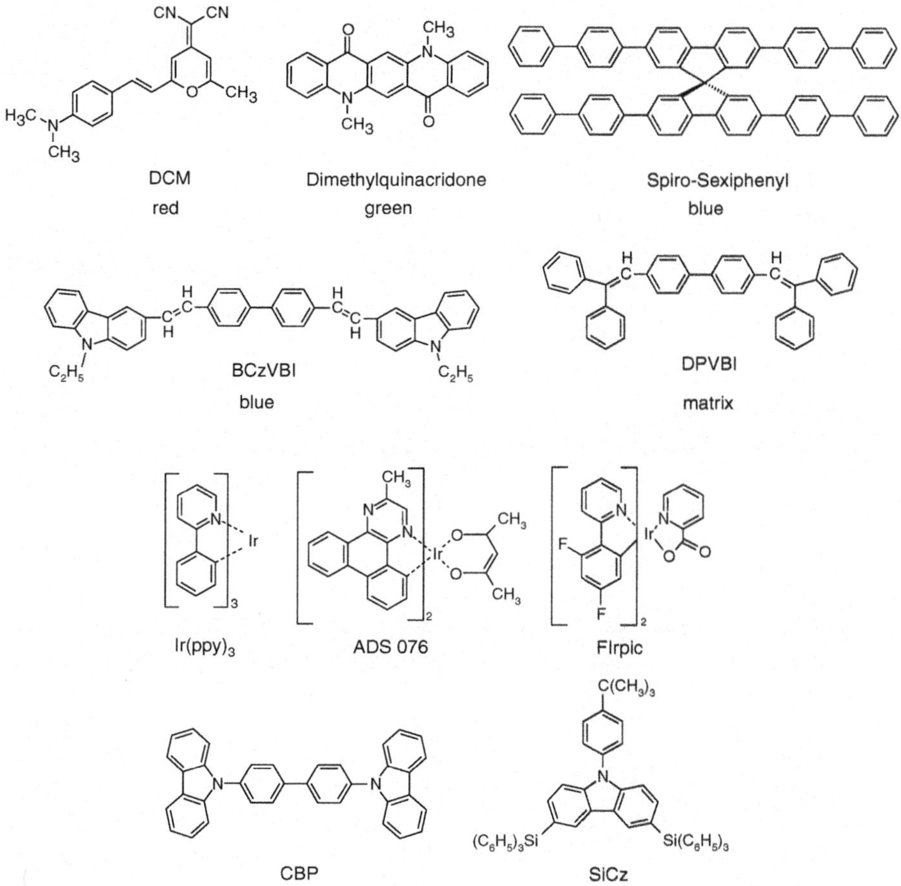

Fig. 7.5 Chemical structures of some common fluorescent (*top*) and phosphorescent (*below*) emitters and matrix materials

phosphorescent emitter Ir(ppy)$_3$, which is used as a dopant in a CBP matrix [8]. CBP has a high S_0-T_1 band gap of 2.55 eV and is well suited as matrix material for the red emitting iridium(III)bis(2-methyldibenzo-[f,h]quinoxaline)(acetylacetonate) (ADS 076) and the green emitter Ir(ppy)$_3$. FIrpic is a phosphorescent greenish-blue emitter. For such emitters, matrix materials with a S_0-T_1 band gap of almost 3 eV like SiCz [19] are necessary to ensure an efficient energy transfer from the host material to the emitter.

The chemical structures of a number of mainchain, highly conjugated polymers used in OLEDs are shown in Fig. 7.6. The electroluminescence of the polymer poly(1,4-phenylenevinylene) (PPV) was reported in 1990 [10]. Since PPV itself is completely insoluble and thin films are only accessible by thermal conversion of a soluble polyelectrolyte precursor, a number of soluble PPV-derivatives have been developed. Among these are MeH-PPV [20] and a number of soluble PPVs

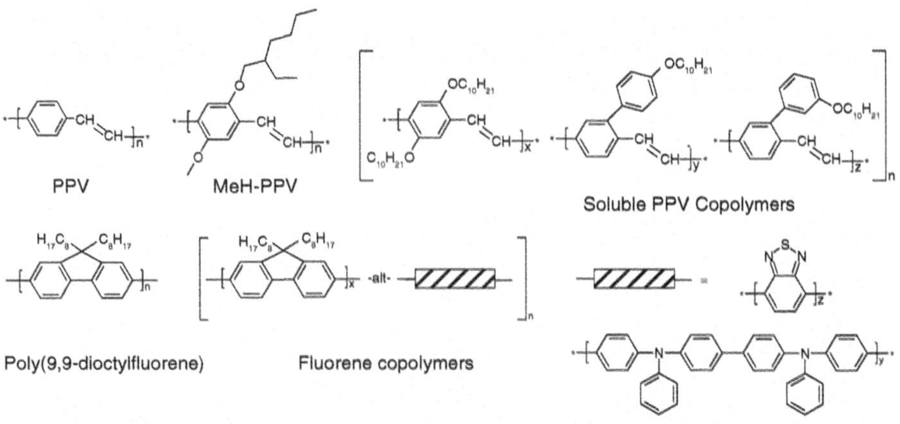

Fig. 7.6 Chemical structures of conjugated polymers frequently used in OLEDs

developed by Covion (Merck Organic Semiconductors) [21]. With PPV emitters the red and the green part of the visible spectrum are accessible. The best known blue emitters are polyfluorene homo- and copolymers [22, 23].

PLEDs are often simpler in construction and are usually fabricated with fewer layers than OLEDs prepared using small molecules and vacuum evaporation. This is due to the fact that the deposition of a second polymer layer often leads to production problems, since the first polymer may be partially dissolved by the solvent used to deposit the second polymer. This problem can be overcome in some cases by the use of orthogonal solvents, which means that the solvent for the second layer is not a solvent for the first polymer deposited. A good example is PEDT/PSS, which is used as aqueous suspension and does not dissolve in common organic solvents, so that other conjugated polymers can be deposited on top of PEDT/PSS. Nevertheless orthogonal solvents are often not available for a pair of conjugated polymers and this makes the preparation of multilayer PLEDs difficult. One alternative for multilayer PLEDs are conjugated polymers with photocrosslinkable units. These polymers are spin coated and then photochemically crosslinked to form an insoluble layer of polymer network, so that the next polymer layer can be coated on top on the first one from solution.

7.3 Liquid Crystals in Polarized OLEDs and PLEDs

Liquid crystal displays (LCDs) are the prevalent flat panel display technology. The LCDs manufactured today range from small, low-cost displays for wristwatches and calculators to very large-area, active-matrix TVs for home use. This gives rise to the question as to whether the huge amount of knowledge of the self-assembling properties of liquid crystals could not be advantageously used in the field of OLED displays. There are two major aspects which make the use of oriented liquid crystalline materials in OLEDs attractive.

The first property of well aligned liquid crystalline films from conjugated materials is their high charge carrier mobility. Mobilities up to 10^{-1} cm^2 V^{-1} s^{-1} have been demonstrated in highly ordered discotic [24] and smectic [25] liquid crystalline phases. The high charge carrier mobility of liquid crystalline polymers [26] and of networks from reactive mesogens [27] are of particular interest for the development of organic field effect transistors (OFETs), see Chap. 8. The second important property is the fact that OLEDs, in which rod-like emitters are aligned parallel to each other, directly emit polarized light. This may lead to large area polarized light sources for applications, such as LCD backlights [28]. Spatial patterning of the polarization direction can also form the basis of three-dimensional displays [29].

The self organisation of liquid crystalline materials has indeed been used to design polarized OLEDs. If the emitters in an OLED are arranged parallel to each other in an LC-monodomain, the device will directly emit linearly polarized light. Polarized OLEDs could be particularly useful as backlights for conventional LCDs because they make a polarizer with its absorptive losses redundant. Such an opportunity was first discussed by Dyrekelev [30], who used mechanical stretching of a bithiophene polymer for the orientation and obtained an OLED with an electroluminescence polarization ratio ($EL_{||}/EL_{\perp}$) of 2.4:1. The method of stretch alignment, however, is not suitable for the manufacturing of OLED displays. Different methods like mechanical alignment, Langmuir Blodgett deposition and liquid crystalline self-organisation have been proposed to align anisotropic molecules in OLEDs and have been described in detail in 1999 by Grell and Bradley [28]. Among these methods the alignment of liquid crystalline molecules to large monodomains has turned out to be most attractive.

Polyfluorenes and fluorene-based low molar mass compounds have attracted a lot of interest for use in emissive devices of polarised emission. This is due to the ability of many highly conjugated, fluorene-based materials to form mesophases in conjunction with the strong blue fluorescence of the fluorene chromophore [31]. Dialkylated polyfluorenes like poly[9,9-di(2-ethylhexyl)]fluorene (PF 2/6) exhibit liquid crystalline phases. Grell et al. showed that the parallel orientation of the liquid crystalline polyfluorenes in an PLED directly leads to the emission of linearly polarized light [32]. In their PLED setup they used rubbed polyimide as orientation layer. Due to the high electrical resistivity the polyimide had to be doped with an aromatic amine to ensure hole transport through the orientation layer. They achieved polarization ratios of 14:1 for the electroluminescence measured parallel and perpendicular to the rubbing direction.

We have described the orientation of poly[2,7-(9,9-dioctyl)fluorene] on top of a rubbed poly(1,4-phenylenevinylene) (PPV) layer (Fig. 7.7) [33, 34]. Since PPV is a conjugated polymer itself the problems with the extremely good insulator polyimide, which is not well suited as a hole transport layer in OLEDs, are not present in such PLEDs. With an OLED setup consisting of ITO, a 30 nm thick rubbed PPV orientation layer, an annealed 70 nm PFO layer and a calcium top electrode we obtained blue electroluminescence with an orientation ratio of 25:1 parallel and perpendicular to the rubbing direction of the PPV. If a polythiophene

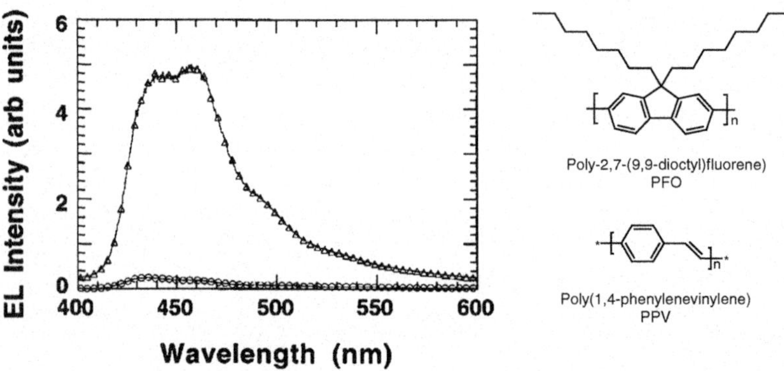

Fig. 7.7 Polarized electroluminescence from an ITO/rubbed PPV/aligned PFO/Ca OLED. Spectra are shown for light polarized parallel (*open triangles*) and perpendicular (*open circles*) to the rubbing direction. Polarization ratio. 25:1

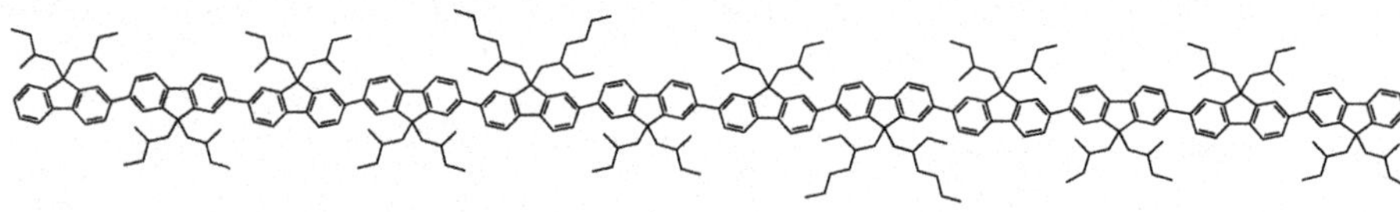

Fig. 7.8 Molecular structure of the fluorene decamer F(MB)10F(EH)2 [41]

is deposited on top of oriented PPV and subsequently rubbed, it is also possible to prepare OLEDs with two orthogonally polarized emitting layers [35]. Neher et al. used a doped photoorientation layer containing azo-groups in polyfluorene OLEDs and reached orientation ratios up to 14:1 [36].

A number of groups synthesized low-molar-mass fluorene model compounds with a different number of fluorene units based on the results using polyfluorenes, e.g., Klaerner et al. described the synthesis of a mixture of oligomers and their separation using HPLC [37]. The synthesis of monodisperse oligofluorenes with up to seven fluorene units by repetitive Suzuki and Yamamoto coupling reactions was reported in 2004 [38]. Fluorene oligomers with up to 12 units were synthesized by Geng et al. [39, 40] and Culligan et al. [41] succeeded in making linearly polarized OLEDs using these oligomers. The fluorene dodecamer F(MB)10F(EH)2, see Fig. 7.8, has a high glass transition temperature (Tg = 123°C) and exhibits a broad nematic mesophase up to the thermal decomposition at 375°C.

OLEDs with polarized emission were fabricated using a the fluorene dodecamer, see Fig. 7.9. Uniaxial molecular alignment was accomplished by spin casting F(MB)10F(EH)2 on top of a rubbed PEDT/PSS conductive alignment layer with subsequent thermal annealing. A peak polarization ratio of 31:1 and a luminance yield of 1.1 cd A^{-1} were obtained from an OLED with an additional electron transport layer and a LiF/MgAg electrode. The concept was extended in 2004 to polarized green and red OLEDs obtained by doping red and green emitting rod-like oligomers into a blue fluorene matrix [42]. Förster energy transfer leads to polarized

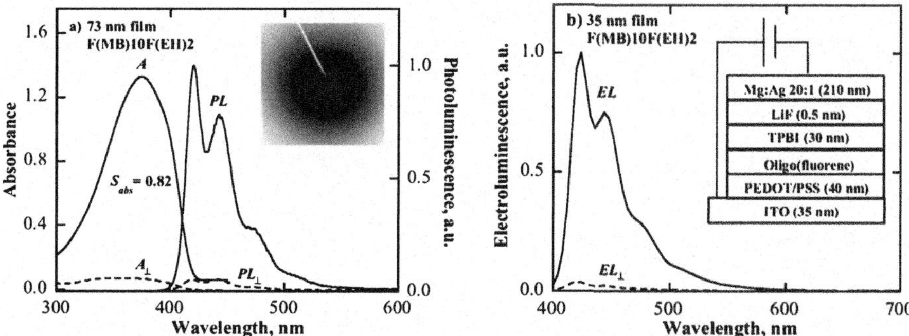

Fig. 7.9 (*Left*) Linearly polarized absorption and photoluminescence spectra (excitation wavelength 370 nm) of an OLED containing a 73 nm thick F(MB)10F(EH)2 film, with an inset showing the electron diffraction pattern. (*Right*) Device structure and polarized EL spectra of an OLED containing a 35 nm thick F(MB)10F(EH)2 film (From Culligan et al. [41])

emission with polarization ratios up to 26:1 and luminance yields up to 6.4 cd A^{-1}. Fluorene oligomers using both electron-donating arylamines and electron-accepting triazine units have been used to fabricate OLEDs with polarized emission [43, 44].

The results discussed above show that from both low molar mass fluorene model compounds and high molecular weight polyfluorenes OLEDs with reasonably high polarization ratios can be fabricated. A number of orientation layers with hole transporting properties have been developed, e.g., rubbed polyimide doped with an aromatic amine, rubbed poly(1,4-phenylenevinylene)(PPV) and rubbed PEDT. In addition to these layers, where orientation is achieved by mechanical forces, photo-orientation layers appear very promising for OLED applications [36, 45], since the orientation is achieved by irradiation with linearly polarized light and mechanical rubbing is not necessary. A further option of photo-orientation layers is their ability to create sets of OLED pixels with orthogonal polarization, which may lead to 3D-OLED-displays.

7.4 Reactive Mesogens in OLED Applications

Besides their liquid crystallinity reactive mesogens exhibit an additional feature. They carry functional groups, such as oxetanes or acrylates, which can be polymerized and densely crosslinked polymer networks are formed. If the crosslinking takes place in the liquid crystalline phase of the material, then the liquid crystalline order is preserved in the resultant polymer network, which is thermally stable up to the decomposition temperature of the polymer network. Furthermore, the liquid crystalline polymer network becomes completely insoluble upon crosslinking, which allows the preparation of multilayer devices using solution processes. In many cases the crosslinking is carried out by a photochemically initiated reaction,

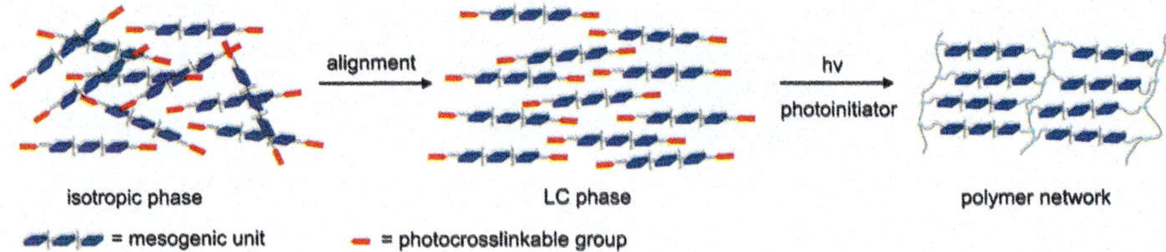

Fig. 7.10 Photochemical crosslinking of reactive mesogens in their liquid crystalline state

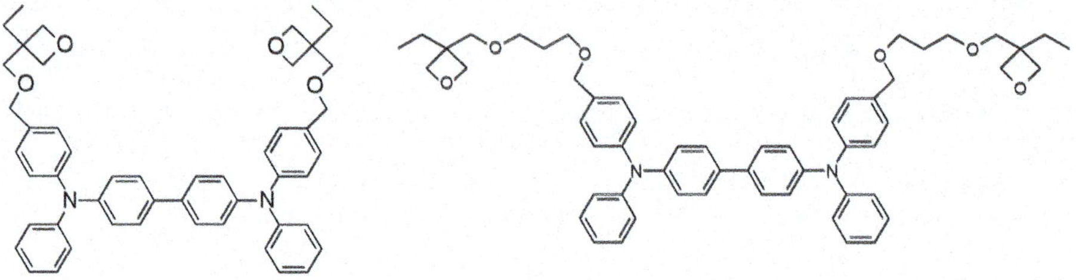

Fig. 7.11 Photocrosslinkable TPD derivatives with cationically polymerisable oxetane units

see Fig. 7.10. If a photomask is used, then the liquid crystalline monomer (reactive mesogen) behaves like a conventional negative photoresist. By dissolving away the non-crosslinked parts of the film, pixel patterns are obtained. A resolution of up to 1 μm is readily achievable [46], which is much smaller compared to that obtained by standard printing techniques.

From 1997 on the first papers on crosslinkable hole transport layers appeared in literature. Li et al. reported a polymethacrylate terpolymer with blue-emitting distyrylbenzene, electron-conducting oxadiazole and photocrosslinkable cinnamoyl repeat units in 1997 [47]. The polymer could be crosslinked by UV-irradiation, but suffered from bleaching of the distyrylbenzene chromophores. Bellman et al. described a series of polynorbonenes with pendant triarylamine groups. The remaining double bonds in the polynorbonene backbone could be crosslinked by UV-irradiation, but the efficiency of the OLEDs decreased after crosslinking [48]. Thermal crosslinking of oligotriarylamines with styrene endgroups [49] and the thermal crosslinking of triphenylamines using silane chemistry [50] have both been reported. In 1999 Bacher et al. described the photopolymerisation of triphenylenes with one, two and three acrylate units [51]. They also successfully demonstrated photopatterning of OLED materials using a photomask. In the same year Bayerl et al. prepared TPD-based hole transport layers with pendant oxetane units, that can be cationically photopolymerised (Fig. 7.11) [52]. The material set with photocrosslinkable oxetane groups has been extended from hole transporting aromatic amines [52, 53] to multifunctional spirobifluorene copolymers and to efficient electrophosphorescent OLEDs with iridium complex emitters [54].

These materials were developed in the first place for an easier OLED processing due to the multilayer capability of the crosslinked polymers. Layer after layer can be successively spincoated without dissolving the crosslinked layer underneath. If a photomask is applied spatially separated pixels are obtained, which is described in the following paragraph.

7.5 Pattern Formation

Although OLEDs prepared by solution-based processes have seen tremendous advances over the years, the pixilation of red, green and blue pixels in the emissive layer still remains one of the key challenges for the production of full colour displays. Various printing techniques, especially ink jet printing, have received much attention in recent years, mostly because of their potential for cost effective production. Nevertheless there are still a number of problems to be solved before printing of conjugated polymers can be used as a routine method to manufacture PLEDs. One problem is that printing often has to be done on prepatterned substrates, which are usually fabricated by conventional lithographic techniques [55]. This adds considerable complexity and cost to the production process.

The use of reactive mesogens and polymers and oligomers with photocrosslinkable units as negative photoresists involves the photo-crosslinking by irradiation through a patterned photo-mask. The exposed parts become insoluble due to polymer network formation, while the non-exposed parts remain soluble and are washed away in the subsequent wet development step, see Fig. 7.12. The

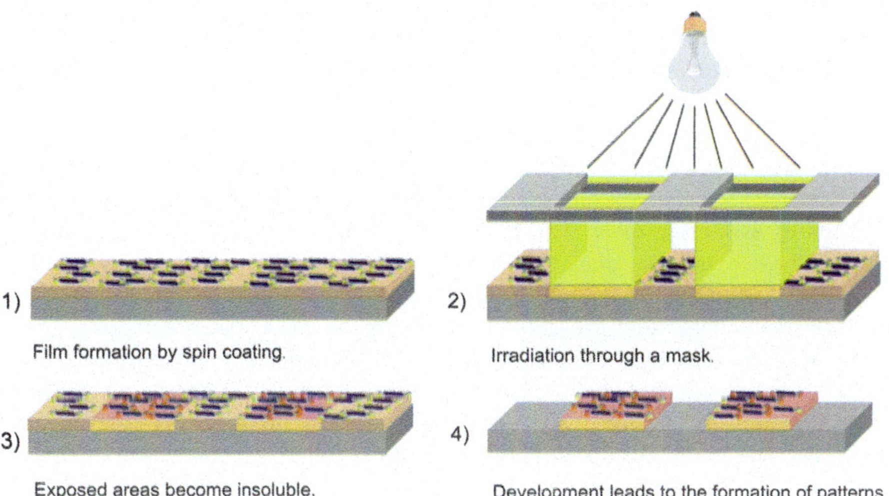

Fig. 7.12 Pattern formation by direct photolithography. (**a**) Formation of a thin film of the reactive mesogen by spin coating. (**b** and **c**) Irradiation through a mask leads to the crosslinking of the exposed parts. (**d**) In the development step the non-crosslinked areas are washed away

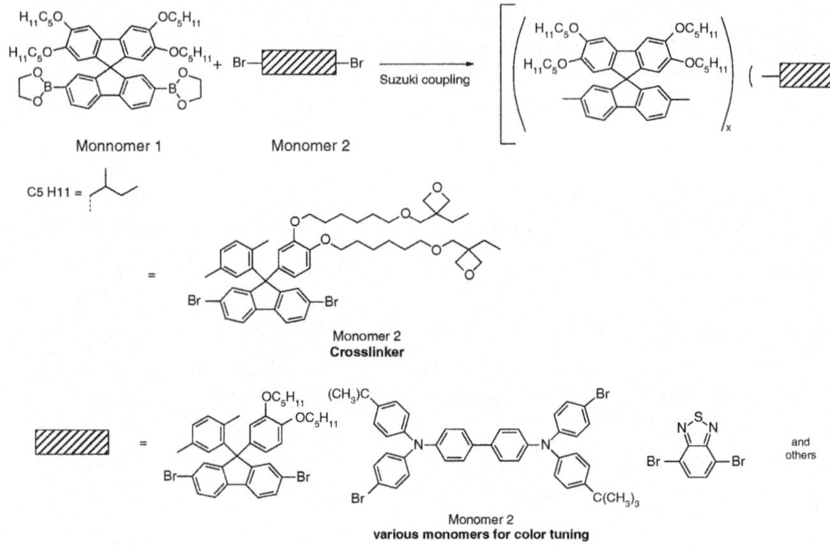

Fig. 7.13 Synthesis of crosslinkable spirobifluorene-co-fluorene copolymers [57]

crosslinkable material itself can serves as an emitting species and/or a charge transport layer. Therefore, it is very important that, during crosslinking, the chemical structure of the material is not altered, i.e., photodegradation is kept to a minimum [56]. Unfortunately, photocrosslinking may sometimes cause a substantial degree of photochemical degradation. Thus, in order to avoid this, the processing conditions have to be very mild in terms of temperature and exposure time.

The best known example for photopatternable OLEDs comes from a collaboration of the groups of Meerholz and Nuyken [57, 58] who reported the synthesis of a number of red-, green- and blue-emitting spiro-copolymers Fig. 7.13, by Suzuki cross coupling of a diborolane (monomer 1) and mixtures of various co-monomers (monomer 2). One of the co-monomers contains two photoactive oxetane units and serves as a crosslinker. The other co-monomers are used for colour tuning. Pixelated OLEDs were made in a multi step process using these polymers as shown in Fig. 7.14. (a) PEDT/PSS and a crosslinkable hole-transport material are first spin-coated and crosslinked on top of an ITO coated substrate; (b). a solution of the blue-light-emitting polymer is then deposited and exposed to UV-light through an aligned shadow mask; (c) the non-crosslinked parts are dissolved in an organic solvent after a soft-curing step; (d) repetition of this procedure for green and red pixels results in parallel stripes of blue-, green-, and red-emitting polymers. The fabrication of this full-colour OLED is completed by evaporating a metal cathode through a shadow mask on top of the organic layers. The idea of crosslinking reactive mesogens for electrooptic devices can be traced back to Broer, who developed thermally stable *passive* optical devices such as optical retarders [59] and efficient reflective polarisers with tuneable absorption wavelength from

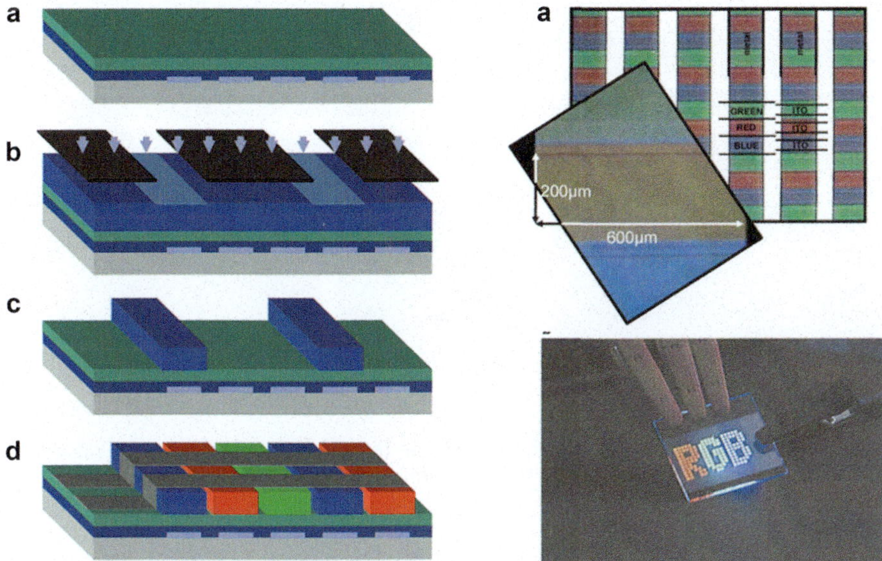

Fig. 7.14 *Left*: Schematic illustration of the direct lithography process. *Top right*: Microscopy image of the completed display taken through the glass substrate. The horizontal polymer stripes are parallel and well aligned with the underlying ITO anode stripes. The metal cathode columns are seen as vertical stripes. (the separating white stripes result from the microscope backlight overexposing the camera.) The inset shows a single RGB triple at a higher magnification. *Bottom right*: Photograph of a RGB OLED device. The dimensions of the glass substrate are 25 × 25 mm (Reprinted from Gather et al. [58])

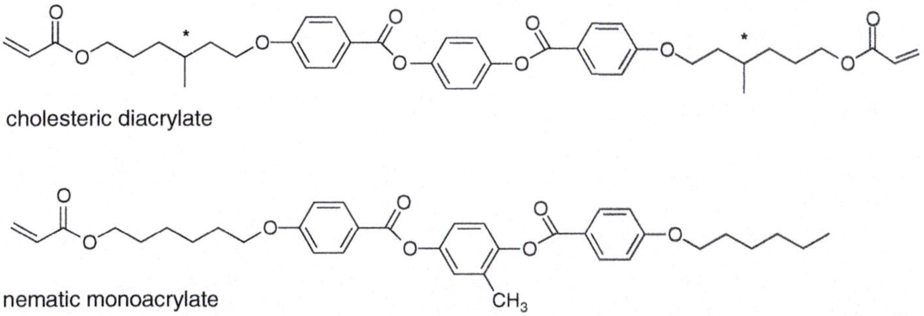

cholesteric diacrylate

nematic monoacrylate

Fig. 7.15 Chemical structure of the cholesteric diacrylate and the nematic monoacrylate used for wide band reflective polarisers [60]

mixtures of a cholesteric diacrylate and a nematic monoacrylate, see Fig. 7.15 [60]. This general approach leads to densely crosslinked, thermally stable, polymer networks and is very promising for the use in multilayer OLEDs.

We have also published a number of papers on photocrosslinkable fluorene containing reactive mesogens and polymers including a series of acrylate functionalized

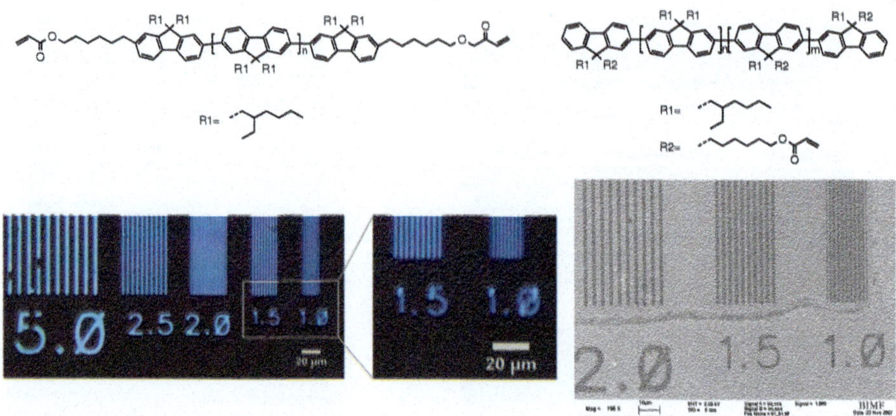

Fig. 7.16 *Top*: Chemical structures of crosslinkable fluorene oligomers. *Bottom*: Fluorescence microscope (*left*) [46] and SEM image (*right*) of a microstructured oligofluorene [62]

fluorene trimers and pentamers and an oligomer [61]. In more recent papers we have focussed on the photolithographic patterning of fluorene oligomers with pendant acrylate units and could demonstrate that a resolution of 1 μm can be readily achieved, see Fig. 7.16 [46, 62]. The first experiments with α,ω-bisacrylates, see Fig. 7.16, top left, show that, although a crosslinking time of only 15 min is needed, some degradation to the polyfluorene system was observed. Thus, we developed a new series of oligomers with tailored properties, see Fig. 7.16, top right. The acrylate content was varied by the introduction of a non-acrylate functionalized co-monomer through a statistical co-oligomerisation reaction. The co-oligomers show similar thermal behaviour, but differ in film formation and crosslinking properties. The exposure time is reduced to 30 s and degradation is no longer observed under these conditions. A lateral resolution of 1 μm was achieved.

7.6 Polarized Emission

One of the first reports on the orientation and subsequent crosslinking of a reactive mesogen for OLED applications with polarised emission has been published by Bacher et al. [63], who synthesized the conjugated bis-stilbene with two polymerizable acrylate groups shown in Fig. 7.17. The reactive mesogen was oriented by heating into the liquid crystalline phase and subsequently thermally crosslinked at 175°C.

We described in 2001 the synthesis of fluorene trimers, pentamers and oligomers with pendant acrylate units [61] (Fig. 7.18). The fluorene reactive mesogens exhibit broad mesophases, for example, the pentamer has a nematic phase between −10 and 123°C, which makes it ideally suited for orientation experiments. A 25 nm thick layer of the pentamer deposited onto rubbed polyimide exhibits a photoluminescence orientation ratio of 25:1. This ratio decreases to 9:1 for an analogous layer of

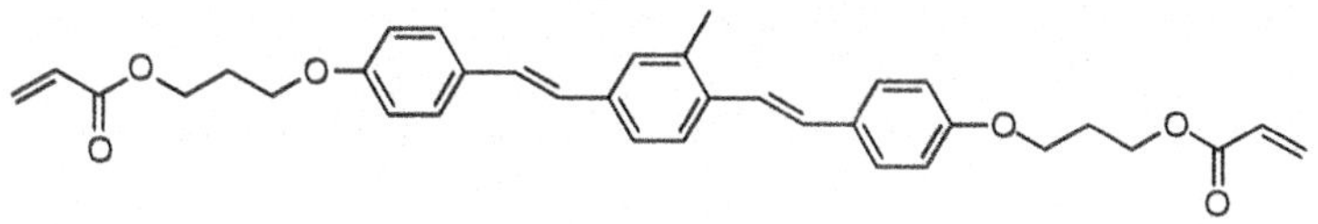

Fig. 7.17 Chemical structure of the conjugated bis-stilbene with pendant acrylate units from [63]

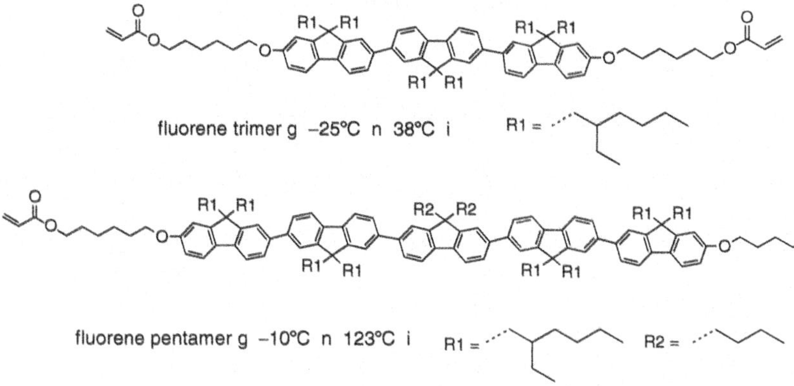

fluorene trimer g –25°C n 38°C i

fluorene pentamer g –10°C n 123°C i

Fig. 7.18 Structure of a fluorene trimer and pentamer with pendant acrylate units [61]

90 nm thickness, which implies the existence of a maximum degree of orientation close to the alignment layer/fluorene interface. If the recombination of electrons and holes can be confined to this zone, high polarization ratios can be expected. This might be the reason for the high polarization ratio of 25:1 that we observed in an OLED with a rubbed poly(1,4-phenylenevinylene) (PPV) orientation layer and a polyfluorene emitter [33].

The conversion of macroscopically oriented reactive mesogens to an intractable polymer network and the possibility of pixel formation by photopatterning was described by the group of Kelly and O'Neill in Hull in 2000 [45]. A reactive mesogen with a fluorene/thiophene core and photocrosslinkable pentadiene end groups was used as shown in Fig. 7.19. The reactive mesogen was oriented by annealing on top of a photoorientation layer consisting of a polymer with coumarin side groups and doped with an aromatic amine to ensure hole transport and subsequently crosslinked by UV-irradiation. A polarization ratio of 10:1 and a brightness of 60 cd m^{-2} were achieved in a prototype OLED.

The synthesis of a number of reactive mesogens with methacrylate and different dienes as polymerizable groups and their polymerization behaviour is described in detail in [64]. It turned out that dienes as shown in Fig. 7.19 require an about 30 times longer irradiation for crosslinking compared to methacrylates, but photodegradation is less pronounced with the diene reactive mesogens compared to methacrylates [65]. The use of acrylate and diene photopolymerizable end groups in the α,ω-positions of mesogens containing fluorene and thiophene units was also reported [66, 67].

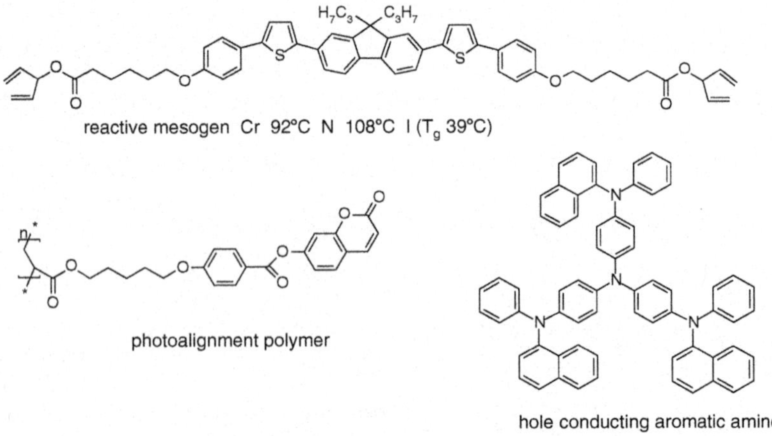

reactive mesogen Cr 92°C N 108°C I (T_g 39°C)

photoalignment polymer

hole conducting aromatic amine

Fig. 7.19 Structures of the reactive mesogen, the photoalignment layer and the hole conducting aromatic amine used for polarised OLEDs [45]

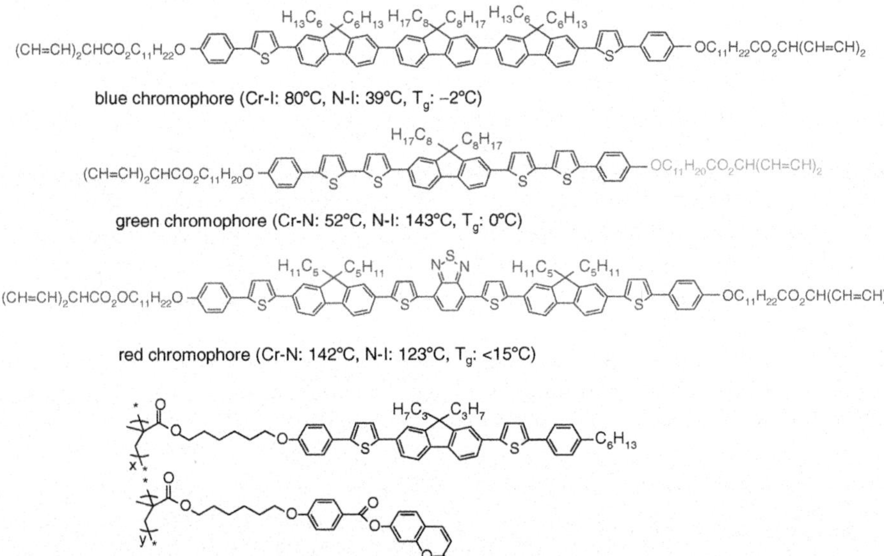

blue chromophore (Cr-I: 80°C, N-I: 39°C, T_g: −2°C)

green chromophore (Cr-N: 52°C, N-I: 143°C, T_g: 0°C)

red chromophore (Cr-N: 142°C, N-I: 123°C, T_g: <15°C)

conducting photoalignment layer (x:y = 1:4, T_g: 69°C)

Fig. 7.20 Chemical structures and phase transition temperatures of the blue, green and red emitting reactive mesogens and the conducting photoalignment layer

The three different reactive mesogens shown in Fig. 7.20 were sequentially deposited and photochemically crosslinked using a photomask to create the red, green and blue areas of the OLED shown in Fig. 7.21. The non-crosslinked parts were removed by washing with chloroform. The green reactive mesogen has

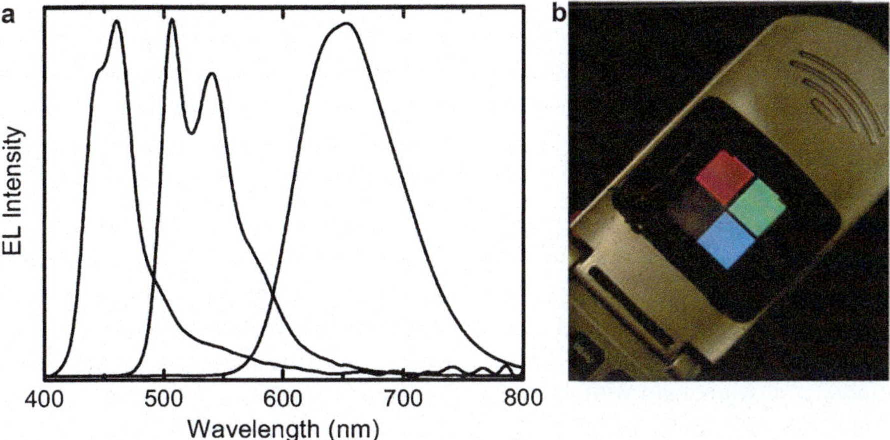

Fig. 7.21 (**a**) Normalized EL spectra of the red, green and blue OLEDs. (**b**) A prototype OLED with a red, green and blue pixel on the same substrate, fabricated by spin coating on a PEDT film covering a patterned ITO substrate. The pixels were defined by irradiation at 325 nm through a mask. Non-irradiated material was removed by washing with chloroform [67]

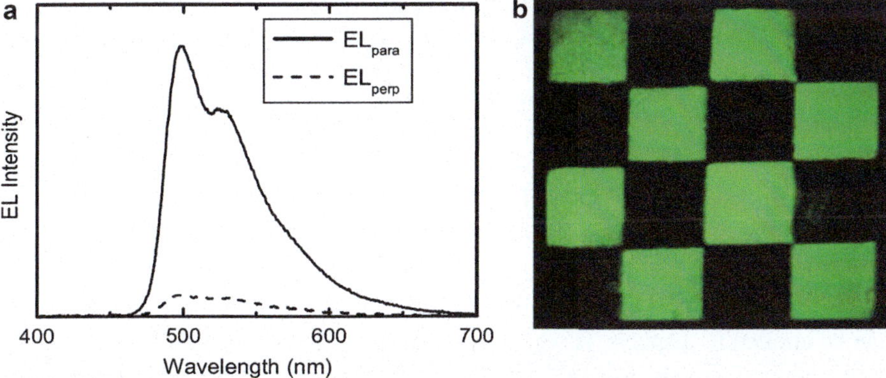

Fig. 7.22 (**a**) Polarized electroluminescence spectrum of an OLED of the photochemically crosslinked green reactive mesogen (Fig. 7.20). The underlying photoorientation layer depicted in Fig. 7.20 was oriented with polarized UV-light. (**b**) Photoluminescence image of a chess pattern of the *green* emitter viewed though a polarizer. In adjacent *squares* the emitter molecules are oriented perpendicular, the orientation direction given by the underlying photoalignment layer. In the *dark regions* the emitter is aligned perpendicular to the polarizer, in the *green regions* the molecules are oriented parallel to the polarizer [67]

also been coated on top of a tailor-made photo-orientation layer [68] containing both charge transporting groups and photo-orienting coumarin units. After thermal alignment an EL polarization ratio of 13:1 is obtained. The photograph shown in Fig. 7.22b shows that polarization patterns are obtained, if the photoalignment layer is patterned with a photomask. The chess pattern was obtained by exposing adjacent squares to orthogonally polarized light. The photoluminescence image was viewed through a polarizer and shows that in the green and in the dark regions the emitter molecules are orthogonally aligned. The viewing of patterned, polarized emission through a pair of orthogonal polarisers creates 3D-effects as each eye sees a different set of pixels. Thus by exploiting the principle of photoalignment, orthogonally oriented pixels can be obtained. In combination with self emitting OLED materials the production of 3D-displays becomes possible. The number of available reactive mesogens has been extended [69–71] and their phase behaviour and transport properties optimised [72].

Kelly et al. recently reported a white light OLED (WOLED) for lighting applications fabricated using liquid crystalline polymer networks [73]. A red emitting liquid crystal was doped into a matrix of a blue reactive mesogen resulting in the emission of white light with the CIE coordinates 0.35, 0.38. Upon crosslinking no colour change was observed. A polarised WOLED was prepared using a rubbed PEDT/PSS/ITO substrate. The reactive mesogen mixture was deposited onto the PEDT orientation layer, then baked at 140°C and photopolymerised with a HeCd laser at 325 nm. After that a hole-blocking layer, an electron-injection layer and an aluminium cathode were vapour deposited on top to complete the WOLED. A device efficiency of 0.4 cd A^{-1} was achieved, which is still too low for practical applications, but is a proof of principle that polarized white electroluminescence from liquid crystalline polymer networks can be obtained. A maximum polarization ratio of 9:1 (electroluminescence) was achieved, see Fig. 7.23.

7.7 Conclusions and Outlook

In this chapter the use of reactive mesogens in polarized OLEDs has been summarized. The major question, which has to be answered before reactive mesogens can be applied in displays, is which advantages do these materials have compared to the state-of-the-art, low-molar-mass compounds and highly conjugated, main chain polymers as light-emitting and charge-transporting, organic semiconductors. There are mainly two aspects of reactive mesogens, which render them attractive for OLED applications.

1. the parallel alignment of calamitic liquid crystals leads to linearly polarized emission.
2. the possibility of pattern formation if a mask is used during the photopolymerisation of reactive mesogens. In this case the material behaves like a negative photoresist.

Large-area, polarized white OLEDs may become attractive as backlights for LCD displays, since the first polarizer which absorbs at least 50% of the incoming light can be omitted. Nevertheless the efficiency of polarized OLEDs from reactive

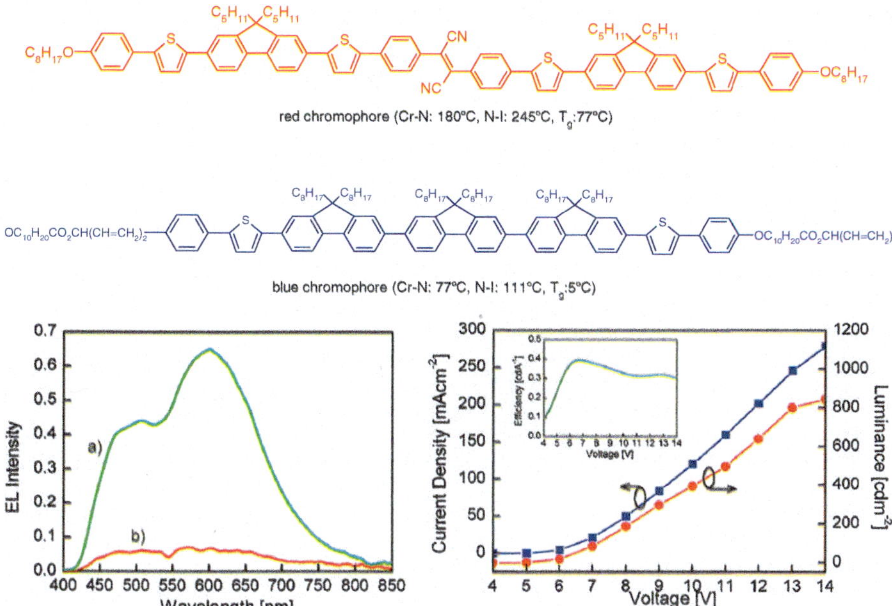

Fig. 7.23 *Top*: Chemical structures of the liquid crystalline emitters used for the white emitting polymer network [73]. *Bottom left*: Electroluminescence spectra of the white crosslinked mixture measured parallel (**a**) and perpendicular (**b**) to the rubbing direction. *Bottom right*: Device characteristics of the WOLED [73]

mesogens has to be increased before they can compete with current light sources, e.g., fluorescent lamps or inorganic high-brightness LEDs. In recent years the efficiency of white OLEDs has rapidly increased and they are now considered as efficient white light sources for large-area lighting. In such OLEDs phosphorescent metal complexes are used as emitters, and it would be an interesting topic to look for rod-like phosphorescent emitters, which can help to increase the energy efficiency of polarized OLEDs. The second important advantage of reactive mesogens is their ability to form high resolution colour pixels for OLED displays, if irradiated through a photomask to form densely crosslinked, completely insoluble films are formed. If both of these advantageous aspects of reactive mesogens, i.e., polarized emission and patterned pixel formation with high resolution, are combined, OLED displays in which two sets of pixels emit with orthogonal polarization can be realized, if an additional photoorientation layer is used. This may pave the way for 3D-OLED displays and shows that the reactive mesogens described in this chapter look promising for future OLED displays.

References

1. Tang, C., vanSlyke, S.A.: Appl. Phys. Lett. **51**, 913 (1987)
2. Chen, C.H., Shi, J., Tang, C.W.: Macromol. Symp. **125**, 1 (1997)
3. Kraft, A., Grimsdale, C., Holmes, A.B.: Angew. Chem. Int. Ed. **37**, 402 (1998)
4. Shirota, Y.: J. Mater. Chem. **10**, 1 (2000)
5. Nalwa, H.S., Rohwer, L.S. (eds.): Organic Light Emitting Diodes. American Scientific Publishers, Valencia (2003)
6. D'Andrade, B.W., Forrest, S.R.: Adv. Mater. **16**, 1585 (2004)
7. Müllen, K., Scherf, U. (eds.): Organic Light Emitting Devices. Wiley VCH, Weinheim (2006)
8. Baldo, M.A., Lamansky, S., Burrows, P.E., Thompson, M.E., Forrest, S.R.: Appl. Phys. Lett. **75**, 4 (1999)
9. Nakayama, T., Hiyama, K., Furukawa, K., Ohtani, H.: Digest of Technical Papers – Society for Information Display International Symposium, vol. 38, p. 1018 (2007)
10. Burroughes, J.H., Bradley, D.D.C., Brown, A.R., Marks, R.N., Mackay, K., Friend, R.H., Burns, P.L., Holmes, A.B.: Nature **347**, 539 (1990)
11. Groenendaal, L.B., Jonas, F., Freitag, D., Pielartzik, H., Reynolds, J.R.: Adv. Mater. **12**, 481 (2000)
12. Borsenberger, P.M., Weiss, D.S. (eds.): Organic Photoreceptors for Xerography. Marcel Dekker Inc, New York (1999)
13. Strohriegl, P., Grazulevicius, J.V.: Adv. Mater. **14**, 1439 (2002)
14. Popovich, Z.D., Aziz, H.: IEEE J. Sel. Top. Quantum Electron. **8**, 362 (2002)
15. Klenkler, R., Aziz, H., Tran, A., Popovic, Z.D., Xu, G.: Org. Electron. **9**, 285 (2008)
16. Salbeck, J.: Ber. Bunsenges. Phys. Chem. **100**, 1667 (1996)
17. Fuhrmann, T., Salbeck, J.: Adv. Photochem. **27**, 83 (2002)
18. Hosokawa, C., Higashi, H., Nakamura, H., Kusumoto, T.: Appl. Phys. Lett. **67**, 3853 (1995)
19. Tsai, M.H., Lin, H.W., Su, H.C., Ke, T.H., Wu, C.C., Fang, F.C., Liao, Y.L., Wong, K.T., Wu, C.I.: Adv. Mater. **18**, 1216 (2006)
20. Braun, D., Heeger, A.J.: Appl. Phys. Lett. **58**, 1982 (1991)
21. Spreitzer, H., Becker, H., Kluge, E., Kreuder, W., Schenk, H., Demandt, R., Schoo, H.: Adv. Mater. **10**, 1340 (1998)
22. Bernius, M.T., Inbasekaran, M., O'Brien, J., Wu, W.: Adv. Mater. **12**, 1737 (2000)
23. Scherf, U., List, E.J.W.: Adv. Mater. **14**, 477 (2002)
24. Adam, D., Schuhmacher, P., Simmerer, J., Haeussling, L., Siemensmeyer, K., Etzbach, K.H., Ringsdorf, H., Haarer, D.: Nature **371**, 141 (1994)
25. Funahashi, M., Hanna, J.: Appl. Phys. Lett. **76**, 2574 (2000)
26. Sirringhaus, H., Wilson, R.H., Friend, R.H., Inbasekaran, M., Wu, W., Woo, E.P., Grell, M., Bradley, D.D.C.: Appl. Phys. Lett. **77**, 406 (2000)
27. McCulloch, I., Zhang, W., Heeney, M., Bailey, C., Giles, M., Graham, D., Shkunov, M., Sparrowe, D., Thierney, S.: J. Mater. Chem. **13**, 2436 (2003)
28. Grell, M., Bradley, D.D.C.: Adv. Mater. **11**, 895 (1999)
29. O'Neill, M., Kelly, S.M.: Ekisho **9**, 9 (2005)
30. Dyrekelev, P., Berggren, M., Inganäs, O., Andersson, M.R., Wennerström, O., Hjertberg, T.: Adv. Mater. **7**, 43 (1995)
31. Grice, A.W., Bradley, D.D.C., Bernuis, M.T., Inbasekaran, M., Wu, W., Woo, E.P.: Appl. Phys. Lett. **73**, 629 (1998)
32. Grell, M., Knoll, W., Lupo, D., Meisel, A., Miteva, T., Neher, D., Nothofer, H.-G., Scherf, U., Yasuda, A.: Adv. Mater. **11**, 671 (1999)
33. Whitehead, K.S., Grell, M., Bradley, D.D.C., Jandke, M., Strohriegl, P.: Appl. Phys. Lett. **76**, 2946 (2000)
34. Jandke, M., Strohriegl, P., Gmeiner, J., Brütting, W., Schwoerer, M.: Synth. Metals **111–112**, 177 (2000)

35. Bolognesi, A., Botta, C., Facchinetti, D., Jandke, M., Kreger, K., Strohriegl, P.: Adv. Mater. **13**, 1072 (2001)
36. Sainova, D., Zen, A., Nothofer, H.-G., Asawapirom, U., Scherf, U., Hagen, R., Bieringer, T., Kostromine, S., Neher, D.: Adv. Funct. Mater. **12**, 49 (2002)
37. Klaerner, G., Miller, R.D.: Macromolecules **31**, 2007 (1998)
38. Jo, J., Chi, C., Höger, S., Wegner, G., Yoon, D.Y.: Chem. Eur. J. **10**, 2681 (2004)
39. Geng, Y., Cullingham, S.W., Trajkovska, A., Wallace, J.U., Chen, S.H.: Chem. Mater. **15**, 542 (2003)
40. Geng, Y., Chen, A.C.A., Ou, J.J., Chen, S.H., Klubek, K., Vaeth, K., Tang, C.W.: Chem. Mater. **15**, 4352 (2003)
41. Culligan, S.W., Geng, Y., Chen, S.H., Klubek, K., Vaeth, K.M., Tang, C.W.: Adv. Mater. **15**, 1176 (2003)
42. Chen, A.C.A., Culligan, S.W., Geng, Y., Chen, S.H., Klubek, K.P., Vaeth, K.M., Tang, C.W.: Adv. Mater. **16**, 783 (2004)
43. Chen, A.C.A., Wallace, J.U., Wei, S.K.-H., Zeng, L., Chen, S.H., Blanton, T.N.: Chem. Mater. **18**, 204 (2006)
44. Chen, A.C.A., Wallace, J.U., Klubek, K.P., Madaras, M.B., Tang, C.W., Chen, S.H.: Chem. Mater. **19**, 4043 (2007)
45. Contoret, A.E.A., Farrar, S.R., Jackson, P.O., Khan, S.M., May, L., O'Neill, M.O., Nicholls, J.E., Kelly, S.M., Richards, G.J.: Adv. Mater. **13**, 971 (2000)
46. Scheler, E., Bauer, I., Strohriegl, P.: Macromol. Symp. **254**, 203 (2007)
47. Li, X.C., Yong, T.M., Grüner, J., Holmes, A.B., Moratti, S.C., Cacialli, F., Friend, R.H.: Synth. Metals **84**, 437 (1997)
48. Bellmann, E., Shaheen, S.E., Thayumanavan, S., Barlow, S., Grubbs, R.H., Marder, S.R., Kippelen, B., Peyghambarian, N.: Chem. Mater. **10**, 1668 (1998)
49. Chen, J.P., Klaerner, G., Lee, J.-I., Markiewicz, D., Lee, V.Y., Miller, R.D., Scott, J.C.: Synth. Metals **107**, 129 (1999)
50. Li, W., Wang, Q., Cui, J., Chou, H., Shaheen, S.E., Jabbour, G.E., Anderson, J., Lee, P., Kippelen, B., Peyghambarian, N., Armstrong, N.E., Marks, T.J.: Adv. Mater. **11**, 731 (1999)
51. Bacher, A., Erdelen, C.H., Paulus, W., Ringsdorf, H., Schmidt, H.W., Schuhmacher, P.: Macromolecules **32**, 4551 (1999)
52. Bayerl, M.S., Braig, T., Nuyken, O., Mueller, C.D., Gross, M., Meerholz, K.: Macromol. Rapid Commun. **20**, 224 (1999)
53. Mueller, C.D., Braig, T., Nothofer, H.G., Arnoldi, M., Gross, M., Scherf, U., Nuyken, O., Meerholz, K.: Chem. Phys. Chem. **1**, 207 (2000)
54. Yang, X., Müller, D.C., Neher, D., Meerholz, K.: Adv. Mater. **18**, 948 (2006)
55. Huang, J., Xia, R., Kim, Y., Wang, X., Dane, J., Hofmann, O., Mosely, A., de Mello, A.J., de Mello, J.C., Bradley, D.D.C.: J. Mater. Chem. **17**, 1043 (2007)
56. Qiang, L., Ma, Z., Zeng, Z., Yin, R., Huang, W.: Macromol. Rapid Commun. **27**, 1779 (2006)
57. Müller, C.D., Falcou, A., Reckefuss, N., Rojahn, M., Wiederhirn, V., Rudati, P., Frohne, H., Nuyken, O., Becker, H., Meerholz, K.: Nature **421**, 829 (2003)
58. Gather, M.C., Koehnen, A., Falcou, A., Becker, H., Meerholz, K.: Adv. Funct. Mater. **17**, 191 (2007)
59. Broer, D.J., Boven, J., Mol, G.N., Challa, G.: Macromol. Chem. **190**, 2255 (1989)
60. Broer, D.J., Lub, J., Mol, G.N.: Nature **378**, 467 (1995)
61. Jandke, M., Hanft, D., Strohriegl, P., Whitehead, K.S., Grell, M., Bradley, D.D.C.: SPIE Proc. **4105**, 338 (2001)
62. Scheler, E., Strohriegl, P.: J. Mater. Chem. **19**, 3207 (2009)
63. Bacher, A., Bentley, P.G., Bradley, D.D.C., Douglas, L.K., Glarvey, P.A., Whitehead, K.S., Turner, M.L.: J. Mater. Chem. **9**, 2985 (1999)
64. Contoret, A.E.A., Farrar, S.R., O'Neill, M., Nicholls, J.E., Richards, G.J., Kelly, S.M., Hall, W.A.: Chem. Mater. **14**, 1477 (2002)
65. O'Neill, M., Kelly, S.M.: Adv. Mater. **15**, 1135 (2003)

66. Aldred, M.P., Eastwood, A.J., Kelly, S.M., Vlachos, P., Contoret, A.E.A., Farrar, S.R., Mansoor, B., O'Neill, M., Tsoi, W.C.: Chem. Mater. **16**, 4928 (2004)
67. Aldred, M.P., Contoret, A.E.A., Farrar, S.R., Kelly, S.M., Mathieson, D., O'Neill, M., Tsoi, W.C., Vlachos, P.: Adv. Mater. **17**, 1368 (2005)
68. Aldred, M.P., Vlachos, P., Contoret, A.E.A., Farrar, S.R., Chung-Tsoi, W., Mansoor, B., Woon, K.L., Hudson, R., Kelly, S.M., O'Neill, M.: J. Mater. Chem. **15**, 3208 (2005)
69. Aldred, M.P., Vlachos, P., Dong, D., Kitney, S.P., Chung-Tsoi, W., O'Neill, M., Kelly, S.M.: Liq. Cryst. **32**, 951 (2005)
70. Aldred, M.P., Eastwood, A.J., Kitney, S.P., Richards, G.J., Vlachos, P., Kelly, S.M., O'Neill, M.: Liq. Cryst. **32**, 1251 (2005)
71. Aldred, M.P., Carrasco-Orozoco, M., Contoret, A.E.A., Dong, D., Farrar, S.R., Kelly, S.M., Kitney, S.P., Mathieson, D., O'Neill, M., Chung Tsoi, W., Vlachos, P.: Liq. Cryst. **33**, 459 (2006)
72. Woon, K.L., Aldred, M.P., Vlachos, P., Mehl, G.H., Stirner, T., Kelly, S.M., O'Neill, M.: Chem. Mater. **18**, 2311 (2006)
73. Liedtke, A., O'Neill, M., Wertmöller, A., Kitney, S.P., Kelly, S.M.: Chem. Mater. **20**, 3579 (2008)

Chapter 8
Liquid Crystals for Organic Photovoltaics

Mary O'Neill and Stephen M. Kelly

8.1 Introduction

The photovoltaic effect involves the generation of a photo-voltage and often a photocurrent on absorption of light in a semiconductor. The device must contain an asymmetry so that the photogenerated electrons and holes separate and travel in different directions to their respective electrodes. In an organic material the absorption of light of energy greater than the exciton energy, given by $E_{ex} = E_{HOMO} - E_{LUMO} - E_{BE}$, transfers an electron from the highest occupied molecular orbital (HOMO) of energy E_{HOMO} to the lowest unoccupied molecular orbital (LUMO) of energy E_{LUMO} and creates an exciton, which is a bound electron hole pair normally localised on a single molecule. The binding energy of the exciton, E_{BE}, is large, some tenths of an eV, so that an electric field is insufficient to dissociate it. Instead the ionisation of the exciton is mostly achieved at an interface between electron-donating and electron-accepting species. Figure 8.1 shows a bilayer photovoltaic device containing an electron accepting material (**A**) overlaying an electron donating medium (**D**), both sandwiched between two dissimilar electrodes, one of which is transparent to transmit the incident light [1].

An energy level diagram of the device is given in Fig. 8.2 to illustrate the principle of operation. The **D** has a low ionisation potential (and thus a high-lying E_{HOMO}) whereas the **A** has a high electron affinity (a low-lying E_{LUMO}). The photogenerated excitons diffuse to the interface and are ionised to generate free

M. O'Neill (✉)
Department of Physics and Mathematics, University of Hull, Hull HU6 7RX, UK
e-mail: m.oneill@hull.ac.uk

S.M. Kelly
Department of Chemistry, University of Hull, Hull HU6 7RX, UK
e-mail: s.m.kelly@hull.ac.uk

R.J. Bushby et al. (eds.), *Liquid Crystalline Semiconductors*, Springer Series in Materials Science 169, DOI 10.1007/978-90-481-2873-0_8,

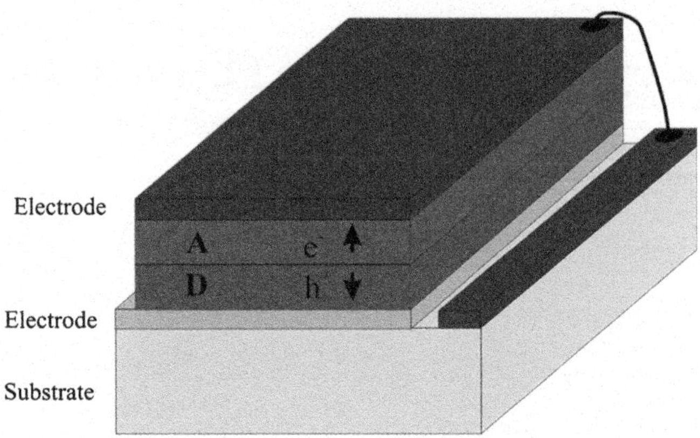

Fig. 8.1 Schematic of a bilayer organic photovoltaic consisting of an electron-accepting material **A** overlying an electron-donating film **D**, sandwiched between two dissimilar electrodes. The hole transporting polymer poly (3,4-ethylenedioxythiophene)/poly(styrene sulfonate) or equivalent is often inserted between the **D** and the anode to assist extraction of holes. The light is incident through the substrate and lower electrode

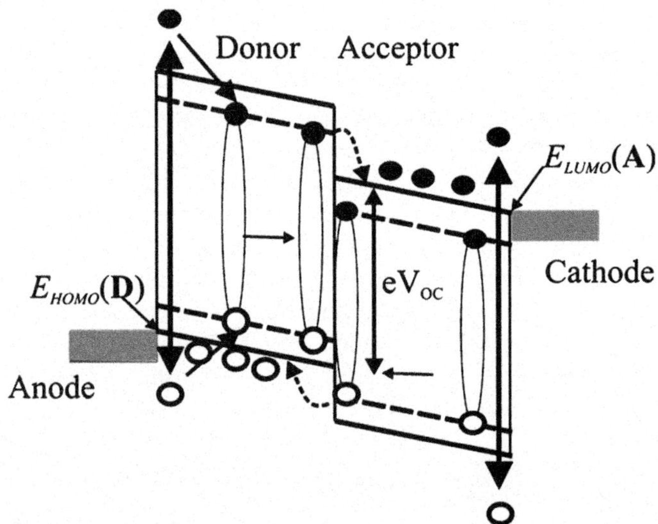

Fig. 8.2 Energy level diagram of a bilayer organic photovoltaic device. E_{HOMO} (**D**) and E_{LUMO} (**A**) denote the HOMO level energy of the **D** and the LUMO level energy of the **A** layer, respectively. Photons with an average photon energy larger than the optical band gap are absorbed on either side of the heterojunction (step 1). The photogenerated electron and hole thermalise and form an exciton (step 2). Excitons diffuse to the heterojunction (step 3) where they dissociate and transfer an electron [hole] into the **A** [**D**] layer (step 4). The electrons and holes are collected at the cathode and anode respectively (step 5). The difference between E_{HOMO} (**D**) and E_{LUMO} (**A**) determines the maximum open circuit voltage (V_{OC}) under illumination

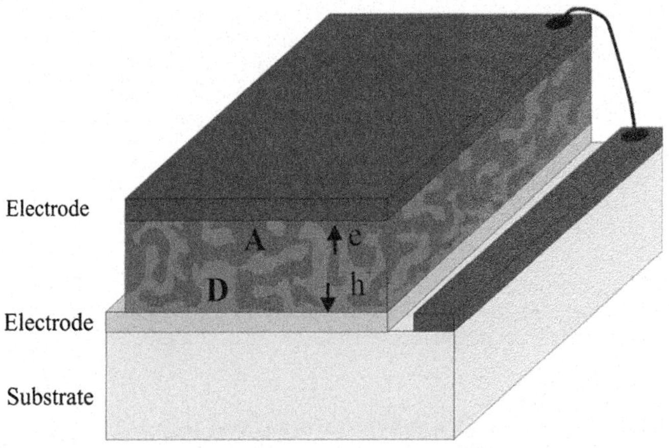

Fig. 8.3 Schematic of a bulk heterojunction organic photovoltaic consisting of a thin film of an electron-accepting material **A** blended with an electron-donating material **D**, sandwiched between two dissimilar electrodes. Photoinduced excitons are dissociated at the distributed interface. Electrons and holes require a continuous pathway through the **A** and **D** material respectively to reach their respective electrodes. The light is incident through the substrate and lower electrode

electrons and holes in the **A** and **D** materials, respectively. Charge separation is energetically favourable when

$$E_{HOMO}\,(\mathbf{D}) - E_{LUMO}\,(\mathbf{A}) < E_{ex} \tag{8.1}$$

for both **D** and **A** materials. The separated carriers drift in the built-in field introduced by dissimilar electrodes and are collected at their respective electrodes.

There are a number of issues that affect the quantum efficiency of the organic photovoltaic device. The absorption coefficient of organic PVs typically peaks at about 0.015 nm^{-1} [2], so that relatively thick films ($\approx$100 nm) are required to absorb all the incident light in a double pass. However the diffusion length of the exciton is of the order of only 10–20 nm before it recombines, so that only excitons generated within about this distance from an interface are ionised. Distributed, rather than single, hetero-interfaces have been frequently used to maximise charge separation whilst maintaining large absorption in so-called bulk heterojunctions devices [3, 4]. These interfaces have been formed in interpenetrating phase-separated blends of electron-donating and electron-accepting polymers, which have nanoscale morphology, see Fig. 8.3. This approach has been extremely successful in producing the most efficient photovoltaics to date, as discussed in more detail below. Alternatively, the co-evaporation of low mass **D**s and **A**s form mixed interfaces. Experimentally the dissociation of the exciton at the interface can be monitored by a photoluminescence quenching experiment.

A second issue is the efficiency of charge generation at the interface. The geminate electron-hole pair, represented by $<\mathbf{D}^{+}\mathbf{A}^{-}|$, at the interface can dissociate

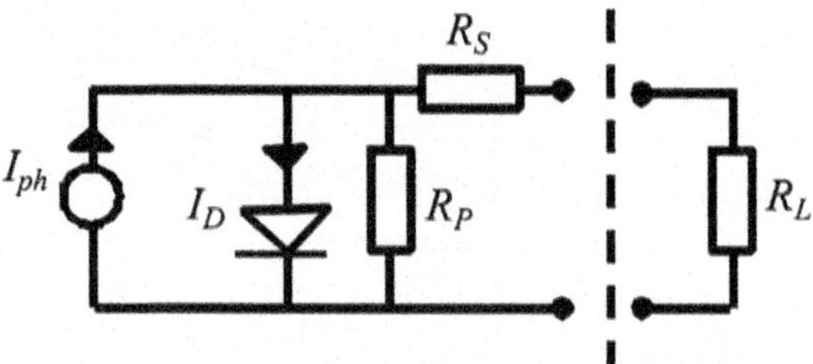

Fig. 8.4 Equivalent circuit for an organic photovoltaic device. The photocurrent (I_{ph}) opposes the diode current (I_D). The series and parallel resistances, R_S and R_P respectively, should be minimised and maximised respectively for optimum performance. R_L denotes the load resistance

to form free electron-hole pairs. Alternatively it can relax to an exciplex state [5], stabilised by the coulombic attraction between the carriers. The recombination of the geminate electron-hole pair or the exciplex may be a significant source of loss. Indeed, the mechanism of charge separation at the interface is not fully understood with disorder and interface dipoles helping charge separation [2]. The separated carriers drift and diffuse to their respective electrodes with efficiency dependent on the **D** and **A** hole and electron mobility, respectively, and the carrier recombination time. Control of the morphology of the phase-separated blend is crucial to optimise charge transport in bulk heterojunction devices. There must be a continuous pathway without dead-ends between the interface and the electrode. Some vertical segregation is desired, with the **D** preferentially located at the anode and the **A** at the cathode, to avoid carrier collection at the "wrong" electrode. The carrier lifetime may decrease, if the phase-separated domains are too finely dispersed, because of the increased probability of recombination between free electrons and holes, which arrive simultaneously on each side of a heterointerface, so-called non-geminate pair recombination. The development of a space charge limits the efficiency in photovoltaics where there are large differences between the mobilities of the holes and electrons [6]. The entire voltage drops across the region next to the electrode where the slower carriers accumulate so that the photocurrent generated in this region is substantially the total current. Finally the efficiency of carrier collection at the electrode can be affected by realignment of the energy levels due to redistribution of charge and/or geometry modifications at the interface.

The organic photovoltaic device can be approximated by the equivalent circuit shown in Fig. 8.4 [2], which is similar to the circuit used to represent inorganic photovoltaics based on *p-n* junctions. The circuit consists of a photocurrent source (I_{ph}) in parallel with a diode, which produces an opposing dark current. The optimum value of the diode's ideality factor n is one, but it is often larger depending on the recombination mechanism. The saturation current I_0 is the current in the dark at reverse bias. The series resistance (R_S), which has to be minimized, results from the finite conductivity of the semiconducting material, the contact resistance between the semiconductors and the electrodes, as well as the resistance associated with electrodes and interconnections. The shunt resistance (R_P), which

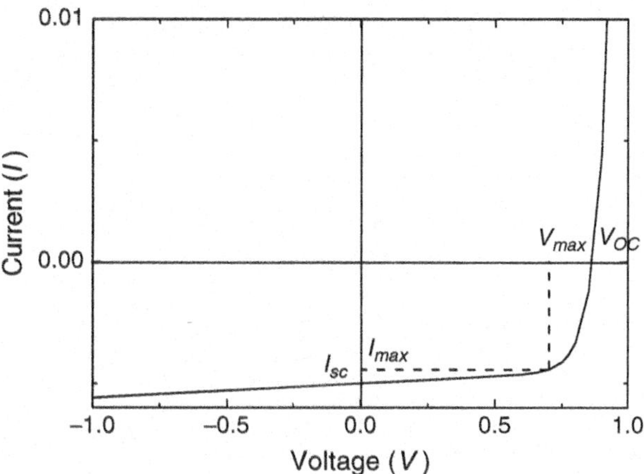

Fig. 8.5 Current-voltage characteristics obtained on solution of Eq. (8.2). V_{OC} and I_{SC} refer to the voltage and current obtained when $I = 0$ and $V = 0$ respectively. I_{max} and V_{max} are the current and voltage values corresponding to maximum power produced

needs to be maximised, results from the loss of carriers *via* possible leakage paths. These paths include collection at the wrong electrode giving continuous pathways for both electrons and holes between the anode and cathode, structural defects such as pinholes in the film or recombination centres. The current-voltage (*I-V*) characteristics for this circuit are described by the Shockley equation [2, 7].

$$I = \frac{1}{1+\frac{R_S}{R_P}}\left[I_0\left\{\exp\left(\frac{V-IR_S}{\frac{nkT}{e}}\right)-1\right\}-\left(I_{ph}-\frac{V}{R_P}\right)\right]$$

$$\approx I_0\left\{\exp\left(\frac{V}{nkT/e}\right)-1\right\}-I_{ph} \tag{8.2}$$

where e denotes the elementary charge, kT the thermal energy, and A the area of the cell. An analysis of Eq. (8.2) shows that R_S is the critical parameter especially when J_{ph} is large. The right hand approximation is valid when $R_S \to 0$ and $R_P \to \infty$. Figure 8.5 shows an example of the *I-V* characteristics obtained on solution of Eq. (8.2). The short-circuit current, I_{SC}, and open-circuit voltage V_{OC} can be obtained from Eq. (8.2) by setting $V = 0$ and $I = 0$ respectively.

$$I_{SC} = \frac{1}{1+\frac{R_S}{R_P}}\left\{I_{ph}-I_0\left[\exp\left(\frac{|I_{SC}|R_S}{\frac{nkT}{e}}\right)-1\right]\right\} \approx I_{ph} \tag{8.3}$$

$$V_{OC} = \frac{nkT}{e}\ln\left\{1+\frac{I_{ph}}{I_0}\left(1-\frac{1-V_{OC}}{I_{ph}R_p}\right)\right\} \approx \frac{nkT}{e}\ln\left(1+\frac{I_{ph}}{I_0}\right) \tag{8.4}$$

Equations (8.2), (8.3) and (8.4) must be solved numerically, except in the limit $R_S \to 0$ and $R_P \to \infty$. As Fig. 8.4 shows, the photovoltaic cell produces an electrical power in the fourth quadrant where I is negative and V positive equal to

$I \times V$. The output voltage V_{max} and current I_{max} is chosen to maximise the output power and the power conversion efficiency η_P, is given by

$$\eta_P = \frac{V_{\max} I_{\max}}{P_{inc}} = \frac{FF \times V_{OC} I_{SC}}{P_{inc}} \tag{8.5}$$

where P_{inc} is the optical power incident to the photovoltaic cell. The fill factor, FF, is the ratio of the maximum power rectangle $V_{max} \times I_{max}$ to the rectangle $V_{OC} \times I_{SC}$. The resistances R_P and R_S can be approximately found from the slopes of the I-V characteristics about $V = 0$ and $V = V_{OC}$ respectively. A rule of thumb is that the value of R_S must be small compared to the characteristic resistance defined as $R_{CH} = V_{OC}/I_{SC}$ while R_P must be large compared to R_{CH}.[2].

For monochromatic illumination the maximum value of I_{SC} is given by

$$I_{SC} = \frac{e\eta_{EQE} P_{inc}}{h\nu} \tag{8.6}$$

where η_{EQE} is the external quantum efficiency defined as the number of carriers collected per incident photon. η_{EQE}, the internal quantum efficiency is the number of carriers collected per incident photon absorbed. In organic photovoltaics, η_{EQE} can be broken down into the product of efficiencies associated with each of the steps. absorption, exciton diffusion, exciton dissociation into free carriers, charge transport, and charge collection. The main motivation in studying organic photovoltaics is to make solar cells. The performance of solar cells for terrestrial applications is characterised using a source that simulates the solar spectrum with standardised illumination conditions denoted by AM 1.5 G. This corresponds to the average intensity of sun light, about 100 mW cm^{-2}, with an angle of incidence of $\theta = 48°$ relative to the normal to the earth's surface. AM denotes the air mass $= 1/\cos\theta$ and G stands for global and refers to direct incident light with a small contribution of diffuse light. Under these conditions I_{SC} becomes

$$I_{SC} = \int_{AM\,1.5} \frac{e\eta_{EQE} P_{inc}(\lambda)\lambda}{hc} d\lambda \tag{8.7}$$

where $\int P_{inc}(\lambda)d\lambda = 100$ mW cm^{-2}.

Ignoring parasitic resistances, Eq. (8.4) shows that V_{OC} increases logarithmically with the photocurrent to a maximum value approximately equal to the energy difference between E_{HOMO}(**D**) and E_{LUMO}(**A**), with corrections for polaron effects [8, 9]. Equation (8.5) would suggest that power conversion would be optimised by designing the molecular structure of the **D** and **A** to maximise V_{OC}. However, Fig. 8.2 shows that there is a trade-off involved. Efficient charge separation at the interface between the **D** and **A** requires a minimum offset between their HOMO and LUMO levels defined as

$$\Delta E_{HOMO} = |E_{HOMO}(\mathbf{D}) - E_{HOMO}(\mathbf{A})|, \tag{8.8}$$

with ΔE_{LUMO} similarly defined, so that Eq. (8.1) is satisfied. It is believed that $\Delta = \Delta E_{HOMO} + \Delta E_{LUMO}$ should be at least 300 meV to overcome the exciton binding energy [10, 11]. η_P becomes lower as Δ increases. For a solar-cell, the absorption spectra of the **D** and **A** should overlap the solar spectrum as much as possible, placing further constraints on the maximum value of V_{OC}.

Since 2009, there have been rapid progress in the development of organic photovoltaic cells, particularly bulk heterojunctions using polymeric **D**s and **A**s, the latter based on fullerene materials, with power conversion efficiencies >7% [12–16]. More recently Konarka reported a certified device with an efficiency of 8.3% [17]. In these devices, "push-pull" copolymers are used as **D**s, see **P-P** shown in Table 8.1 as an example, which have alternating electron-withdrawing and electron-donating components to increase the double bond character between the units. This stabilises the quinoidal form of the polymer resulting in a low band gap whilst maintaining a sufficiently low E_{HOMO} to preserve a high V_{OC}. These co-polymers replace the previously used polymers, such as regioregular poly(3-hexylthiophene), (P3HT), which absorb at shorter wavelengths. Various processing strategies, such as the use of mixed solvents, have also improved morphological control of the phase-separated blends. The most commonly used **A** is the highly soluble fullerene derivative, phenyl-C61-butyric acid methyl ester (PC61BM), but the use of the larger fullerene [6,6]-phenyl C71 butyric acid methyl ester $PC_{71}BM$ can give better performance [18]. The performance of organic photovoltaics using non-fullerene based **A**s, such as perylene derivatives, has been disappointing, possibly because of poorer phase separation and recombination *via* charge-transfer states. *N,N′*-di-(1-ethylpropyl)-3,4,9,10-perylene*bis*(dicarboximide) (PDI) and *N,N′*-di-(9,9-fluoren-2-yl)-3,4,9,10-perylene*bis*(dicarboximide) (PDIF) are examples of perylene **A**s. The chemical structures of the materials listed here are shown in Table 8.1.

8.2 Discotic Photovoltaics

8.2.1 Devices with Perylene-Based Acceptors

As discussed in Chaps. 3 and 4, (columnar) discotic liquid crystals are oriented in columns separated by molten aliphatic chains and, consequently, they can conduct charge efficiently along the channels in one dimension. The organization of the different phases is described elsewhere [19, 20] and the efficiency of charge transport can be directly related to the short intermolecular spacing and order of different types of mesophase, with few exceptions [21]. For example, hole mobility is higher in ordered, rather than disordered, columnar phases and even higher in helically-ordered phases where molecular rotation is suppressed about the columnar axis [22]. Some mesomorphic derivatives of hexabenzocoronene, for example hexaphenyl-substituted hexabenzocoronene (HBC_{12}, see Table 8.2 for chemical structures of all discotic materials discussed here) have hole mobilities

Table 8.1 Non LC compounds

P-P	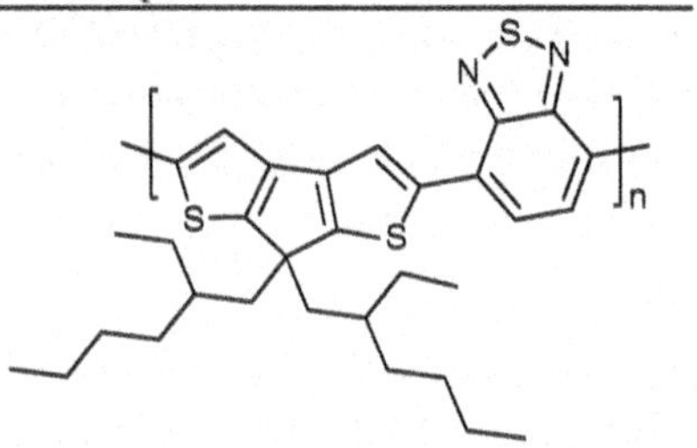
P3HT	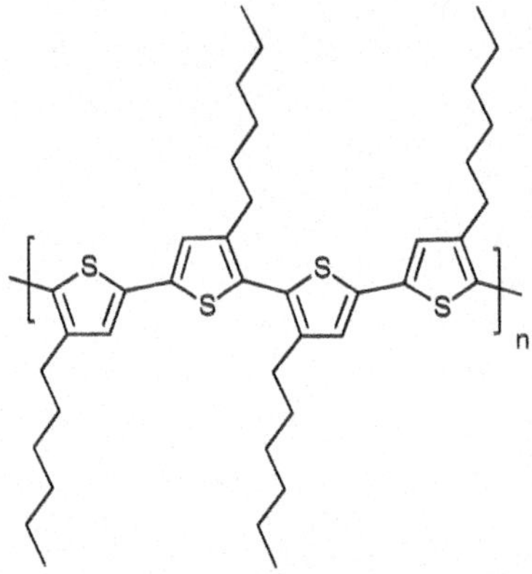 P3HT. $t_g \approx 67°C$; Cr-I $\approx 238°C$
PC61BM	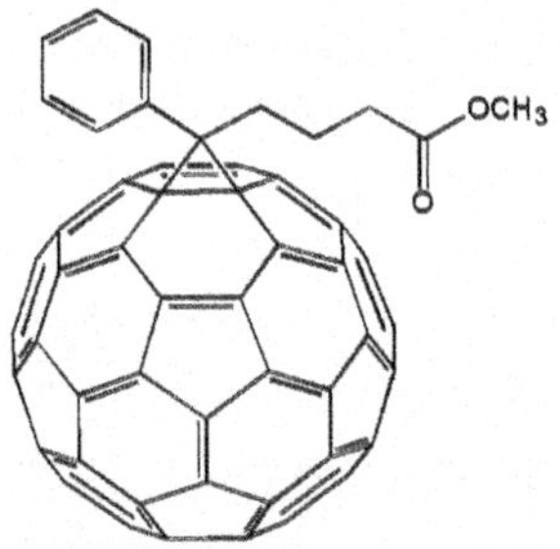
PC71BM	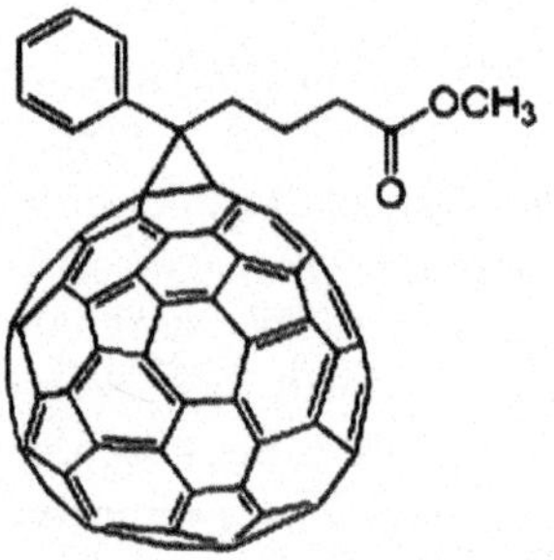

(continued)

Table 8.1 (continued)

PDI

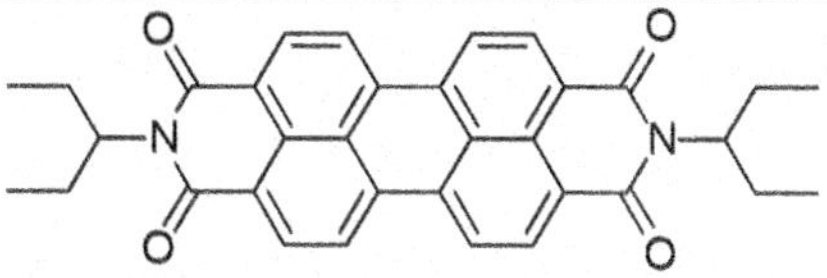

Cr-I = 435°C

PDIF

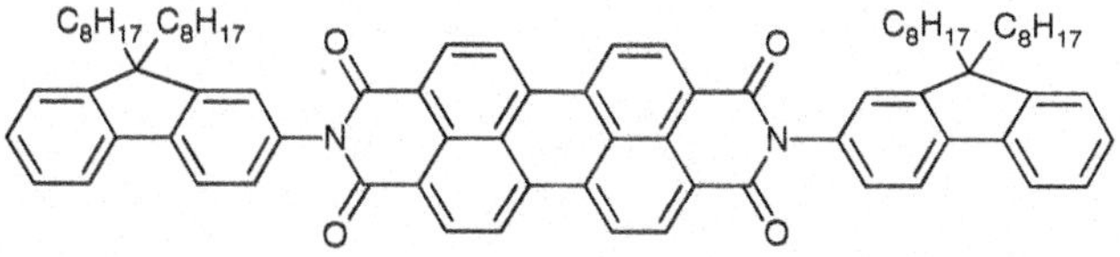

C_8H_{17} C_8H_{17} C_8H_{17} C_8H_{17}

Cr-I = 274°C (forms a glass on rapid cooling in thin films)

Table 8.2 Columnar liquid crystals

HBC_{12}

HBC ≡ 12

HBC_{mn}

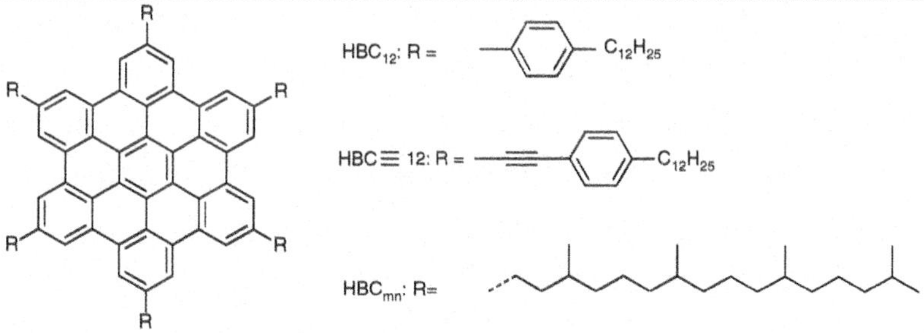

HBC_{12}: R = —C₆H₄—$C_{12}H_{25}$

HBC ≡ 12: R = —C≡C—C₆H₄—$C_{12}H_{25}$

HBC_{mn}: R=

$HBC_{12.}$ Cr- Col_1 < -100°C; Col_1- Col_2 = 18°C; Col_2-Col_3 = 83°C; Col_3- I > 400°C
HBC-$C_{4\text{-}16.}$ Cr-36°C, Col_{ho} 231°C, I

$DG_{14,10}$

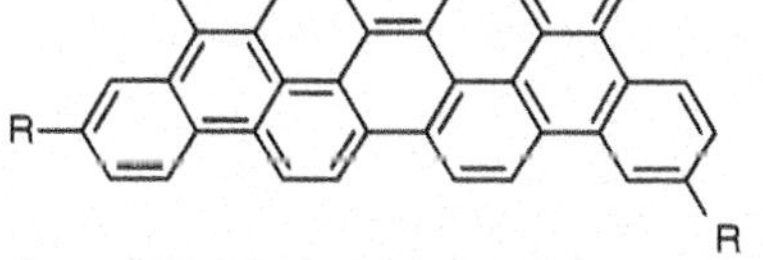

R = $-CH_2CH(C_{10}H_{13})(C_{12}H_{17})$
Cr-Col1 < -70°C; Col_1-Col_2 = 24°C; Col_2- I > 500°C

(continued)

Table 8.2 (continued)

6 F-HBC

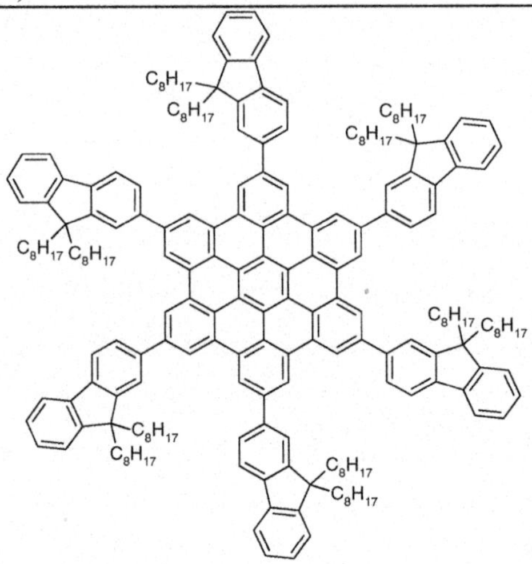

sym2F-HBC

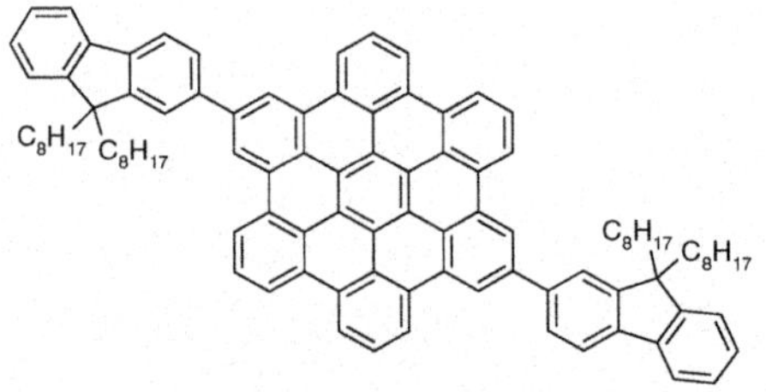

asym2F-HBC

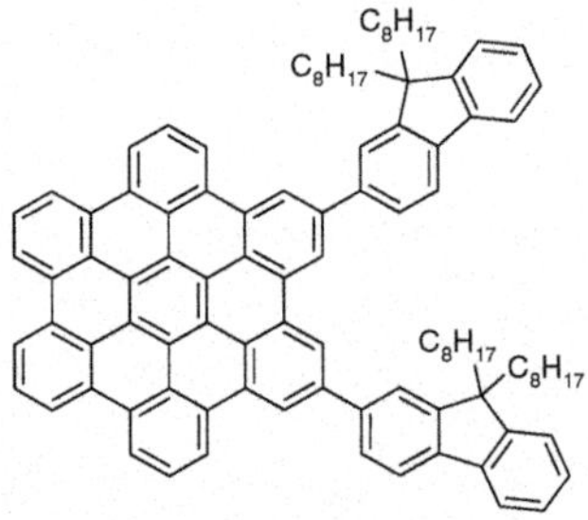

HBC-F-PCBM

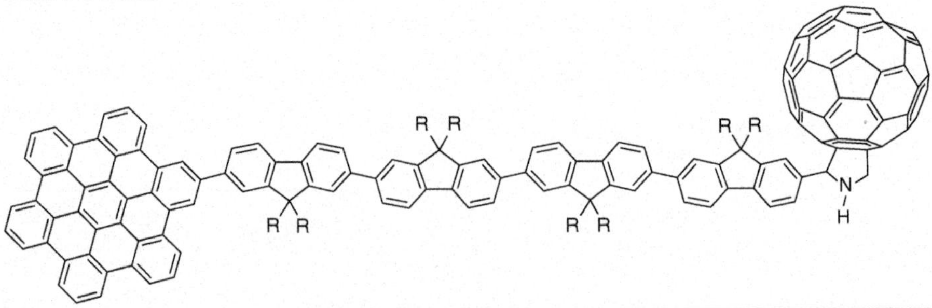

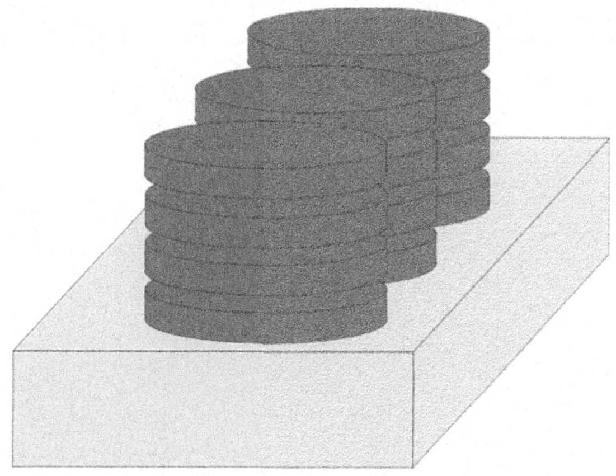

Fig. 8.6 Illustration of face-on alignment of discotic molecules giving homeotropic alignment of the columns

above 0.2 $cm^2 V^{-1} s^{-1}$ in the room temperatures discotic phase [23]. However, these measurements give the value of the local mobility only and the reported value of the charge mobility is much lower ($\mu \leq 0.02\ cm^2 V^{-1} s^{-1}$) in field-effect transistor devices [24, 25]. This behaviour is mainly due to the presence of poor contacts as well as the difficulty in obtaining uniform macroscopic alignment and in optimising order in solution-processed devices. Photovoltaic operation requires the transport of charges from the active layer to the electrodes in a direction perpendicular to the substrate so that homeotropic alignment of the columns is required, as illustrated in Fig. 8.6. Hexabenzocoronene and its derivatives have low ionisation potentials, 5.0–5.4 eV [26, 27], small intermolecular separation ≈3.5 Å, along the columns and so are suitable as **D** materials in photovoltaic devices. Their main disadvantage is the poor overlap of their absorption spectra, often peaking at wavelengths <400 nm, with the solar spectrum.

HBC_{12} is soluble because of its six flexible alkyl side-chains, which also induce low-temperature LC phases. An early breakthrough was the demonstration of impressive photovoltaic performance from a phase-separated blend of HBC_{12} mixed with *N*,*N′*-di-(1-ethylpropyl)-3,4,9,10-perylene-*bis*(dicarboximide) (PDI) as an **A** [27]. Figure 8.7 shows that E_{HOMO} and E_{LUMO} of the **D** and **A** respectively are well aligned to favour charge separation at the **D-A** interface [27, 28]. A 1.1 blend was deposited by spin-coating and the components were vertically segregated by evaporation of the solvent to produce a thin film with an **A** rich surface coating overlying a **D** rich layer. This spontaneous self-assembly gives a distributed interface for separation of electrons and holes into the top and lower layers respectively. Maximum values of η_{EQE} and η_P were 34% and 2% respectively at the peak absorption wavelength and with a small incident irradiance <1 mW cm^{-2}. However, performance degraded drastically at higher intensities and subsequently was found to be equal to 0.22% under 1 sun illumination at AM 1.5 G [29].

Further studies were carried out to investigate what limits the performance of the HBC_{12}-PDI bulk heterojunction device [29–31]. The intermolecular separation

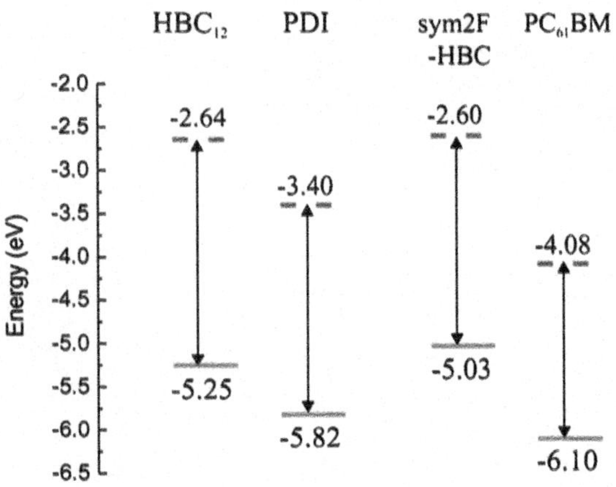

Fig. 8.7 E_{HOMO} (*solid lines*) and E_{LUMO} (*dashed line*) for different D-A pairs used for photovoltaic devices

between the **D**s in the blend was controlled by varying the length of the aliphatic side chains of the coronene, HBC_n. with $n = 6, 8, 12$ and 16. A further compound was investigated, HBC ≡ 12, which has an additional triple bond between the coronene core and the periphery substituents, see Table 8.2, resulting in cubic rather than hexagonal packing [29]. Incomplete photoluminescence quenching indicates that charge generation at the HBC ≡ 12-**A** interface is incomplete. Radiative recombination from a charge transfer state is also observed. Photoluminescence quenching is higher for HBC compounds with less bulky side-chains indicating that these provide more intimate contact between the **D** and **A** at the interface and better charge separation. It is also expected that there is better intercolumnar charge transport in materials with shorter side-chains. This is important since channels are not continuous across the whole heterojunction so that intercolumnar hops are required for carriers to reach the electrodes. Annealing improves the crystallinity of blend components and so enhances mobility. However, annealing in the mesophase also leads to high surface roughness. This results in the formation of large crystals of the perylene dye, which penetrate the thin film. This problem is significantly reduced by annealing the blend, when capped with an elastomeric stamp made from polydimethylsiloxane (PDMS), which adheres spontaneously to organic surfaces [30]. Alternatively surface roughness is minimised by annealing after deposition of the top electrode giving enhanced performance [29]. However, the annealed device shows poor performance at high irradiance. This is attributed to space charge effects resulting from unequal hole and electron mobility. Indeed the electron mobility of perylene compounds is known to vary by many orders of magnitude depending on the film morphology. Incomplete phase separation may also lead to greater non-geminate pair recombination at higher incident intensities. It is suggested that improved homeotropic alignment is required to fully exploit the potential for fast charge transport in these devices. Swallow-tailed hexabenzocoronenes having branched aliphatic side-chains, HBC_{mm}, $m = 4, 8, 12$, were also blended with PDI in bulk heterojunction devices [32]. Photovoltaic performance is best for the molecule with the shortest side-chain.

Triangle-shaped discotic graphenes with three swallow-tailed alkyl substituents, DG_{mn}, (m,n = 10, 6 and 14, 10) have been used in photovoltaics [33]. Both of these materials exhibit a columnar liquid crystalline phase over an extremely wide temperatures range (from −70°C to >+500°C). The compound $DG_{14,10}$ shows a second, more-ordered phase above 24°C. Self-healing occurred for this compound only, on annealing at 120°C. This reduces macroscopic domain boundaries giving larger domains. It also enhances the local order of the molecules and removes defects resulting in improved OPV performance in blends with PDI. $DG_{10,6}$ and $DG_{14,10}$ have E_{HOMO} values of 5.1 and 5.2 eV respectively. They were individually blended with PDI and annealed at 120°C. In both cases a 'homogeneous' bicontinuous phase is deduced from the scanning electron microscopy images across the thickness with cylindrical PDI crystals evenly dispersed within the liquid crystalline matrix of $DG_{10,6}$ and $DG_{14,10}$. Despite their similar morphology the performance of the photovoltaic containing $DG_{14,10}$ is substantially better than that with $DG_{10,6}$. Under 1 sun illumination conditions η_{EQE} at a wavelength of 490 nm is equal to 22% and 11% respectively. The better performance from $DG_{14,10}$ devices is attributed to its improved intracolumnar packing upon annealing, which enhances charge-carrier transport and increases the exciton diffusion length. Under the same conditions, HBC_{mn}-PDI devices (m,n = 6,2, 10,6 and 14,2) shows peak values of η_{EQE}, = 12%.

8.2.2 Devices with Fullerene-Based Acceptors

The HBC materials discussed above are on the periphery of solubility and a more soluble group of compounds, 6F-HBC, sym2F-HBC and asym2F-HBC, see Table 8.1, were synthesised by replacing the aliphatic side-chains with 9,9-dioctylfluorene groups. 6F-HBC has six fluorene substituents [26]. In the sym2F-HBC the two fluorene groups are in the 1,4 positions whereas they occupy the 2,3-positions in asym2F-HBC. The bulky fluorene groups reduces π–π stacking in 6F-HBC and wide angles X-ray scattering measurements could not reveal any intermolecular spacing showing high disorder. In contrast, both compounds with only two fluorenyl substituents reveal a typical well-ordered columnar discotic liquid crystalline organization with intermolecular distances of 3.5 Å and liquid crystalline organisation obtained between −100 and +200°C proving pronounced mesophase stability. Bulk heterojunction solar cells were fabricated with each of the three materials blended with $PC_{61}BM$. Figure 8.7 shows that E_{HOMO} and E_{LUMO} of the **D** and **A**, respectively, are well aligned to favour charge separation at the **D-A** interface. This is confirmed by the complete quenching of fluorescence for all three blends. A TiO_x hole blocking layer was inserted between the blend and the cathode [34]. This oxide film also acts as an optical spacer to ensure that the optical mode peaks in the centre of the heterojunction layer. As expected the performance of cells incorporating 6F-HBC is much worse than those incorporating either of the other two **D**s. These devices improve substantially on annealing (after deposition of the TiO_x cap). A record η_{EQE} of 39% is obtained at a wavelength of 400 nm for the device with the symmetric **D**. Even more impressive is the high fill factor of 65% and V_{OC} of 0.9 V obtained after annealing at 150°C for 15 s. The η_P value of 1.5%, measured under simulated AM 1.5 G conditions, is extremely promising considering the high band-gap of the **D** and **A** materials. The good performance is attributed to good phase separation on an appropriate spatial scale. It is possible that thermal annealing increases molecular ordering (crystallinity) of the individual phase domains (both **D** and **A**) and improves the interfacial contact between the layers in the device. Another possible reason for good performance is that the blend has similar hole and electron mobility. For 1.2 blends of sym2F-HBC and $PC_{61}BM$ in an OFET, the hole carrier mobility is 2.8×10^{-4} cm^2 V^{-1} s^{-1} while the electron mobility is 1.2×10^{-4} cm^2 V^{-1} s^{-1}. Note however, that the mobility in an OFET device may not be the same as that in a photovoltaic, where the charge carrier moves perpendicular to the substrate.

A hybrid molecule incorporating a HBC **D** component separated from a fullerene **A** moiety by a fluorene spacer, HBC-F-PCBM shown in Table 8.2, was used in a single layer PV device capped with a TiO_x film [35]. The material shows ambipolar behaviour with similar, albeit low ($2–3 \times 10^{-6}$ cm^2 V^{-1} s^{-1}) electron and hole mobility. A low fill-factor of 0.31 and η_P of 0.2% were obtained implying that performance is limited by charge recombination and the inability to avoid continuous pathways for electrons and holes in both directions across the device.

8.3 Nematic Materials for PVs

8.3.1 Reactive Mesogens for Distributed Bilayer Devices

In 1972 the photovoltaic effect was first demonstrated in devices with nematic liquid crystals by means of ionic conduction [36]. Although electronic charge transport was widely researched in these materials [37, 38], it was not until 2006 that electronic conduction was first applied to photovoltaics in nematics [39]. A novel approach based on reactive mesogens was used to create a **D-A** bilayer with a distributed interface. Reactive mesogens are polymerisable equivalents of small molecule LCs, but with two additional polymerisable groups, one at each end of a flexible aliphatic spacer attached to the aromatic core. Chapters 2 and 5 discusses charge transport in these materials. Figure 8.8 illustrates the photopolymerisation of such molecules.

The semiconducting aromatic core has orientational, but no positional order. A thin film is formed by solution processing, such as spin or drop casting or ink-jet printing. The nematic molecules adopt a planar alignment with the long molecular axis aligned in the substrate plane [40]. They can be uniaxially aligned using rubbing or photoalignment techniques resulting in very high order parameters particularly for materials with extended molecular cores [41–43]. Polymerisation and crosslinking occur either by the thermal or photoinduced generation of free

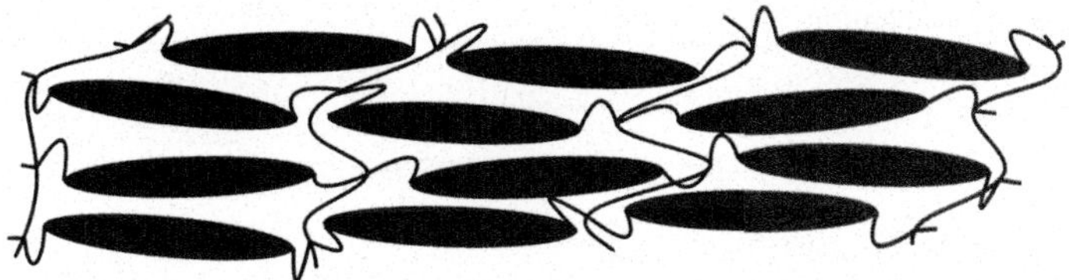

Fig. 8.8 Schematic of a crosslinked polymer network. The nematic reactive mesogens have photopolymerisable groups at each end of an extended aromatic core with semiconducting properties. The mesogens are deposited as a thin film by solution processing. They are polymerised and crosslinked either thermally or on irradiation with ultraviolet light

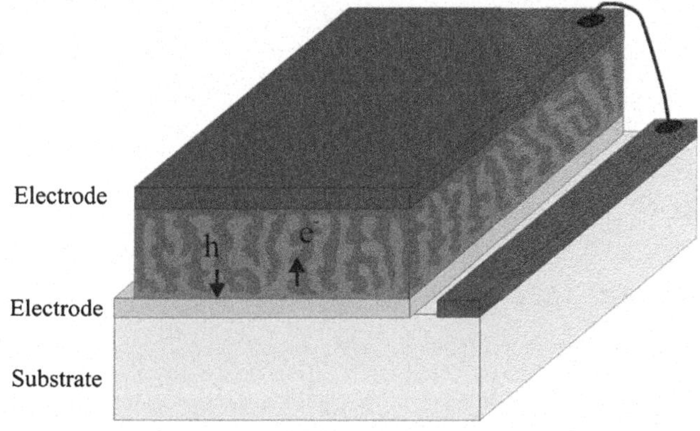

Fig. 8.9 Schematic of ideal organic photovoltaic with a distributed interface between a diffuse **D** layer and an overlying **A** layer

radicals using ultraviolet light or by ionic photoinitation [42, 44–46]. The thin film becomes insoluble so that multilayer devices are easily formed. This is a key advantage since the deposition of multiple polymer layers using conventional solution processing can prove problematic, because layers may mix when the solvents used for spin coating upper layers dissolve the underlying film deposited previously. Photopolymerisation offers the further advantage of pixellation by photolithography, unexposed regions are simply removed by washing in the original spin-casting solvent.

A liquid crystal composite approach was used to provide a distributed interface to vertically separate **D** and **A** films in an OPV device [39, 47]. The concept is illustrated in Fig. 8.9. A nematic polymer network with a porous surface with sub-micron scaled grooves and electron-donating properties was prepared by photopolymerising a thin film containing a blend of the reactive mesogen, FT1, which has a fluorene-thiophene structure, and a non-polymerisable analogue, compound FT2. Photopolymerisation leads to phase separation of the polymerised and the non-polymerised materials, the latter of which is removed by washing in a suitable solvent [39]. The chemical structures of the nematic materials discussed here are

Table 8.3 Nematic calamitic liquid crystals

FT1

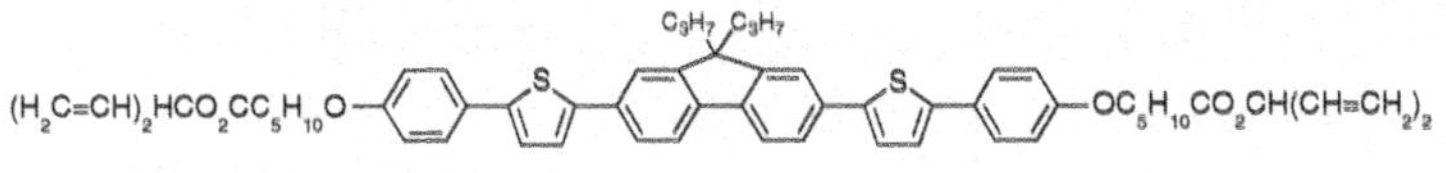

t_g = 39°C; Cr-N = 92°C, N-I = 108°C

FT2

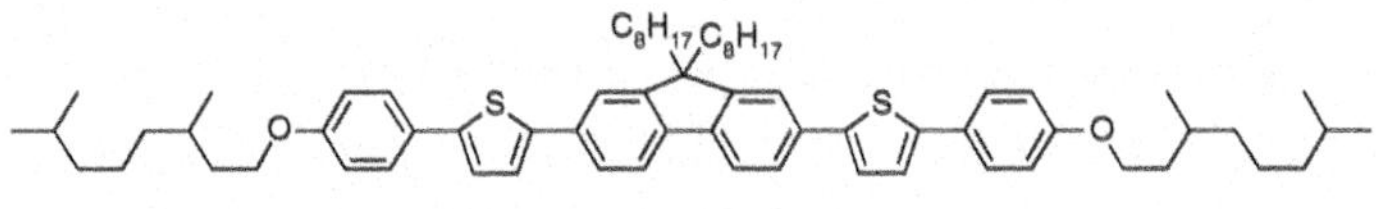

t_g = -4; Cr-N = 72°C; N-I = 61°C (monotropic)

FT3

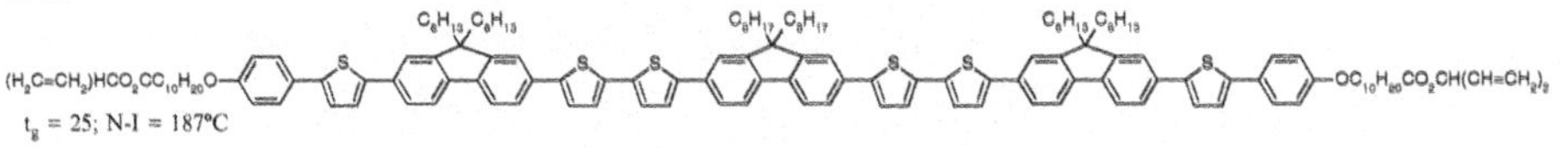

t_g = 25; N-I = 187°C

FT4

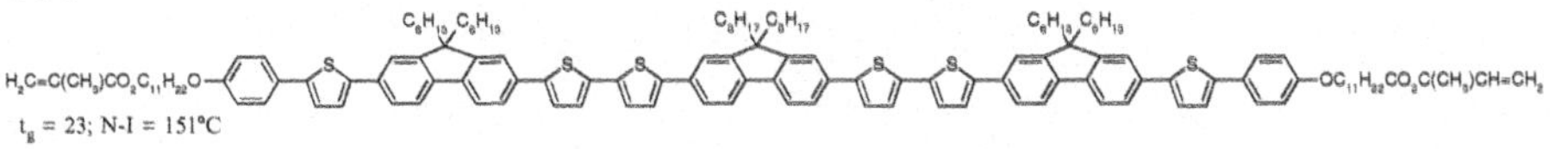

t_g = 23; N-I = 151°C

FT5

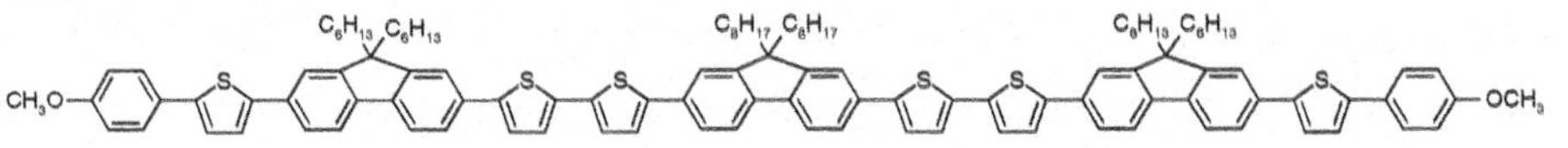

t_g = 80; N-I = 218°C

shown in Table 8.3. Figure 8.10a, b shows the rms surface-roughness-amplitude of such a porous, polymerized, lower layer, obtained from Atomic Force Microscopy, before and after washing in toluene. The roughness increases in amplitude by over a factor of four on washing giving a mean peak-to-valley roughness of 6.5 nm. The structured film is crosslinked, which gives improved thermal stability. Another key parameter is its in-plane structure characterised by spatial frequencies obtained by Fourier analysis. The concept assumes that the morphological structure of the interface is retained when the porous **D** layer is in-filled with an **A** layer. Assuming an exciton diffusion length of 10 nm, we require the in-plane spatial frequencies to be 0.05 nm^{-1} to ensure that all excitons reach an interface before recombination. However, the power spectral density for the washed layer peaks at a frequency of only 0.005 nm^{-1} extending to 0.04 nm^{-1}. An electron-accepting, perylene PDIF was deposited by spin-casting to in-fill the grooves to give a smooth surface with low roughness as confirmed by Fig. 8.10c. The resulting PV device has a η_P of 0.6% and η_{EQE} of 6% on illumination with a wavelength of 500 nm and an irradiance of 45 mW cm^{-2}. A large V_{OC} of 1.1 V is obtained. This approaches

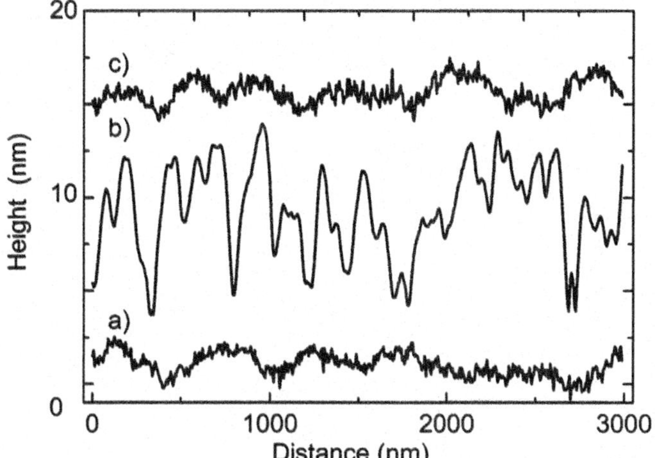

Fig. 8.10 A cross section of surface morphology obtained using AFM of the thin-film blend of the polymerisable compound FT1 and its non-polymerisable homologue FT2 following (**a**) photopolymerisation, (**b**) washing of the photopolymerised blend to remove the non-crosslinked molecules, and (**c**) deposition of an overlying **A**, PDIF. The plots are displaced upwards for clarity. Adapted from Carrasco-Orozco et al. [39]

the maximum value for the **D-A** combination $= E_{HOMO}(\mathbf{D}) - E_{LUMO}(\mathbf{A}) = 1.37$ V [39]. The relatively poor performance was attributed to the imperfect interface. The roughness is insufficiently deep and too coarse so that a relatively small fraction of excitons dissociate. A structured surface of greater amplitude was obtained by polymerising and washing a film of the nematic **D**, FT3, blended with PDIF [47]. The amplitude distribution peaks at 13.3 nm and extends from 7 to 19.5 nm (FWHM). Unfortunately, the in-plane spatial features are about three times coarser than above. The resulting bilayer device with an overlying layer of PDIF shows a η_P of up to 0.8% at the peak of the absorption spectrum. The performance of the devices was investigated as a function of incident irradiance using the equivalent circuit illustrated in Fig. 8.4. Figure 8.11 shows the resulting current-voltage characteristics. The simulated characteristics required the inclusion of a blocking contact and an intensity dependent R_P in order for the measured data to be accurately simulated. The modeled characteristics are shown as solid lines for the two highest input irradiances. The intensity dependent R_P (varying between $5 \times 10^6\ \Omega\text{cm}^2$ and $1.4 \times 10^3\ \Omega\text{cm}^2$) is attributed to continuous pathways for electrons across the device. R_S is independent of irradiance, but was two orders of magnitude higher than the best PV devices ($\approx 5\ \Omega\text{cm}^2$). More recently bilayer devices with η_P of 0.9% were fabricated from a photopolymerisable **D**, FT4, where the diene photoreactive moiety of FT3 is replaced with a methacrylate group [48]. The **A** PDIF was used in the devices.

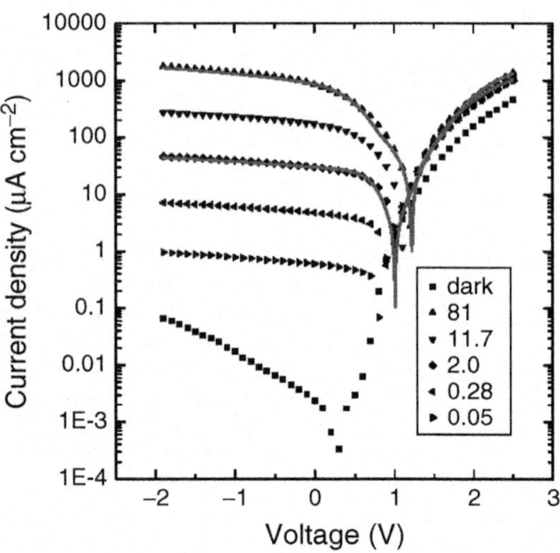

Fig. 8.11 The current density versus voltage of the photovoltaic device incorporating a thin film of PDIF infilling the network formed from a blend of FT3 and PDIF. The *inset* labels the irradiance in mW cm^{-2} of the input light source of wavelength 475 nm. The *solid lines* show the simulated plots using an equivalent circuit for input irradiances of 81 and 2 mW cm^{-2}

Table 8.4 Performance parameters for the PV device using a FT5. PDIF blend with ratio 1.2 by weight annealed at 120°C, on varying the irradiance at 466 nm [50]

Light intensity (mW cm^{-2})	V_{oc} (V)	J_{sc} mA cm^{-2}	P_{max} mW cm^{-2}	FF	EQE (%)	PCE (%)
22	1	0.71	0.25	0.36	8.6	1.1
15	1	0.50	0.18	0.37	8.9	1.3
9.3	1	0.32	0.13	0.39	9.3	1.4
2.6	0.8	0.10	0.03	0.32	10.3	1.0

8.3.2 Nematic Molecules in PV Blends

Nematic liquid crystalline **D**s with a fluorene-thiophene structure, e.g., FT3, were blended with different perylene-based **A**s including PDIF having similar electron affinities, but different thermotropic phases, to form single-layer OPV devices [49]. Interestingly, blends of equal weights of the **D** and **A** retain supercooled nematic glassy phases at room temperature after annealing, although the **A** is crystalline. Best PV results are obtained when a non-polymerisable donor **D**, FT5, with the same aromatic core as FT1 is blended with PDIF [50]. Table 8.4 shows that the maximum η_P and η_{EQE} are 1.4% and 10.3% respectively. The corresponding η_{IQE} of 53% is a promising result and shows that performance is limited by low absorption in an insufficiently thick blend.

8.4 Liquid Crystalline Polymers for Photovoltaics

An excellent demonstration of the benefits of the self-assembly properties of liquid crystalline phases lies in the recent development of liquid crystalline, semiconducting, thiophene polymers and copolymers, many of which contain

Table 8.5 Liquid crystalline mainchain polymers

PBTTT	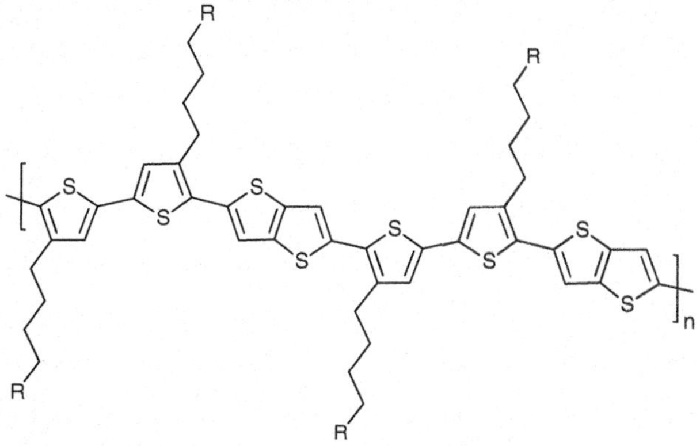
	Cr-SmX ≈ 171°C; SmX-I ≈ 251°C [R = C_nH_{2n+1}]

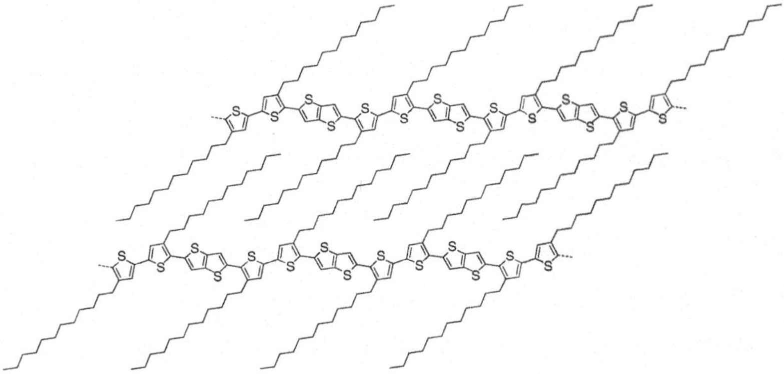

Fig. 8.12 Schematic of the packing of polymer chains of a sanditic liquid crystal with uniform side chain interdigitation

thienothiophene groups [51–57], see for example poly(2,5-bis(3-alkylthiophen-2-yl)thieno[3,2-b]thiophene) (pBTTT) which is shown in Table 8.5. This shows exceptionally high field-effect values of mobility up to 1 $cm^2\ V^{-1}\ s^{-1}$. Such polymers exhibit a mesophase above room temperature associated with a lower side-chain density than the polymer P3HT, where each thiophene ring is substituted with an alkyl group. The latter material is regularly used in **D-A** blends with PCBM giving η_P up to 5.2% [58]. The lower density allows interdigitation, as illustrated in Fig. 8.12, and the mesophase corresponds to a highly-ordered smectic or sanditic phase of π-stacked backbones separated by melted and interdigitated aliphatic, non-conducting side-chains [59, 60]. On annealing to the mesophase, the backbones can freely move and reorganise into a more ordered state, whilst removing defects. Upon cooling, the side chains recrystallise to capture and maintain the high level of layer order. Consequently OFET performance improves on annealing above the mesophase transition temperature. P3HT has no corresponding mechanism capable

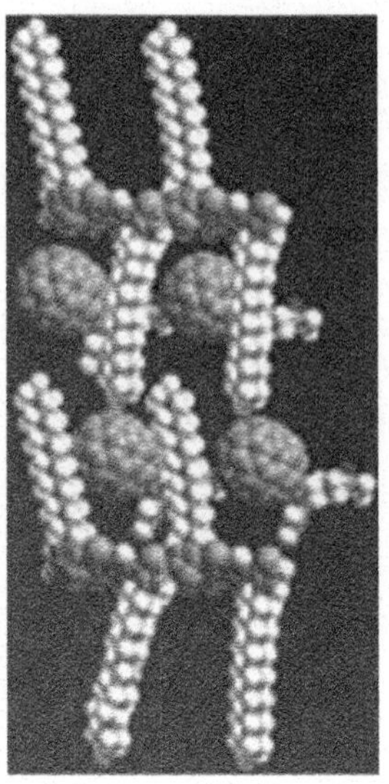

Fig. 8.13 Simulation of bimolecular crystal containing molecules of $PC_{71}BM$ intercalated between monomer units of pBTTT Reprinted with permission from [64] Copyright © 2009 WILEY-VCH

of improving the degree of order. The field effect mobility experiment probes in-plane charge transport at the interface of a polymer thin film and is affected by the local molecular arrangement. Time-of-flight (TOF) studies probe the charge transport through the bulk of the film perpendicular to the substrate plane. The TOF mobility of pBTTT is $\approx 10^{-5}$ cm^2 V^{-1} s^{-1}, orders of magnitude less than that obtained using the field-effect method [61]. However, the mobility improves by an order of magnitude due to increased order following prolonged annealing. This result is consistent with the view that the polymer chains align in the plane. The intermolecular separation between the layers is low so the perpendicular charge transport proceeds by an indirect percolation route.

To date there has been very little research to examine whether thioenothiophene polymers prove to be as useful for OPVs as they are for OFETs. pBTTT has a relatively broad absorption band extending to 650 nm and shows extremely well-ordered, phase-separated domains in a blend with PCBM. The power conversion efficiency of the blends is $>1\%$ under standard conditions [62]. This performance is significantly lower than for blends of PCBM with other polymer **D**s, which have a less ordered morphology. This may be due to high geminate recombination losses and poor interconnectivity between the domains. Much improved performance was obtained in blends of pBTTT with the electron acceptor $PC_{71}BM$ [63]. This has a larger diameter than PCBM and so packs efficiently when intercalating between the side chains of pBTTT forming a bimolecular crystal, as illustrated in Fig. 8.13 [64]. The bimolecular crystal, having a 1.1 ratio of the pBTTT monomer unit and $PC_{71}BM$, shows good hole transport but suppressed electron transport because of poor interconnectivity between the $PC_{71}BM$ molecules. Improved electron transport requires a higher fraction of $PC_{71}BM$, giving a phase-separated blend of a bimolecular crystal and an amorphous $PC_{71}BM$ [64, 65]. Hence, the most efficient blended pBTTT. $PC_{71}BM$ solar cells have a high fraction 75% of $PC_{71}BM$ by weight with η_{EQE} equal to 50%, corresponding to η_P of 2.5%. The fullerene derivative, bisPC_{71}BM is too large to effectively intercalate with pBTTT. As a result PV device performance was optimised for a 1.1 blend [63]. However, photoluminescence is not completely quenched in the blend and the performance is not as good as for the pBTTT. PC_{71}BM blend. The side-chains of P3HT are too closely packed to allow intercalation of fullerene **A**s. The temperature dependent solubility of pBTTT allows a bilayer PV device to be constructed, each layer consisting of an interpenetrating electron donor-acceptor network. pBTTT is the **D** in the lower layer and the **D** in the upper layer is chosen to increase the spectral coverage of the device [66].

8.5 Anisotropic Luminescent Solar Concentrator

A novel application of liquid crystals to solar cells is the use of anisotropic fluorescent dyes and mesogens forming a guest-host system to control the direction of emitted light in a luminescent solar concentrator [67]. A luminescent solar concentrator allows sunlight to penetrate the top surface of an inexpensive large-area waveguide (plastic or glass). The light is absorbed by embedded dye molecules and re-emitted at a longer wavelength. A fraction of the re-emitted light is trapped in the waveguide by total internal reflection and coupled from the edges of the waveguide onto a solar cell of small area. Figure 8.14 illustrates the principle of the anisotropic concentrator.

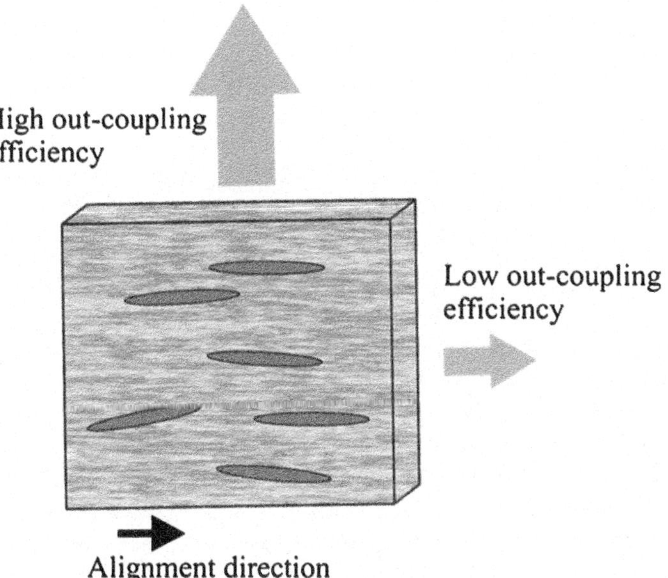

Fig. 8.14 Illustration of principle of luminescent solar concentrator with. The solar cell is not included in the illustration and would butt-up to the edge to maximise efficiency of light collection

The guest-host matrix forms the waveguide and is uniaxially aligned. The transition dipole moment of the rod-shaped dye molecules is aligned along their long axes. The polarisation direction of the emitted light will be similarly aligned. Light is a transverse wave so that more light is coupled out of the two edges parallel to the alignment direction than from the edges, which are perpendicularly oriented. Out-coupling from the parallel edge direction is higher by 60% than that from the perpendicular edge and the output from the favoured edge of the aligned system exceeds the output of any edge of an isotropic system by up to 30%. This improvement is relatively low, because of mode mixing in the waveguide.

8.6 Discussion

As discussed in Sect. 8.1, there have been many developments in the performance of solar cells based on non-liquid crystalline, main-chain polymers blended with fullerene leading to state of the art values of $\eta_P = 8\%$. The performance of any of the liquid crystal containing devices discussed here is far from the state of the art. However, we must remember that reaching the current state of the art required painstaking research over many years and the effort devoted to liquid crystalline materials for photovoltaics is orders of magnitude less. There has also been enormous progress in the understanding of the various parameters and trade-offs required to optimise performance, as discussed in many excellent reviews [2, 11, 15, 68], and similar strategies could be applied to liquid crystalline materials for improved performance. The main motivation for using liquid crystals is to exploit their self-assembly properties to control the morphology of the **D-A** composite. For nematic reactive mesogens, there is the added incentive that polymerisation and crosslinking should lock-in the optimised morphology providing long term stability.

The best solar cell incorporating discotic liquid crystals is the device containing sym2F-HBC blended with PCBM, with $\eta_{EQE} = 39\%$ with monochromatic irradiation and $\eta_P = 1.5\%$ under AM 1.5 G conditions. $\Delta = \Delta E_{HOMO} + \Delta E_{LUMO}$, as defined in Eq. (8.8), should be about 0.3 eV to optimise efficiency of both charge separation and photon energy loss. As Fig. 8.7 shows, band-alignment is far from optimum in this blend with Δ equal to 2.55 eV. Improved performance would require a lowering of E_{HOMO} and a red-shift in the absorption of the discotic, a challenge for the synthetic chemist. Despite the non-optimised energy levels, the performance of the best cells is very encouraging. This is related to the effective use of annealing to control the molecular ordering (crystallinity) of the individual phase domains (both **D** and **A**). Interestingly, as also shown in Fig. 8.7, better band alignment is found for discotics **D**s combined with PDI **A**s. However, the photovoltaic performance of these blends is substantially worse than mixtures of discotic **D**s and fullerene **A**s, a result reflecting trends already seen in devices with polymer **D**s. Sample processing also emerges as a critical factor, which influences device performance. For example, annealing in the liquid crystalline phase, although very beneficial to modify phase separation, can cause problems. Enhanced surface roughness and large, phase-separated domains are also found in annealed blends of coronenes and perylenes. Much smaller domains and improved photovoltaic performance are obtained by annealing after deposition of a cap or the top electrode. Interestingly, a small portion (3 wt.%) of the (columnar) discotic liquid crystal, 2,3,6,7,10,11-hexaacetoxytriphenylene has been used as an additive to improve by 30% the performance of well-established regioregular P3HT.$PC_{61}BM$ bulk-heterojunction OPVs [69]. The improved performance was attributed to the strong self-assembling ability of the discotic, which improves the crystallinity of the P3HT. This increases the hole mobility by an order of magnitude resulting in improved balance of the electron and hole mobilities.

More effort is required to realise the potential of the distributed bilayer photovltaic concept based on nematic reactive mesogens. Many of the push-pull **D** polymers, which have red-shifted absorption spectra are rod-shaped. Similar modifications of chemical strucutres could be used to shift the energy levels of nematic reactive mesogens. More effective phase separation of the polymerisable and nonpolymerisable components during photopolymerisation is the key to obtaining a **D** layer with a sufficiently deep and laterally fine, nanostructured, interface after washing. The combined use of solvent vapour annealing and ultraviolet irradiation, whereby the film is saturated with vapour during photopolymerisation is a new tool to control the degree and spatial scale of phase separation.

8.7 Conclusions

We have described how the self assembly properties of discotic, nematic and sanditic liquid crystals can be applied to organic photovolatics. A range of hexabenzocoronene molecules with room temperature discotic phases have excellent electron-donating properties and combine with perylene and, preferably, fullerene-based **A**s to make very promising distributed heterojunction photovoltaics. Annealing in the liquid crystalline phase effectively controls phase separation of the blend. A suitable application would be near UV-blue photodetection. The photopolymerisation of nematic, reactive mesogens is a novel way to make a bilayer photovoltaic with a distributed interface, although better control of the spatial scale of the distributed is required. A new class of thienothiophene polymers with high temperature sanditic phases, which were originally designed for charge transport in field-effect transistors, have also been applied to photovoltaics. Interestingly intercalation of the **A** $PC_{71}BM$ between the ordered sidechains of the polymer gives a bimolecular crystal aiding hole transport. Research on the application of liquid crystals to photovoltaics is relatively young and improved synthesis and cleaning, morphology control and processing is required to achieve state-of-the-art.

References

1. Tang, C.W.: Two-layer organic photovoltaic cell. Appl. Phys. Lett. **48**, 183 (1986)
2. Kippelen, B., Brédas, J.L.: Organic photovoltaics. Energy Environ. Sci. **2**(3), 251–261 (2009)
3. Halls, J.J.M., Walsh, C.A., Greenham, N.C., Marseglla, E.A., Friend, R.H., Moratti, S.C., Holmes, A.B.: Efficient photodiodes from interpenetrating polymer networks. Nature **376**(6540), 498–500 (1995)
4. Yu, G., Gao, J., Hummelen, J.C., Wudl, F., Heeger, A.J.: Polymer photovoltaic cells. Enhanced efficiencies via a network of internal donor-acceptor heterojunctions. Science **270**(5243), 1789–1791 (1995)
5. Morteani, A.C., Sreearunothai, P., Herz, L.M., Friend, R.H., Silva, C.: Exciton regeneration at polymeric semiconductor heterojunctions. Phys. Rev. Lett. **92**(24), 247402 (2004)
6. Mihailetchi, V.D., Wildeman, J., Blom, P.W.M.: Space-charge limited photocurrent. Phys. Rev. Lett. **94**(12), 126602.1–126602.4 (2005)
7. Sze, S.M.: Semiconductor Devices. Physics and Technology, 2nd edn. Wiley, New York (2002)
8. Rand, B.P., Burk, D.P., Forrest, S.R.: Offset energies at organic semiconductor heterojunctions and their influence on the open-circuit voltage of thin-film solar cells. Phys. Rev. B Condens. Matter Mater. Phys. **75**(11), 115327 (2007)
9. Scharber, M.C., Mühlbacher, D., Koppe, M., Denk, P., Waldauf, C., Heeger, A.J., Brabec, C.J.: Design rules for donors in bulk-heterojunction solar cells – towards 10% energy-conversion efficiency. Adv. Mater. **18**(6), 789–794 (2006)
10. Brédas, J.L., Beljonne, D., Coropceanu, V., Cornil, J.: Charge-transfer and energy-transfer processes in Ï€-conjugated oligomers and polymers. A molecular picture. Chem. Rev. **104**(11), 4971–5003 (2004)
11. Dennler, G., Scharber, M.C., Brabec, C.J.: Polymer-fullerene bulk-heterojunction solar cells. Adv. Mater. **21**(13), 1323–1338 (2009)
12. Park, S.H., Roy, A., Beaupré, S., Cho, S., Coates, N., Moon, J.S., Moses, D., Leclerc, M., Lee, K., Heeger, A.J.: Bulk heterojunction solar cells with internal quantum efficiency approaching 100%. Nat. Photon. **3**(5), 297–303 (2009)
13. Chen, H.Y., Hou, J., Zhang, S., Liang, Y., Yang, G., Yang, Y., Yu, L., Wu, Y., Li, G.: Polymer solar cells with enhanced open-circuit voltage and efficiency. Nat. Photon. **3**(11), 649–653 (2009)
14. Liang, Y., Xu, Z., Xia, J., Tsai, S.T., Wu, Y., Li, G., Ray, C., Yu, L.: For the bright future-bulk heterojunction polymer solar cells with power conversion efficiency of 7.4%. Adv. Mater. **22**(20), E135–E138 (2010)
15. Brabec, C.J., Gowrisanker, S., Halls, J.J.M., Laird, D., Jia, S., Williams, S.P.: Polymer-fullerene bulk-heterojunction solar cells. Adv. Mater. **22**(34), 3839–3856 (2010)
16. Liang, Y., Yu, L.: A new class of semiconducting polymers for bulk heterojunction solar cells with exceptionally high performance. Acc. Chem. Res. **43**(9), 1227–1236 (2010)
17. http.//www.konarka.com/index.php/newsroom/konarka- news/. http.//www.konarka.com/index.php/newsroom/konarka- news/. Accessed 18 Apr 2011
18. Hummelen, J.C., Knight, B.W., Lepeq, F., Wudl, F., Yao, J., Wilkins, C.L.: Preparation and characterization of fulleroid and methanofullerene derivatives. J. Org. Chem. **60**(3), 532–538 (1995)
19. Funahashi, M.: Development of liquid-crystalline semiconductors with high carrier mobilities and their application to thin-film transistors. Polym. J. **41**(6), 459–469 (2009)
20. Pisula, W., Zorn, M., Chang, J.Y., Müllen, K., Zentel, R.: Liquid crystalline ordering and charge transport in semiconducting materials. Macromol. Rapid Commun. **30**(14), 1179–1202 (2009)
21. Tsao, H.N., Pisula, W., Liu, Z., Osikowicz, W., Salaneck, W.R., Müllen, K.: From ambi- to unipolar behavior in discotic dye field-effect transistors. Adv. Mater. **20**(14), 2715–2719 (2008)
22. Adam, D., Schuhmacher, P., Simmerer, J., Häussling, L., Siemensmeyer, K., Etzbach, K.H., Ringsdorf, H., Haarer, D.: Fast photoconduction in the highly ordered columnar phase of a discotic liquid crystal. Nature **371**(6493), 141–143 (1994)

23. Van De Craats, A.M., Warman, J.M., Fechtenkötter, A., Brand, J.D., Harbison, M.A., Müllen, K.: Record charge carrier mobility in a room-temperature discotic liquid-crystalline derivative of hexabenzocoronene. Adv. Mater. **11**(17), 1469–1472 (1999)
24. Pisula, W., Menon, A., Stepputat, M., Lieberwirth, I., Kolb, U., Tracz, A., Sirringhaus, H., Pakula, T., Müllen, K.: A zone-casting technique for device fabrication of field-effect transistors based on discotic hexa-perihexabenzocoronene. Adv. Mater. **17**(6), 684–688 (2005)
25. Xiao, S., Myers, M., Miao, Q., Sanaur, S., Pang, K., Steigerwald, M.L., Nuckolls, C.: Molecular wires from contorted aromatic compounds. Angew. Chem. Int. Ed. **44**(45), 7390–7394 (2005)
26. Wong, W.W.H., Birendra Singh, T., Vak, D., Pisula, W., Yan, C., Feng, X., Williams, E.L., Chan, K.L., Mao, Q., Jones, D.J., Chang-Qi, M., Müllen, K., Bauerle, P., Holmes, A.B.: Solution processable fluorenyl hexa-peri-hexabenzocoronenes in organic field-effect transistors and solar cells. Adv. Funct. Mater. **20**(6), 927–928 (2010)
27. Schmidt-Mende, L., Fechtenkötter, A., Müllen, K., Moons, E., Friend, R.H., MacKenzie, J.D.: Self-organized discotic liquid crystals for high-efficiency organic photovoltaics. Science **293**(5532), 1119–1122 (2001)
28. Shin, W.S., Jeong, H.H., Kim, M.K., Jin, S.H., Kim, M.R., Lee, J.K., Lee, J.W., Gal, Y.S.: Effects of functional groups at perylene diimide derivatives on organic photovoltaic device application. J. Mater. Chem. **16**(4), 384–390 (2006)
29. Hesse, H.C., Weickert, J., Al-Hussein, M., Dassel, L., Feng, X., Mullen, K., Schmidt-Mende, L.: Discotic materials for organic solar cells. Effects of chemical structure on assembly and performance. Solar Energy Mater. Solar Cells **94**(3), 560–567 (2010)
30. Schmidtke, J.P., Friend, R.H., Kastler, M., Müllen, K.: Control of morphology in efficient photovoltaic diodes from discotic liquid crystals. J. Chem. Phys. **124**(17), 174704 (2006)
31. Schmidt-Mende, L., Fechtenkötter, A., Müllen, K., Friend, R.H., MacKenzie, J.D.: Efficient organic photovoltaics from soluble discotic liquid crystalline materials. Physica E **14**(1–2), 263–267 (2002)
32. Li, J., Kastler, M., Pisula, W., Robertson, J.W.F., Wasserfallen, D., Grimsdale, A.C., Wu, J., Müllen, K.: Organic bulk-heterojunction photovoltaics based on alkyl substituted discotics. Adv. Funct. Mater. **17**(14), 2528–2533 (2007)
33. Feng, X., Liu, M., Pisula, W., Takase, M., Li, J., Müllen, K.: Supramolecular organization and photovoltaics of triangle-shaped discotic graphenes with swallow-tailed alkyl substituent. Adv. Mater. **20**(14), 2684–2689 (2008)
34. Kim, J.Y., Kim, S.H., Lee, H.H., Lee, K., Ma, W., Gong, X., Heeger, A.J.: New architecture for high-efficiency polymer photovoltaic cells using solution-based titanium oxide as an optical spacer. Adv. Mater. **18**(5), 572–576 (2006)
35. Wong, W.W.H., Vak, D., Singh, T.B., Ren, S., Yan, C., Jones, D.J., Liaw, I.I., Lamb, R.N., Holmes, A.B.: Ambipolar hexa-peri-hexabenzocoronene-fullerene hybrid materials. Org. Lett. **12**(21), 5000–5003 (2010)
36. Kamei, H., Ozawa, T., Katawama, Y.: Photovoltaic effect in the nematic liquid crystal. Jpn. J. Appl. Phys. **11**, 1385 (1972)
37. Funahashi, H., Hanna, J.: Fast hole transport in a new calamitic liquid crystal of 2-(4'-heptyloxyphenyl)-6-dodecylthiobenzothiazole. Phys. Rev. Lett. **78**, 2184 (1997)
38. Farrar, S.R., Contoret, A.E.A., O'Neill, M., Nicholls, J.E., Richards, G.J., Kelly, S.M.: Nondispersive hole transport of liquid crystalline glasses and a cross-linked network for organic electroluminescence. Phys. Rev. B **66**(12) (2002). doi:12510710.1103/PhysRevB.66.125107
39. Carrasco-Orozco, M., Tsoi, W.C., O'Neill, M., Aldred, M.P., Vlachos, P., Kelly, S.M.: New photovoltaic concept. Liquid-crystal solar cells using a nematic gel template. Adv. Mater. **18**(13), 1754 (2006). doi:10.1002/adma.200502008
40. Droege, S., Khalifah, M.S.A., O'Neill, M., Thomas, H.E., Simmonds, H.S., Macdonald, J.E., Aldred, M.P., Vlachos, P., Kitney, S.P., Loebbert, A., Kelly, S.M.: Grazing incidence X-ray diffraction of a photoaligned nematic semiconductor. J. Phys. Chem. B **113**(1), 49–53 (2009)
41. Aldred, M.P., Vlachos, P., Contoret, A.E.A., Farrar, S.R., Chung-Tsoi, W., Mansoor, B., Woon, K.L., Hudson, R., Kelly, S.M., O'Neill, M.: Linearly polarised organic light-emitting diodes (OLEDs). Synthesis and characterisation of a novel hole-transporting photoalignment copolymer. J. Mater. Chem. **15**(31), 3208–3213 (2005). doi:10.1039/b50603j

42. Contoret, A.E.A., Farrar, S.R., Jackson, P.O., Khan, S.M., May, L., O'Neill, M., Nicholls, J.E., Kelly, S.M., Richards, G.J.: Polarized electroluminescence from an anisotropic nematic network on a non-contact photoalignment layer. Adv. Mater. **12**(13), 971 (2000)
43. Woon, K.L., Contoret, A.E.A., Farrar, S.R., Liedtke, A., O'Neill, M., Vlachos, P., Aldred, M.P., Kelly, S.M.: Material and device properties of highly birefringent nematic glasses and polymer networks for organic electroluminescence. J. Soc. Inf. Disp. **14**(6), 557–563 (2006)
44. Aldred, M.P., Eastwood, A.J., Kelly, S.M., Vlachos, P., Contoret, A.E.A., Farrar, S.R., Mansoor, B., O'Neill, M., Tsoi, W.C.: Light-emitting fluorene photoreactive liquid crystals for organic electroluminescence. Chem. Mater. **16**(24)), 4928–4936 (2004). doi:10.1021/cm0351893
45. Contoret, A.E.A., Farrar, S.R., O'Neill, M., Nicholls, J.E.: The photopolymerization and cross-linking of electroluminescent liquid crystals containing methacrylate and diene photopolymerizable end groups for multilayer organic light-emitting diodes. Chem. Mater. **14**(4), 1477–1487 (2002). doi:10.1021/cm011111f
46. McCulloch, I., Zhang, W., Heeney, M., Bailey, C., Giles, M., Graham, D., Shkunov, M., Sparrowe, D., Tierney, S.: Polymerisable liquid crystalline organic semiconductors and their fabrication in organic field effect transistors. J. Mater. Chem. **13**(10), 2436–2444 (2003)
47. Tsoi, W.C., O'Neill, M., Aldred, M.P., Kitney, S.P., Vlachos, P., Kelly, S.M.: Distributed bilayer photovoltaics based on nematic liquid crystal polymer networks. Chem. Mater. **19**((23), 5475–5484 (2007). doi:10.1021/cm071727q
48. Lei, C., O'Neill, M., Myers, S.A., Kelly, S.M., Kitney, S.: To be published
49. Lei, C., Al Khalifah, M.S., O'Neill, M., Aldred, M.P., Kitney, S.P., Vlachos, P., Kelly, S.M.: Calamitic liquid crystal blends for organic photovoltaics. In: Proceedings of SPIE – The International Society for Optical Engineering, Bellingham, USA (2008)
50. Al Khalifah, M.S.: Nematic liquid crystals for nanostructured organic photovoltaic. PhD thesis, The University of Hull (2010)
51. Ong, B.S., Wu, Y., Liu, P., Gardner, S.: High-performance semiconducting polythiophenes for organic thin-film transistors. J. Am. Chem. Soc. **126**(11), 3378–3379 (2004)
52. McCulloch, I., Heeney, M., Bailey, C., Genevicius, K., MacDonald, I., Shkunov, M., Sparrowe, D., Tierney, S., Wagner, R., Zhang, W., Chabinyc, M.L., Kline, R.J., McGehee, M.D., Toney, M.F.: Liquid-crystalline semiconducting polymers with high charge-carrier mobility. Nat. Mater. **5**(4), 328–333 (2006)
53. McCulloch, I., Heeney, M., Chabinyc, M.L., Delongchamp, D., Kline, R.J., Cölle, M., Duffy, W., Fischer, D., Gundlach, D., Hamadani, B., Hamilton, R., Richter, L., Salleo, A., Shkunov, M., Sparrowe, D., Tierney, S., Zhang, W.: Semiconducting thienothiophene copolymers. Design, synthesis, morphology, and performance in thin-film organic transistors. Adv. Mater. **21**(10–11), 1091–1109 (2009)
54. Heeney, M., Bailey, C., Genevicius, K., Shkunov, M., Sparrowe, D., Tierney, S., McCulloch, I.: Stable polythiophene semiconductors incorporating thieno[2,3-6]thiophene. J. Am. Chem. Soc. **127**(4), 1078–1079 (2005)
55. Hamadani, B.H., Gundlach, D.J., McCulloch, I., Heeney, M.: Undoped polythiophene field-effect transistors with mobility of 1 $cm^2\ V^{-1}\ s^{-1}$. Appl. Phys. Lett. **91**(24), 243512 (2007)
56. Li, Y., Wu, Y., Liu, P., Birau, M., Pan, H., Ong, B.S.: Poly(2,5-bis(2-thienyl)-3,6-dialkylthieno[3,2-b]thiophene)s-high-mobility semiconductors for thin-film transistors. Adv. Mater. **18**(22), 3029–3032 (2006)
57. Kim, D.H., Lee, B.L., Moon, H., Kang, H.M., Jeong, E.J., Park, J.I., Han, K.M., Lee, S., Yoo, B.W., Koo, B.W., Kim, J.Y., Lee, W.H., Cho, K., Becerril, H.A., Bao, Z.: Liquid-crystalline semiconducting copolymers with intramolecular donor-acceptor building blocks for high-stability polymer transistors. J. Am. Chem. Soc. **131**(17), 6124–6132 (2009)
58. Irwin, M.D., Buchholz, D.B., Hains, A.W., Chang, R.P.H., Marks, T.J.: p-Type semiconducting nickel oxide as an efficiency-enhancing anode interfacial layer in polymer bulk-heterojunction solar cells. Proc. Natl. Acad. Sci. U.S.A. **105**(8), 2783–2787 (2008)
59. Delongchamp, D.M., Kline, R.J., Jung, Y., Lin, E.K., Fischer, D.A., Gundlach, D.J., Cotts, S.K., Moad, A.J., Richter, L.J., Toney, M.F., Heeney, M., McCulloch, I.: Molecular basis of mesophase ordering in a thiophene-based copolymer. Macromolecules **41**(15), 5709–5715 (2008)

60. DeLongchamp, D.M., Kline, R.J., Lin, E.K., Fischer, D.A., Richter, L.J., Lucas, L.A., Heeney, M., McCulloch, I., Northrup, J.E.: High carrier mobility polythiophene thin films. Structure determination by experiment and theory. Adv. Mater. **19**(6), 833–837 (2007)
61. Baklar, M., Barard, S., Sparrowe, D., Wilson, R.M., McCulloch, I., Heeney, M., Kreouzis, T., Stingelin, N.: Bulk charge transport in liquid-crystalline polymer semiconductors based on poly(2,5-bis(3-alkylthiophen-2-yl)thieno[3,2-b]thiophene). Polymer Chem. **1**(9), 1448–1452 (2010)
62. Hwang, I.W., Kim, J.Y., Cho, S., Yuen, J., Coates, N., Lee, K., Heeney, M., McCulloch, I., Moses, D., Heeger, A.J.: Bulk heterojunction materials composed of poly(2,5-bis(3-tetradecylthiophen-2-yl)thieno[3,2-b;]thiophene). Ultrafast electron transfer and carrier recombination. J. Phys. Chem. C **112**(21), 7853–7857 (2008)
63. Cates, N.C., Gysel, R., Beiley, Z., Miller, C.E., Toney, M.F., Heeney, M., McCulloch, L., McGehee, M.D.: Tuning the properties of polymer bulk heterojunction solar cells by adjusting fullerene size to control intercalation. Nano Lett. **9**(12), 4153–4157 (2009)
64. Mayer, A.C., Toney, M.F., Scully, S.R., Rivnay, J., Brabec, C.J., Scharber, M., Koppe, M., Heeney, M., McCulloch, I., McGehee, M.D.: Bimolecular crystals of fullerenes in conjugated polymers and the implications of molecular mixing for solar cells. Adv. Funct. Mater. **19**(8), 1173–1179 (2009)
65. Miller, N.C., Gysel, R., Miller, C.E., Verploegen, E., Beiley, Z., Heeney, M., McCulloch, I., Bao, Z., Toney, M.F., McGehee, M.D.: The phase behavior of a polymer-fullerene bulk heterojunction system that contains bimolecular crystals. J. Polym. Sci., Part B: Polym. Phys. **49**(7), 499–503 (2011)
66. Sun, Q., Park, K., Dai, L.: Liquid crystalline polymers for efficient bilayer-bulk-heterojunction solar cells. J. Phys. Chem. C **113**(18), 7892–7897 (2009)
67. Verbunt, P.P.C., Kaiser, A., Hermans, K., Bastiaansen, C.W.M., Broer, D.J., Debije, M.G.: Controlling light emission in luminescent solar concentrators through use of dye molecules aligned in a planar manner by liquid crystals. Adv. Funct. Mater. **19**(17), 2714–2719 (2009)
68. Brabec, C.J., Heeney, M., McCulloch, I., Nelson, J.: Influence of blend microstructure on bulk heterojunction organic photovoltaic performance. Chem. Soc. Rev. **40**(3), 1185–1199 (2011)
69. Jeong, S., Kwon, Y., Choi, B.D., Ade, H., Han, Y.S.: Improved efficiency of bulk heterojunction poly(3-hexylthiophene).[6,6]- phenyl- C61 -butyric acid methyl ester photovoltaic devices using discotic liquid crystal additives. Appl. Phys. Lett. **96**(18), 183305 (2010)

Chapter 9
Liquid Crystals for Organic Field-Effect Transistors

Mary O'Neill and Stephen M. Kelly

9.1 Introduction

Organic semiconductors are becoming a viable alternative to amorphous silicon for a range of thin film transistor devices. Indeed, plastic and organic electronics are considered disruptive technologies enabling new applications including intelligent or interactive packaging, RFID tags, e-readers, flexible power sources and lighting panels. The future success of the industry depends on the availability of high performance, solution-processable, materials for low cost manufacturing as well as low voltage device operation. The organic field effect transistor (OFET) is the fundamental building block of plastic electronics and is used to amplify and switch electronic signals. The key figure of merit for OFETs is the charge carrier mobility, (μ), the hole/electron velocity per unit field. Carrier transport occurs by hopping via π-π interactions between sites, which may be traps, single molecules, several repeat units of a polymer chain or even a number of delocalised chain segments in high-mobility conjugated mainchain polymers. There has been excellent progress in the development of solution processed organic semiconductors for OFETs [1–3]. The state-of-the-art performance is now equivalent to that of amorphous silicon.

M. O'Neill (✉)
Department of Physics and Mathematics, University of Hull, Hull, HU6 7RX UK
e-mail: m.oneill@hull.ac.uk

S.M. Kelly
Department of Chemistry, University of Hull, Hull, HU6 7RX UK
e-mail: s.m.kelly@hull.ac.uk

R.J. Bushby et al. (eds.), *Liquid Crystalline Semiconductors*, Springer Series in Materials Science 169, DOI 10.1007/978-90-481-2873-0_9,

9.2 Principle of Organic Field-Effect Transistors

Transistors are organic semiconductor devices used to amplify and switch electronic signals. In an organic field-effect transistor (OFET) a voltage applied to the gate electrode controls the current between the source and drain electrodes [4, 5]. The construction of a typical OFET is shown in Fig. 9.1. An insulator is placed between the gate and the organic semiconducting thin film. The source and drain electrode are separated by a semiconducting channel of length L and width W . Ideally, when no gate voltage, V_G, is applied, the conductance of the semiconductor film is extremely low, because there are no mobile charge carriers, i.e., the device is "off". The application of V_G gives an electric field across the dielectric, which induces mobile charges in the semiconductor film. These move in response to the voltage, V_D, applied between the source and drain, i.e., the transistor is "on". Figure 9.1 is an example of a bottom-gate, top-contact device configuration. The source and drain electrodes may be deposited between the dielectric and semiconducting thin film giving a bottom-gate, bottom-contact device. Alternatively the layers can be applied sequentially in the order: source and drain electrodes, semiconducting thin film, insulator and gate electrode giving a top-gate configuration. Figure 9.1 also show the circuit used to test the OFET. The source electrode is placed at ground. A hole drain current, I_D, is measured when a negative V_G is applied, as illustrated in Fig. 9.1. The electron drain current is measured when a positive V_G is applied. V_G must be greater than a threshold value V_T, before current flows because the contacts may not be ohmic and charge-carrier traps must first be filled. Figure 9.2 shows the output characteristic curve of the transistor, i.e., a plot of I_D versus V_D for different values of V_G. Two regions of the characteristic curve can be differentiated: I_D increases approximately linearly for low values of V_D, is when $V_D << (V_G - V_T)$. When $(V_G - V_T) > V_D$, I_D saturates. The physics underlying these characteristics is beyond the scope of this volume and is well explained elsewhere [5].

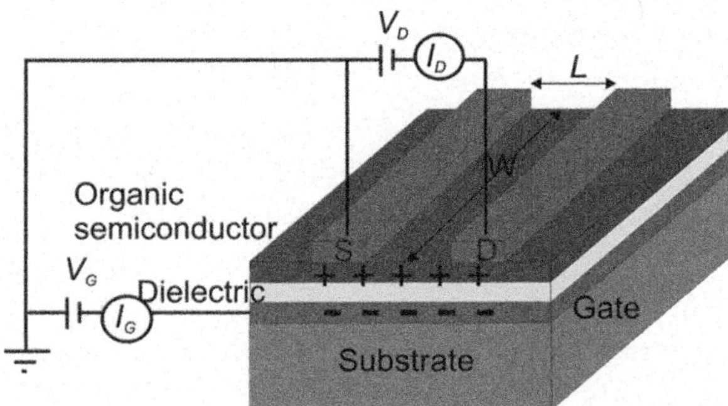

Fig. 9.1 Schematic of an OFET. The gate electrode is deposited on a substrate. The dielectric layer and p-type organic semiconducting film are deposited sequentially on top. The source S and drain D electrodes are photo lithographically patterned to define a channel of width W and length L. The gate and drain voltages are V_G and V_D respectively with corresponding notation for current I

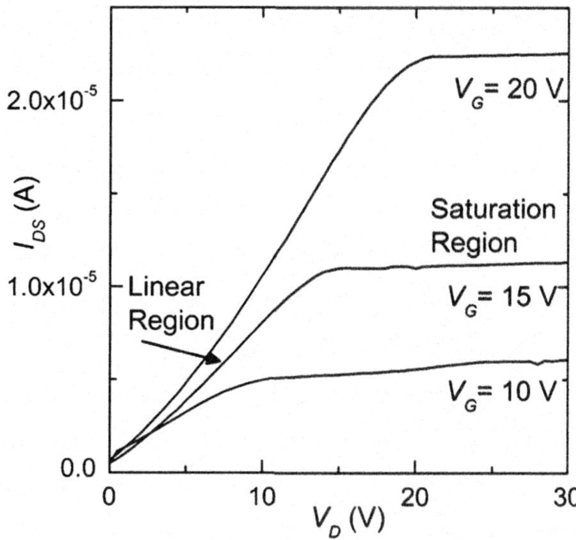

Fig. 9.2 Output characteristics of a typical OFET showing I_D versus V_D plotted for different values of V_G. At low voltages, there is an approximate linear change of I_D with V_D. The current saturates at higher voltages

The charge-carrier mobility μ_{FET} is a key parameter obtained from the characteristics. It can be determined from the linear region using the equation

$$I_D = \frac{WC}{L}\mu_{FET}\left[(V_G - V_T)\,V_D - {V_D}^2\right] \tag{9.1}$$

where C is the capacitance per unit area of the insulator. Alternatively, keeping V_D constant in the saturation region, μ_{FET} can be calculated from a plot of I_D versus $(V_G - V_T)^2$ using the equation

$$I_D = \frac{WC}{L}\mu_{FET}(V_G - V_T)^2 \tag{9.2}$$

It is desired to maximise the magnitude of μ_{FET} so that the operating voltage and power consumption of the transistor can be minimised. Another important parameter is the on/off ratio, equal to I_D (on)/I_D (off) which should be sufficiently large to define the states "0" and "1" in electronic circuits.

9.3 Polymer Liquid Crystals for OFETs

9.3.1 Lamellar Polymers with High Temperature Smectic Phases

Before 2005 the bench-mark prototype semiconducting polymer was regioregular poly(3-hexylthiophene), compound **1** (P3HT), which shows hole mobility values of up to 0.3 cm^2 V^{-1} s^{-1} [2, 6]. Its chemical structure is shown in Table 9.1. Polymer **1**

Table 9.1 Chemical structures and phase transition temperatures of some liquid crystalline polymers

1	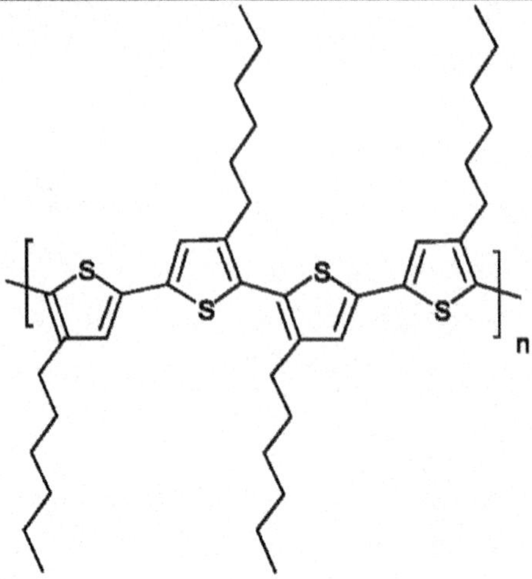 P3HT: $t_g \approx 67\ ^{\circ}C$; Cr-I ≈ 238 °C
2	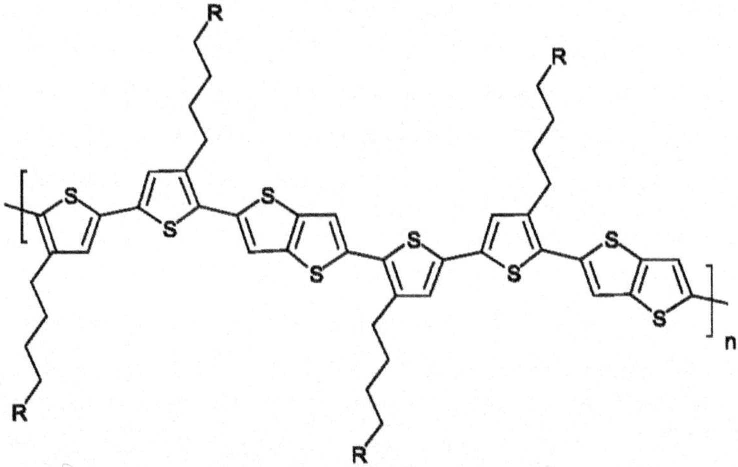 PBTTT: Cr-SmX ≈ 171 °C; SmX-I ≈ 251 °C [$R = C_6H_{13}$]
3	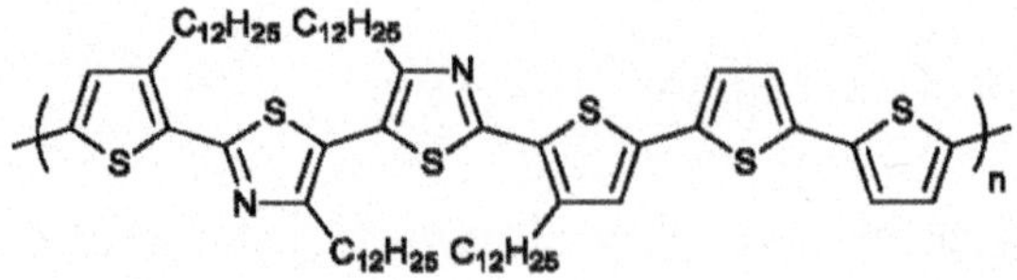 Sanditic polymer

(continued)

Table 9.1 (continued)

4	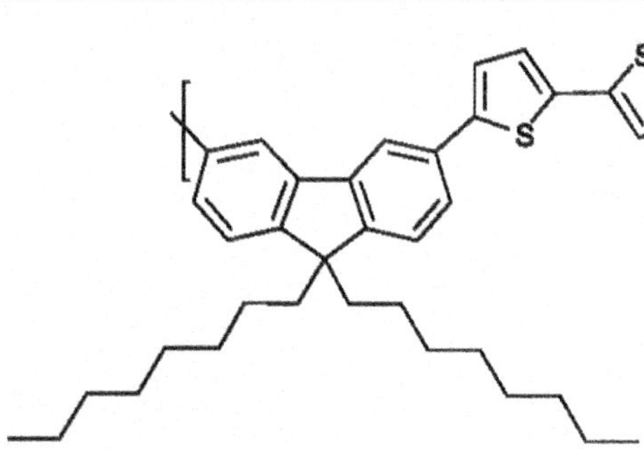 F8T2: Cr-N ≈ 265 °C; N-I > 300 °C
5	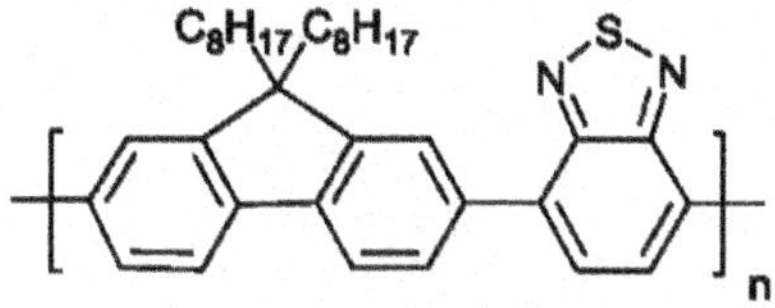 F8BT: t_g ≈ 125-135 °C; Cr-N ≈ 258-283 °C; N-I > 300 °C

has a microcrystalline, lamellar structure, as shown in Fig. 9.3, consisting of closely-spaced layers of stacked, highly conjugated, aromatic backbones separated by layers of solubilising and insulating side chains, which facilitate large π-π coupling between adjacent polymers in the x direction because of the small separation between them. The insulating side-chains inhibit charge transport in the y direction. In-plane carrier transport is enhanced when the polymers adopt an an edge-on orientation on the surface as shown in the Fig. 9.3 and charge transport between segments of the same conjugated polymer chain is fast. However, the observed values of the charge carrier mobility can vary by orders of magnitude depending on the molecular weight and microcrystalline morphology of the polymer [7–9]. This variation may be due to poor interconnectivity between microcrystalline domains.

An excellent demonstration of the benefits of the self-assembly properties of liquid crystalline phases lies in the recent development of liquid crystalline, semiconducting, thiophene polymers and copolymers, many of which contain thienothiophene groups [11–17], see polymer **2**, poly(2,5-bis(3-alkylthiophen-2-yl)thieno [3,2-b]thiophene) (pBTTT), for example, which exhibits exceptionally high field-effect values of mobility up to 1 $cm^2\ V^{-1}\ s^{-1}$. Such polymers possess a mesophase above room temperature associated with a lower side-chain density than polymer **1**, where each thiophene ring is substituted with an alkyl group. The lower density allows interdigitation, as illustrated in Fig. 9.4, and the mesophase corresponds to a highly-ordered smectic or saniditic phase of π-stacked back-bones separated by melted and interdigitated aliphatic, non-conducting side-chains

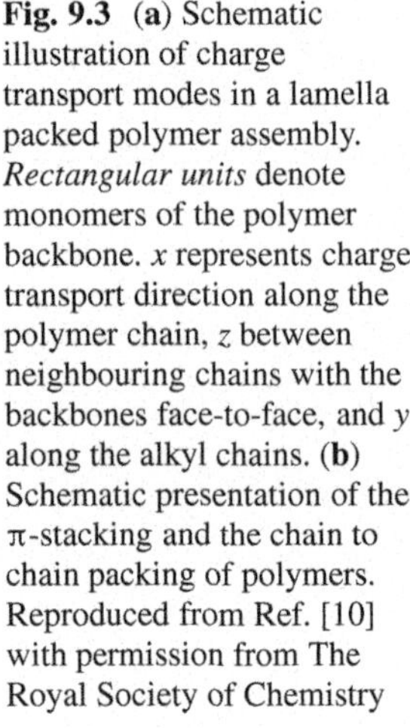

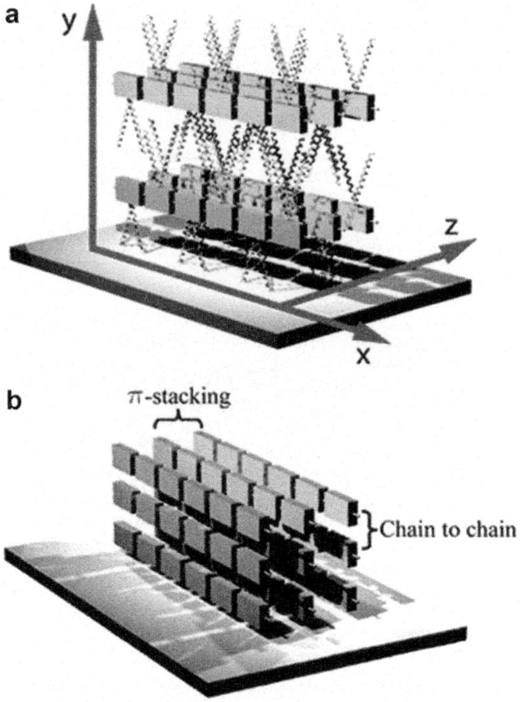

Fig. 9.3 (**a**) Schematic illustration of charge transport modes in a lamella packed polymer assembly. *Rectangular units* denote monomers of the polymer backbone. x represents charge transport direction along the polymer chain, z between neighbouring chains with the backbones face-to-face, and y along the alkyl chains. (**b**) Schematic presentation of the π-stacking and the chain to chain packing of polymers. Reproduced from Ref. [10] with permission from The Royal Society of Chemistry

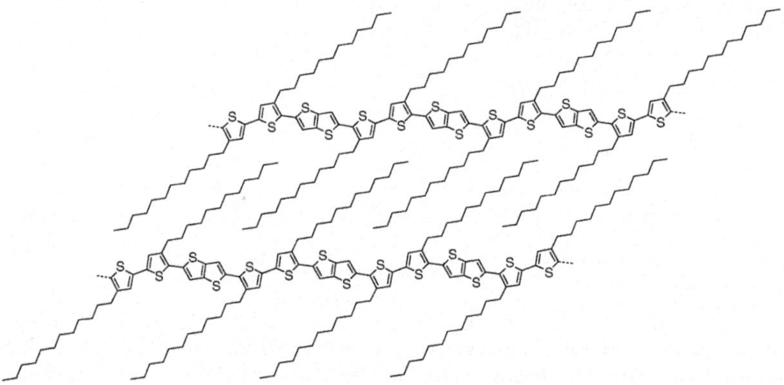

Fig. 9.4 Schematic of the packing of polymer chains of a saniditic liquid crystal with uniform side chain interdigitation

[18, 19]. The OFET performance improves on annealing above the mesophase transition temperature. Polymer **1** has no corresponding mechanism capable of improving the degree of order.

Figure 9.5 shows how the order of a saniditic liquid crystal polymer **3** changes with annealing. The top and middle three images show grazing incidence X-ray diffraction patterns and AFM images of the surfaces, respectively. The left-hand

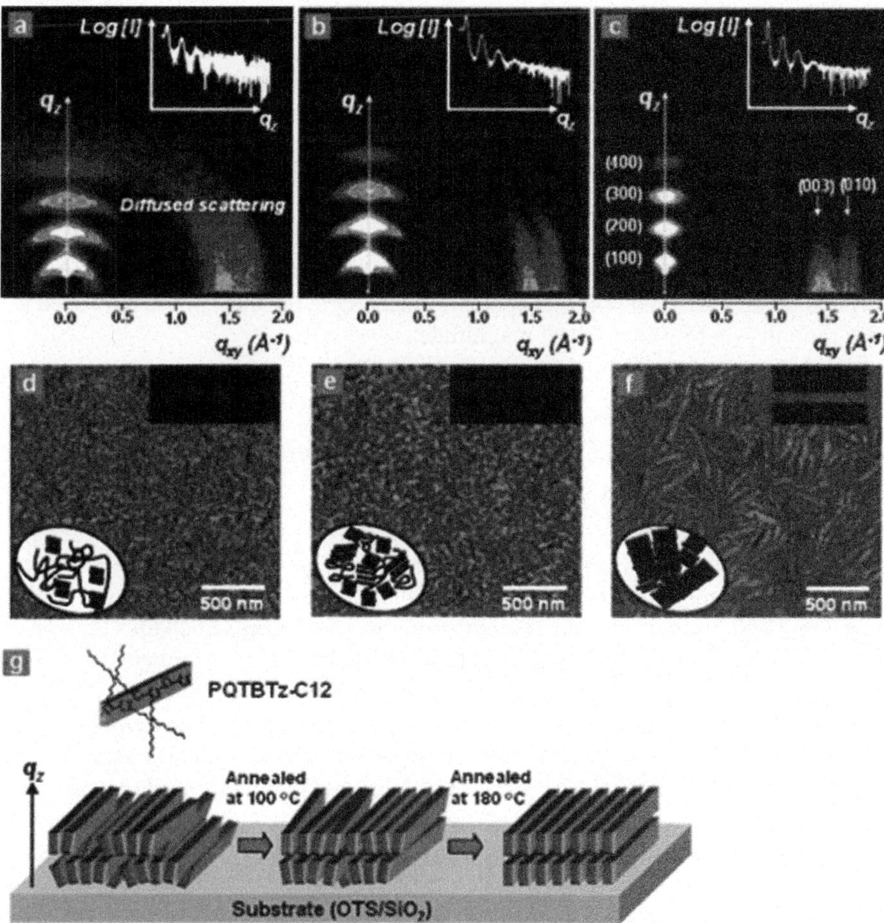

Fig. 9.5 Effects of thermal annealing on molecular orientation and morphological features of the saniditic liquid crystal 3. 2D grazing incidence X-ray diffraction patterns of the thin films spin-coated onto OTS-treated SiO_2 substrates for various annealing temperatures: (**a**) as-spun, (**b**) 100°C, (**c**) 180°C. The data are plotted versus the scattering vector q. (The *insets* in (**a**), (**b**), and (**c**) show the log-scale X-ray intensity profiles along the q_z axis, and the crystallographic assignments of the peaks are labelled.) AFM topographs of the thin films spin-coated onto the OTS-treated SiO_2 substrates for various annealing temperatures: (**d**) as-spun, (**e**) 100°C, (**f**) 180°C. The insets in (**d**), (**e**), and (**f**) show the polarized optical microscope images (*right-up*) and schematic features of self-organization of the polymer chains for various annealing temperatures (*left-down*). (**g**) Schematic representation of liquid-crystalline domains in thin films for various annealing temperatures, where q_z is the surface normal direction (Reprinted with permission from [17]. Copyright (2009) American Chemical Society)

images are taken from spin-cast films whilst the middle and right hand images correspond to films annealed at 100°C and 180°C, respectively, the former below the mesophase transition temperature. The bottom set of cartoons illustrate improved crystallisation on annealing. On annealing to the mesophase, the backbones can freely move and reorganise into a more-ordered state, whilst removing defects. Upon cooling, the side chains recrystallise to capture and maintain the high level of layer order. The (*h*00) set of spots in the X-ray images correspond to lamellar stacking due to the alkyl side-chains with a *d* spacing of 21 Å. The X-ray patterns in Fig. 9.5 provide evidence of improved order following annealing in the mesophase; in particular the narrow spots in the out-of-plane, q_z, direction showing perfect edge-on orientation. The corresponding AFM image, Figure 9.5, shows polycrystalline structures with rod-shape grains, which arise from the self-organization of the perpendicularly oriented polymer chains. Annealing in the mesophase improved the FET mobility of **3** by an order of magnitude.

9.3.2 Oriented OFETs Using Liquid Crystalline Semiconductors

As discussed in detail in Chaps. 2 and 5, electronic transport in organic semiconductors occurs by hopping between sites, which may be traps, single molecules, several repeat units of a polymer chain or even a number of delocalised chain segments in high-mobility conjugated mainchain polymers [20]. For highly conjugated polymers and extended oligomers, the hopping may be intramolecular, between different sites on the same chain or intermolecular between sites on neighbouring chains. The charge transport is anisotropic since the rates of intramolecular and intermolecular hopping are not equal. The first application of liquid crystals to OFETs exploited the anisotropic carrier transport properties of the liquid crystalline polyfluorene co-polymer **4** when uniformly oriented [21]. The polymer **4** exhibits a nematic phase above 265°C, as shown in Table 9.1. A top-gate, bottom-drain device configuration was used. The source and drain electrodes were deposited onto a rubbed polyimide alignment layer. The copolymer was then deposited and annealed in the liquid crystalline phase. The nematic ordering was maintained in a glassy state by rapid quenching to prevent recrystallisation and the dielectric layer and gate electrode were then sequentially deposited. The FET mobility was 2×10^{-2} cm^2 V^{-1} s^{-1} when the quenched glass was oriented along the transport direction between the source and drain. The mobility was an order of magnitude less for perpendicular orientation. This result shows that the rate of intra-chain transport of holes is very much greater than the hopping rate between chains and aligning the chains along the channel minimises the number of inter-chain hops. Similar results were found for a uniformly oriented oligofluorene and polyfluorene [22].

The lamellar liquid crystalline polymer **2** has a high-temperature phase transition at about 251°C. Heating above this phase transition produces films with a smectic-like, chain-extended conformation in which the polymer chains assemble into crystalline nano-ribbons with a characteristic width of 70–80 nm, consistent with the length of the fully extended polymer chain backbone. These ribbons, separated by grain boundaries, extend over several micrometers in the direction of π–π stacking

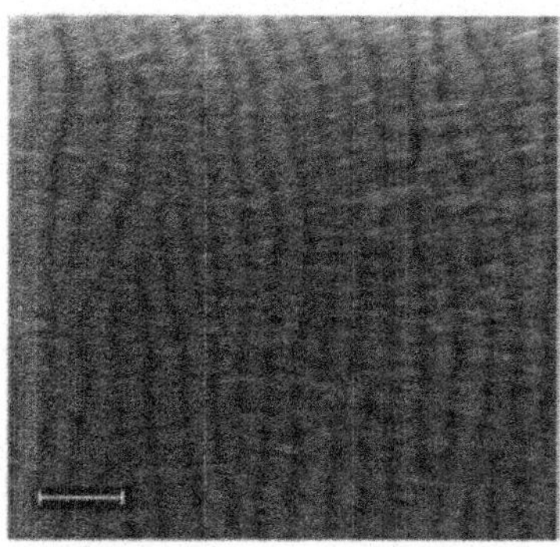

Fig. 9.6 SEM image of a uniaxially aligned thin film of 2 produced by zone casting. The scale denotes a length of 200 nm. The extended polymer chains are aligned in the *horizontal directions* with smectic-like organisation (Reproduced with pernmission from [24] Copyright @ 2011 WILEY-VCH)

[23]. A zone casting method, followed by annealing at 275°C, was used to uniaxially align the nano-ribbons, producing films with structures shown in Fig. 9.6 [24]. This involved dispensing a solution of the polymer through a linear-shaped nozzle onto a linearly translating substrate, whose speed was carefully controlled, as was the rate at which the solution was dispensed by the plunger. The temperatures of the substrate and the solution were also optimised to control the drying speed, in order to prevent nucleation in solution and to induce uniaxial alignment of the polymer near the drying front.

The mobility of the aligned polymer was anisotropic having a value of $0.13\ cm^2\ V^{-1}\ s^{-1}$ parallel to the polymer chains, and a factor of five less in the perpendicular orientation. The mobility along the chain direction does not reflect directly the fast, intrachain, charge transport along the polymer backbone, but is limited by disordered grain boundaries between the chain-extended ribbons occurring on a 60–80 nm length scale. Along the ribbons, a close inspection of Fig. 9.6 shows that domain boundaries are also encountered on a 20–30 nm length scale. In these grain boundaries charges encounter a relatively favourable parallel chain orientation. The measured mobility anisotropy of about 5 is close to the ratio of grain sizes encountered in the two transport directions.

9.3.3 *Polymer Light-Emitting Field Effect Transistors*

Polymers, such as **4**, with a nematic phase generally exhibit mobility mobility values typically at least an order of magnitude less than that of the sanidic materials discussed in Sect. 9.3.1. However, as discussed in more detail in Chaps. 2 and 5, the former class of materials are often efficient light-emitters and, as such, they are promising candidates for ambipolar, light-emitting, field-effect transistors (LEFETs). These devices can form electron and hole accumulation layers simultaneously within the channel between the source and drain electrodes, as illustrated in Fig. 9.7. Charge recombination occurs at the interface between the two regions and thus light emission is observed. LEFETs are currently the subject

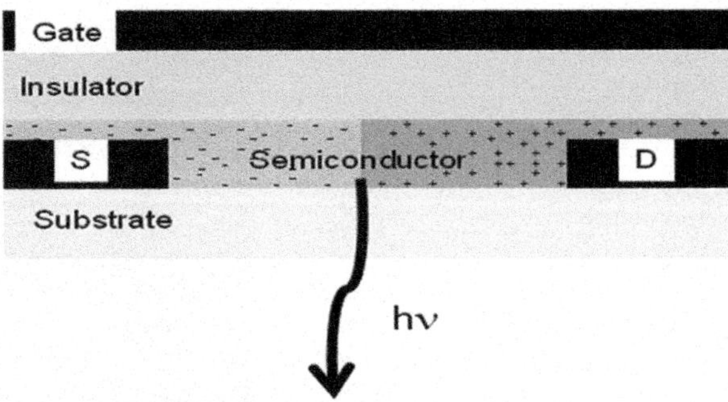

Fig. 9.7 Schematic of the structure of a top gate LEFET. Under appropriate driving conditions, the source and drain electrodes inject charge carriers of different sign so that both electrons and holes accumulate in the channel. Carrier recombination and light emission occurs at the intersection between the accumulation regions (Reproduced with pernmission from [29] Copyright @ 2011 WILEY-VCH)

of a lot of interest because they are compatible with waveguide transmission of the emitted light for optoelectronic integrated circuits. High current densities can be obtained, which generate an attractive architecture for the potential realization of electrically-pumped organic lasers. The liquid crystalline poly(9,9-dioctylfluorene-*alt*-benzothiadiazole) **5** (F8BT), possesses HOMO and LUMO energies with similar barriers for electron- and hole-injection from gold electrodes and so has been widely used in LEFETs [25–27]. In most cases the transistors' characteristics are optimised following annealing from the liquid crystaline phase, although the resulting film is polycrystalline at room temperature [27]. However, high-temperature annealing results in high surface roughness, which increases the losses of the optical wave-guide. Modification of the gold electrodes with a 1-decanethiol, self-assembled monolayer (SAM) significantly improves performance of the low-temperature (120°C) devices, allowing efficient charge injection into F8BT films deposited on top [25]. The emitted light from a LEFET based on polymer **5** was coupled efficiently into the resonant mode of a passive Ta_2O_5 waveguide by placing the recombination zone of the LEFET directly above the waveguide ridge [28]. The waveguide provides distributed feedback by means of a grating and optically pumped lasing is obtained. Interestingly, low temperature annealing was used to give an amorphous, rather than a polycrystalline, film and so minimise light scattering. A LEFET was constructed with polymer **5** uniaxially-aligned in a monodomain [26]. The location of the emission along the channel between the source and drain is dependent on the applied voltage between them. The width of the emission zone is 5–10 times greater than that when the polymer is aligned parallel to the current than for perpendicular alignment, implying a more than 100 times smaller recombination rate constant than expected for Langevin recombination. This result suggests that anisotropic mobility affects the probability of exciton formation, decreasing it parallel to the aligned chains and increasing it in the perpendicular direction.

9.4 Columnar Liquid Crystals in OFETs

Discotic liquid crystals are oriented in columns separated by molten aliphatic chains and consequently they can conduct charge efficiently along the columns in one dimension. The organization of the different phases is described in Chaps. 1 and 2 as well as elsewhere [30, 31] and the efficiency of charge transport can be directly related to the short intermolecular spacing and order of different types of mesophase, as discussed in detail in Chaps. 2 and 3. Some mesomorphic derivatives of hexabenzocoronene, possess values of hole mobility above 0.2 $cm^2\ V^{-1}\ s^{-1}$ in the discotic phase at room temperature, while others exhibit values above 1 $cm^2\ V^{-1}\ s^{-1}$ in a crystalline phase at elevated temperatures [32]. However, these measurements give the value of the local mobility only and the reported value of the charge mobility is much lower ($\leq$0.02 $cm^2\ V^{-1}\ s^{-1}$) in actual OFET devices [33, 34]. This behaviour is mainly due to the presence of poor contacts as well as the difficulty in obtaining uniform macroscopic alignment and in optimising order in solution-processed devices. In-plane transport is required between the source and drain electrodes of OFETS, so that the columns must have an edge-on, or homogeneous configuration, as illustrated in Fig. 9.8. Zone casting was used to provide a homogeneously aligned film of the hexabenzocoronene **6**, whose structure is shown in Table 9.2, giving one of the best performing discotic columnar OFET devices with

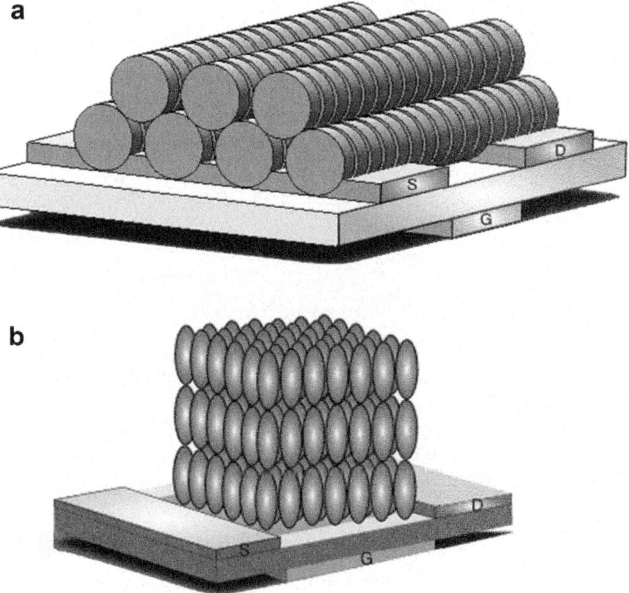

Fig. 9.8 Schematic illustration of optimum orientation of (**a**) columnar and (**b**) smectic liquid crystals for OFET operation (Reproduced with pernmission from [29] Copyright @ 2011 WILEY-VCH)

Table 9.2 Chemical structures of compounds with columnar phases

6

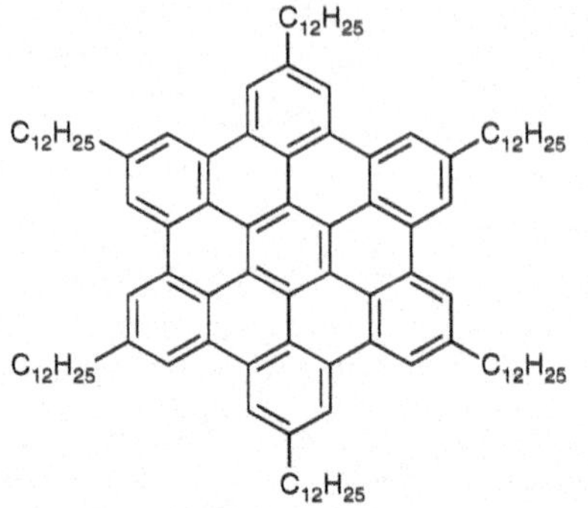

$C_{12}H_{25}$ $C_{12}H_{25}$ $C_{12}H_{25}$ $C_{12}H_{25}$ $C_{12}H_{25}$ $C_{12}H_{25}$

HBC-C12: Tg-Col_l = 42°C; Col_l- I = 107°C

7

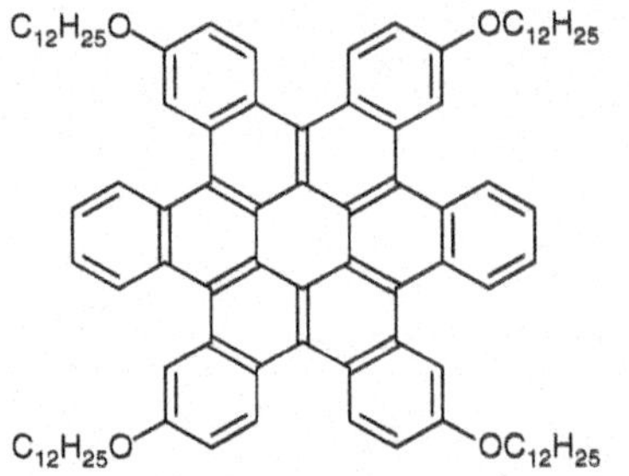

$C_{12}H_{25}O$ $OC_{12}H_{25}$ $C_{12}H_{25}O$ $OC_{12}H_{25}$

HBC-C_{O12}: Cr-Col_l = 91°C; Col_l- I = 285°C

8

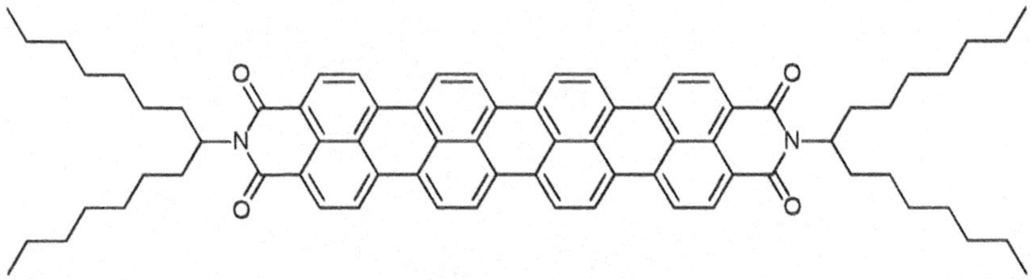

O N O O N O

SWQDI: Col_P- Col_{ho} = 188°C; Col_{ho}- I > 500°C

a hole mobility of 0.01 $cm^2\ V^{-1}\ s^{-1}$ [33, 35]. Electron and X-ray diffraction confirm that the zone-cast layer is crystalline with an edge-on orientation to the substrate. Vertically adjacent layers are tilted at ±45° to the substrate with a herringbone arrangement. The discotic hexabenzocoronene **7** has a non-planar core, which arises as a result of steric congestion attributable to its peripheral phenyl groups [34]. Homogeneous alignment was realised by spin-casting the material onto a SAM-modified SiO_2 insulator and OFETs with a mobility of 0.02 $cm^2\ V^{-1}\ s^{-1}$were obtained. It is suggested that non-planarity may improve the mobility, because the π-surfaces of contorted molecules can approach each other and arrange themselves in very different ways. Ambipolar, organic field-effect transistors (OFETs) are of special interest owing to their application in complementary-like circuits or light-emitting field-effect transistors. Swallow-tailed, quarterrylene-tetracarboxdiimide (SWQDI) compounds with discotic columnar superstructures, see **8**, were utilised as an active layer in ambipolar thin film transistors [36]. The work function of gold lies approximately in the middle of the band-gap of **8**. An OFET with a drop-cast active layer has a roughly equal magnitude of electron and hole mobility of $\approx 1 \times 10^{-3}$ cm^2 V^{-1} s^{-1}. Interestingly, hole transport disappears and electron mobility is reduced on annealing to a more ordered film.

9.5 Low-Mass Smectic Liquid Crystals in OFETs

Low-molar-mass, smectic liquid crystals self-organise in layers with order in two dimensions. Silane-based surfactants are generally used to give the homeotropic alignment required to provide in-plane transport between the electrodes of OFETs, see Fig. 9.8. There are a number of smectic phases as illustrated in Chaps. 1 and 2 as well as elsewhere [30, 31]. In the smectic A and C phases the packing in the layers is fluid-like, while the more-ordered, crystalline smectics, with hexatic, hexagonal or herring-bone packing, facilitate greater π-π overlap, as discussed in more detail in Chaps. 1 and 2 as well as elsewhere. Many smectic liquid crystals show ambipolar transport [37]. Some high-performance, thin-film transistors with a mobility of up to 0.4 cm^2 V^{-1} s^{-1} have active layers of thermally-evaporated thiophene-phenylene oligomers, for example compound **9**, whose chemical structure is shown in Table 9.3, with smectic phases at high temperatures [38]. However, the liquid crystalline properties are not directly utilised in devices. Crystal grain boundaries can be avoided by annealing and reorganization in the fluid smectic phase, followed by slow cooling through more-ordered smectic phases by the process of paramorphosis [39, 40]. Hence, a monodomain crystalline sample of compound **10** was prepared by solution processing and annealing on a rubbed polyimide layer, as shown in Fig. 9.9. This contrasts with the leaf-like multi-domain crystals that are produced by spin-casting onto an untreated substrate. An OFET mobility of 0.02 cm^2 V^{-1} s^{-1} was obtained for the monodomain sample, which is an order of magnitude larger than that found using multi-domain samples of the same material [39]. The mobility is anisotropic, possibly because the tilting of the molecules in the layers reduces the π-π overlap in one direction. Paramorphosis does not occur, if the crystalline phase has a significantly different molecular arrangement to that of the mesophases. Interestingly, a thiophene-naphthalene-based mesogen, compound **11**, retains its layered structure in a polycrystalline phase at room temperature in a thermally deposited OFET, giving a high carrier mobility of 0.14 cm^2 V^{-1} s^{-1} [41]. However, its mobility decreases following annealing in the smectic phase at 100°C, possibly because of cracking of the organic layer caused by differential degrees of expansion of the layer and the substrate. A mesogen **12**, with a room temperature smectic phase, showed a time-of-flight carrier mobility of 0.07 and 0.2 cm^2 V^{-1} s^{-1} for holes and electrons respectively [42]. A solution-processed

Table 9.3 Chemical structures of compounds with smectic phases

9

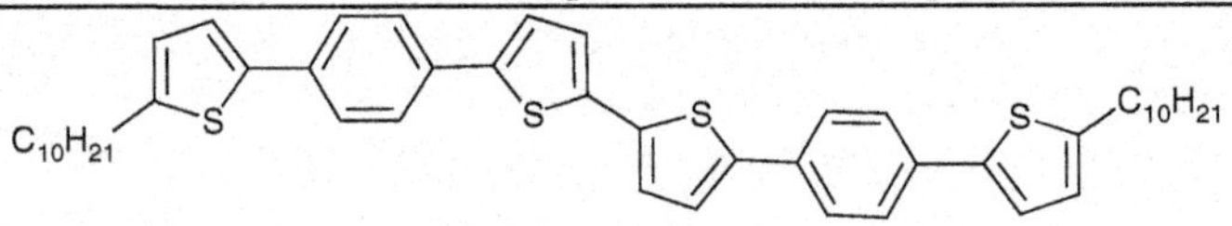

Dec-(TPhT)$_2$-Dec: Cr-SmX = 316°C; SmX-I = 351°C

Dec-(TPhT)$_2$-Dec: Cr-SmX = 316°C; SmX-I = 351°C

10

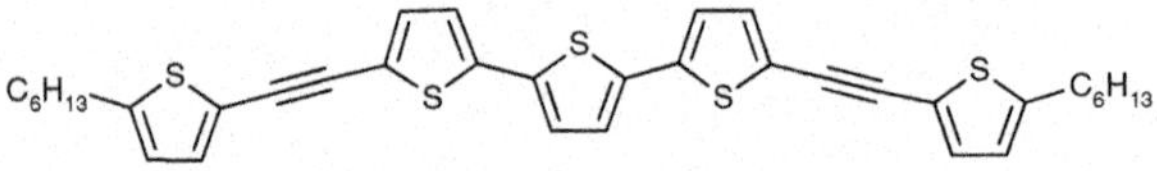

Cr-SmC = 92°C; SmC-SmA = 111°C; SmA-N = 115°C; N-I = 180°C

11

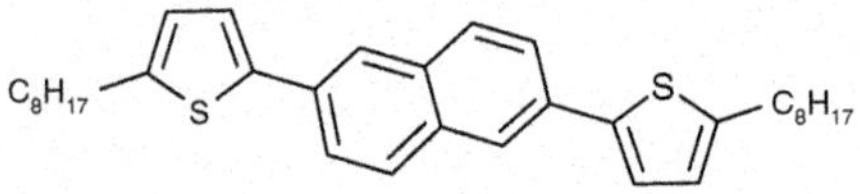

8-TNAT-8: Cr-SmX = 95°C; SmX-I = 182°C

12

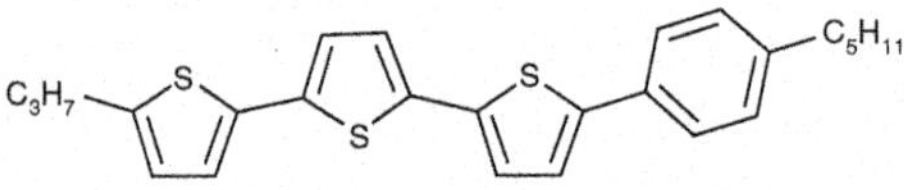

Cr-SmX < –50°C; SmX–I = 57°C

13

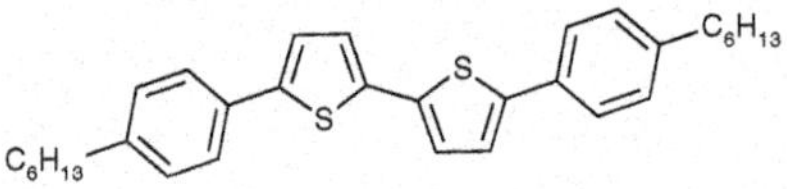

DH-PTTP: Cr-SmC 232°C; SmF-SmB 139°C; SmB-SmC = 150°C; SmC-SmA = 235°C; SmA-I = 239°C

14

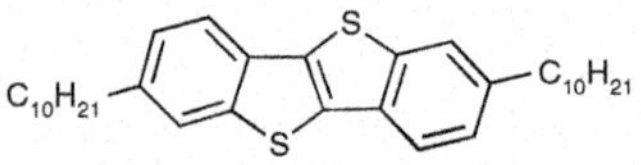

C_{10}-BTBT: Cr-SmC 232°C; SmF-SmB 139°C; SmB-SmC = 150°C; SmC-SmA = 235°C; SmA-I = 239°C

15

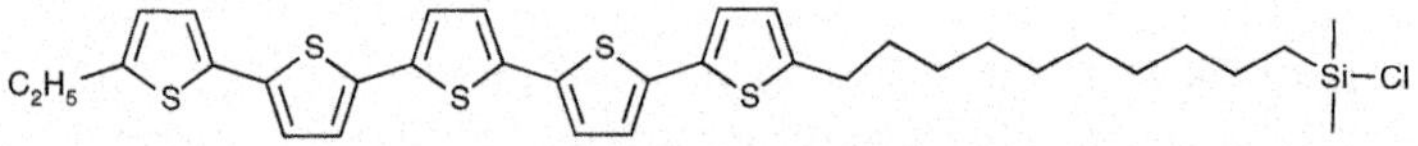

8-TTP-8: Cr-SmC 232°C; SmG-SmF 139°C; SmF-SmC = 150°C; SmC-I = 235°C

(continued)

Table 9.3 (continued)

16	$C_8H_{17}O$ ... OC_8H_{17} Doo-P2TP
17	C_2H_5 ... Si–Cl
18	Cr-SmG = 75°C; SmG-I = 174°C

Fig. 9.9 Light transmission micrographs of two films of compound **10** (polarizer in the *horizontal*, analyzer in the *vertical* direction): (**a**) untreated, as-spun film; (**b**) monodomain with the rubbing direction of the PI layer parallel to the analyzer; (**c**) rubbing direction at 22.5° with respect to the analyzer; (**d**) rubbing direction at an angle of 45° with respect to the analyzer (Reprinted with permission from [39]. Copyright (2006) American Chemical Society)

film of compound **12** exhibited hole mobility values of up to 0.04 $cm^2 V^{-1} s^{-1}$ in an OFET, when well aligned by a silane-based, self-assembled monolayer after annealing [43, 44]. A similar result was obtained on a 3 % strained polyimide substrate, which demonstrates the suitabilty of smectic compounds for flexible plastic electronics [45]. Bottom-contact OFETs with a mobility value of up to 0.03 $cm^2 V^{-1} s^{-1}$ have been fabricated using a very simple process involving neither solvents nor vacuum-deposition. A liquid-crystalline organic semiconductor, **13**, was melted directly onto a SiO_2 dielectric layer treated with a surfactant to promote homeotropic alignment. A surfactant-treated cover-slide was placed on top of the powder before melting to prevent dewetting. The melting and cooling conditions were optimised to maximise charge mobility, which was significantly larger than that obtained for thermally evaporated films.

Recently some solution processed smectic materials have shown extremely high values of OFET mobility. Compound **14** has a benzothienobenzothiophene core and exhibits a smectic A (SmA) phase in the temperature range from 95 to 122°C. It has a mobility up to 0.86 $cm^2 V^{-1} s^{-1}$ when spin-coated at room temperature and annealed at 80°C. Similarly processed compounds with the same backbone but with alkyl chains of different lengths showed mobilities as high as 2.75 $cm^2 V^{-1} s^{-1}$ [46], the mobility tending to increase with alkyl chain length. The extremely high mobility is attributed to excellent molecular overlap resulting from the herringbone monoclinic crystal structure, which enhances the sulfur–sulfur interaction, and the hydrophobic interactions effected by long alkyl chains [47]. For films deposited by solution processing at room temperature, recrystallization takes place on the substrate during solvent evaporation and the inhomogeneity of the films is not fully improved by post-thermal annealing. The wetting and homogeneity of thin films of **14** was substantially improved by spin-coating when in the liquid crystalline phase, resulting in extremely high mobilities up to 3 $cm^2 V^{-1} s^{-1}$ [48]. Similarly the mobility of the terthiophene oligomer **15** increased by over an order of magnitude to 0.1 $cm^2 V^{-1} s^{-1}$by spin-casting in the smectic phase [48].

Photoalignment has been used to homogeneously align the phenylene-thiophene compound **16**, which possesses a high-temperature smectic phase [49, 50]. The magnitude of the mobility was enhanced, when the direction of electronic transport was parallel to the direction of the intermolecular hops (perpendicular to the molecular rods) [49]. An OFET hole mobility of 0.05 $cm^2 V^{-1} s^{-1}$ was obtained, compared to a value of 0.003 $cm^2 V^{-1} s^{-1}$ measured in the orthogonal direction. This is the opposite result to that obtained for the mainchain polymer, discussed in Sect. 9.3.2 above. The result for **16** is not surprising since intramolecular hopping, which is faster than intermolecular hopping in polymers, is a meaningless concept for so short a molecular core and intermolecular transport is faster within a smectic layer rather than between layers.

Self-assembled monolayer FETs (SAMFETs), where the semiconductor is a monolayer spontaneously formed by self-assembly on the dielectric layer, is a promising technology for the mass production of organic electronics. However, the presence of defects limit charge transport, so that devices with channels of only submicron length work well. SAMFETs with long-range intermolecular π-π coupling in the monolayer were obtained by dense packing with liquid-crystalline compounds, for example compound **17** [51]. These materials consist of a π-conjugated mesogenic core separated by a long aliphatic chain from a monofunctionalised anchor group. The resulting SAMFETs exhibit a bulk-like carrier mobility up to 0.04 $cm^2\ V^{-1}\ s^{-1}$, large current modulation and high reproducibility.

OFETs require organic semiconductors with a high glass transition temperature for long-term stability. An alternative way to ensure stability makes use of the *in-situ* crosslinking of a thin film of reactive mesogens, which are low-molar-mass liquid crystals with at least two polymerisable groups often separated from the molecular core of the liquid crystal by flexible aliphatic spacer units [52–56]. Reactive mesogens often exhibit a low viscosity and can be designed to exhibit mesophases at low temperatures, so they can be easily ordered or macroscopically aligned in thin films at or near room temperature. The order is then permanently fixed by cross-linking of adjacent molecules either by irradiation with ultraviolet light or thermal polymerisation. The use of reactive mesogens in OLEDs and photovoltaics is discussed in Chaps. 7 and 8, respectively. Chapter 5 discusses charge transport in these materials. Compound **18** is an example of a photopolymerisable smectic liquid crystal with non-conjugated diene polymerisable terminal groups decoupled by aliphatic chains from the semiconducting aromatic core. Thermal or photopolymerisation links the reactive groups together. In this case each molecule has two reactive groups so a highly crosslinked, insoluble and intractable polymer network of mesogenic units is obtained. Different types of polymerisable groups, such as methacrylate, acrylate, diene and oxetane reactive groups, have been used. Scheme 9.1 shows the general structures of the polymer backbones formed on crosslinking. Photopolymerisation offers the further possibility of pixellation by photolithography. Unexposed and, therefore, non-crosslinked, regions are still soluble in organic solvents and are simply removed by washing in the original spin-casting solvent. OFETs fabricated from both crosslinked and uncrosslinked films of smectic reactive mesogens with reactive end-groups, such as **18**, have been studied [57–59]. Relatively low values of mobility are found, $<0.005\ cm^2\ V^{-1}\ s^{-1}$, possibly because of disorder induced by the long terminal aliphatic chains. On crosslinking the mobility decreases for most of the compounds studied by at least a factor of five, possibly because the layer structure of the smectic mesogens is disrupted, when the end groups of the molecules form a covalently bonded polymer backbone. However, this result may not be general as time-of-flight mobility values, $>10^{-2}\ cm^2\ V^{-1}\ s^{-1}$, were found to be similar for smectic reactive mesogens with oxetane reactive end-groups before and after polymerisation [60]. Crosslinking was also shown not to degrade charge transport in an OFET based on star-shaped molecules and in a nematic reactive mesogen with diene photoreactive endgroups studied with time-of-flight measurements [61, 62].

Scheme 9.1 Polymerisable groups, acryalte, methacrylate, diene and oxetane groups from top to bottom, and the corresponding polymer backbones formed on polymerisation of these monomers

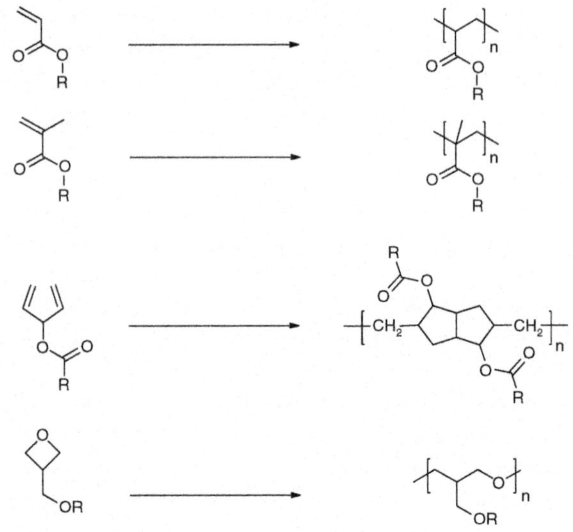

9.6 Conclusions

There have been many major advances in liquid crystalline organic semiconductors in recent years including the ability of saniditic liquid crystals to allow very high values of the charge carrier mobility in polymer-based OFETs with state-of-the art performance [63]. The high values of the mobility results from re-ordering of both the aliphatic layers and the conjugated polymer backbone into close-packed structures. The side-chains can interdigitate between vertically adjacent lamellae and facilitate thereby the formation of large domains with a high degree of molecular order. The use of more-disordered, liquid-crystalline, semiconducting polymers in LEFETs is a promising development, although more investigation is needed on why recombination is inhibited in aligned samples. Although the time-of-flight charge carrier mobility of columnar and smectic materials can have very high values, $>0.1\ cm^2\ V^{-1}\ s^{-1}$, the OFET mobility values are mostly significantly lower particularly in solution-processed devices. This is a critical issue, since solution processing is compatible with high-volume, large-scale manufacturing. An exception is the recent work from the Hiroshima and Hanna groups who obtained state-of-the-art mobilities $>1\ cm^2\ V^{-1}\ s^{-1}$ from materials having the benzothienobenzothiophene core and high temperature smectic phases. The mobility and film quality improved by spin-casting in the smectic phase. Thus, sample processing has emerged as a critical factor, which influences electronic and optoelectronic device properties. Zone and drop casting are alternative solution-based methods to produce highly ordered thin films. An extremely high FET mobility was obtained from a dip-cast sample of a semiconducting polymer, which is not liquid crystalline, following annealing [64]. A similar approach would allow the anisotropic properties of liquid crystalline polymers to be best exploited. Although initial efforts to obtain monodomain OFETs are promising, more work is needed to further exploit this important capability of liquid crystals, especially as defects are known to limit charge transport in devices. Very often the liquid crystalline phases of mesomorphic organic semiconductors exist at high temperatures and annealing at these temperatures may well not be compatible with efficient and cost-effective manufacturing processes. Alternatively, solvent vapour annealing may be carried out at lower temperatures using solvents with high boiling points as plasticisers [65].

In conclusion, the application of self-organised materials to organic electronics is maturing well. Some liquid crystalline organic semiconductors are used in devices with state-of-the-art performance, whilst others, though not optimized, give novel device properties or alternative processing opportunities. As the research fields of organic electronics become more sophisticated, more complex devices will be required, so that the ability to pattern and fix self-assembled, liquid-crystalline organic semiconductors will be become increasingly important.

References

1. Coropceanu, V., Cornil, J., da Silva Filho, D.A., Olivier, Y., Silbey, R., Brédas, J.L.: Charge transport in organic semiconductors. Chem. Rev. **107**(4), 926–952 (2007)
2. Sirringhaus, H.: Device physics of solution-processed organic field-effect transistors. Adv. Mater. **17**(20), 2411–2425 (2005). doi:10.1002/adma.200501152
3. Zaumseil, J., Sirringhaus, H.: Electron and ambipolar transport in organic field-effect transistors. Chem. Rev. **107**(4), 1296–1323 (2007)
4. Allard, S., Forster, M., Souharce, B., Thiem, H., Scherf, U.: Organic semiconductors for solution-processable field-effect transistors (OFETs). Angew. Chem. Int. Ed. **47**(22), 4070–4098 (2008)
5. Newman, C.R., Frisbie, C.D., da Silva, D.A., Bredas, J.L., Ewbank, P.C., Mann, K.R.: Introduction to organic thin film transistors and design of n-channel organic semiconductors. Chem. Mater. **16**(23), 4436–4451 (2004). doi:10.1021/cm049391x
6. Sirringhaus, H., Brown, P.J., Friend, R.H., Nielsen, M.M., Bechgaard, K., Langeveld-Voss, B.M.W., Spiering, A.J.H., Janssen, R.A.J., Meijer, E.W., Herwig, P., De Leeuw, D.M.: Two-dimensional charge transport in self-organized, high-mobility conjugated polymers. Nature **401**(6754), 685–688 (1999)
7. Chang, J.F., Clark, J., Zhao, N., Sirringhaus, H., Breiby, D.W., Andreasen, J.W., Nielsen, M.M., Giles, M., Heeney, M., McCulloch, I.: Molecular-weight dependence of interchain polaron delocalization and exciton bandwidth in high-mobility conjugated polymers. Phys. Rev. B Conden. Matter Mater. Phys. **74**(11) (2006)
8. Kline, R.J., McGehee, M.D.: Morphology and charge transport in conjugated polymers. Polym. Rev. **46**(1), 27–45 (2006)
9. Kline, R.J., McGehee, M.D., Toney, M.F.: Highly oriented crystals at the buried interface in polythiophene thin-film transistors. Nat. Mater. **5**(3), 222–228 (2006)
10. Tsao, H.N., Mullen, K.: Improving polymer transistor performance via morphology control. Chem. Soc. Rev. **39**(7), 2372–2386 (2010)
11. Ong, B.S., Wu, Y., Liu, P., Gardner, S.: High-performance semiconducting polythiophenes for organic thin-film transistors. J. Am. Chem. Soc. **126**(11), 3378–3379 (2004)
12. McCulloch, I., Heeney, M., Bailey, C., Genevicius, K., MacDonald, I., Shkunov, M., Sparrowe, D., Tierney, S., Wagner, R., Zhang, W., Chabinyc, M.L., Kline, R.J., McGehee, M.D., Toney, M.F.: Liquid-crystalline semiconducting polymers with high charge-carrier mobility. Nat. Mater. **5**(4), 328–333 (2006)

13. McCulloch, I., Heeney, M., Chabinyc, M.L., Delongchamp, D., Kline, R.J., Cölle, M., Duffy, W., Fischer, D., Gundlach, D., Hamadani, B., Hamilton, R., Richter, L., Salleo, A., Shkunov, M., Sparrowe, D., Tierney, S., Zhang, W.: Semiconducting thienothiophene copolymers: design, synthesis, morphology, and performance in thin-film organic transistors. Adv. Mater. **21**(10–11), 1091–1109 (2009)
14. Heeney, M., Bailey, C., Genevicius, K., Shkunov, M., Sparrowe, D., Tierney, S., McCulloch, I.: Stable polythiophene semiconductors incorporating thieno[2,3-6]thiophene. J. Am. Chem. Soc. **127**(4), 1078–1079 (2005)
15. Hamadani, B.H., Gundlach, D.J., McCulloch, I., Heeney, M.: Undoped polythiophene field-effect transistors with mobility of 1 cm^2 V^{-1} s^{-1}. Appl. Phys. Lett. **91**(24), P243512 (2007)
16. Li, Y., Wu, Y., Liu, P., Birau, M., Pan, H., Ong, B.S.: Poly(2,5-bis(2-thienyl)-3,6-dialkylthieno[3,2-b]thiophene)s-high-mobility semiconductors for thin-film transistors. Adv. Mater. **18**(22), 3029–3032 (2006)
17. Kim, D.H., Lee, B.L., Moon, H., Kang, H.M., Jeong, E.J., Park, J.I., Han, K.M., Lee, S., Yoo, B.W., Koo, B.W., Kim, J.Y., Lee, W.H., Cho, K., Becerril, H.A., Bao, Z.: Liquid-crystalline semiconducting copolymers with intramolecular donor-acceptor building blocks for high-stability polymer transistors. J. Am. Chem. Soc. **131**(17), 6124–6132 (2009)
18. Delongchamp, D.M., Kline, R.J., Jung, Y., Lin, E.K., Fischer, D.A., Gundlach, D.J., Cotts, S.K., Moad, A.J., Richter, L.J., Toney, M.F., Heeney, M., McCulloch, I.: Molecular basis of mesophase ordering in a thiophene-based copolymer. Macromolecules **41**(15), 5709–5715 (2008)
19. DeLongchamp, D.M., Kline, R.J., Lin, E.K., Fischer, D.A., Richter, L.J., Lucas, L.A., Heeney, M., McCulloch, I., Northrup, J.E.: High carrier mobility polythiophene thin films: structure determination by experiment and theory. Adv. Mater. **19**(6), 833–837 (2007)
20. Chang, J.F., Sirringhaus, H., Giles, M., Heeney, M., McCulloch, I.: Relative importance of polaron activation and disorder on charge transport in high-mobility conjugated polymer field-effect transistors. Phys. Rev. B Conden. Matter Mater. Phys. **76**(20), 205204 (2007)
21. Sirringhaus, H., Wilson, R.J., Friend, R.H., Inbasekaran, M., Wu, W., Woo, E.P., Grell, M., Bradley, D.D.C.: Mobility enhancement in conjugated polymer field-effect transistors through chain alignment in a liquid-crystalline phase. Appl. Phys. Lett. **77**(3), 406–408 (2000)
22. Yasuda, T., Fujita, K., Tsutsui, T., Geng, Y., Culligan, S.W., Chen, S.H.: Carrier transport properties of monodisperse glassy-nematic oligofluorenes in organic field-effect transistors. Chem. Mater. **17**(2), 264–268 (2005)
23. Delongchamp, D.M., Kline, R.J., Jung, Y., Germack, D.S., Lin, E.K., Moad, A.J., Richter, L.J., Toney, M.F., Heeney, M., McCulloch, I.: Controlling the orientation of terraced nanoscale "ribbons" of a poly(thiophene) semiconductor. ACS Nano **3**(4), 780–787 (2009)
24. Lee, M.J., Gupta, D., Zhao, N., Heeney, M., McCulloch, I., Sirringhaus, H.: Anisotropy of charge transport in a uniaxially aligned and chain-extended, high-mobility, conjugated polymer semiconductor. Adv. Funct. Mater. **21**(5), 932–940 (2011)
25. Gwinner, M.C., Khodabakhsh, S., Giessen, H., Sirringhaus, H.: Simultaneous optimization of light gain and charge transport in ambipolar light-emitting polymer field-effect transistors. Chem. Mater. **21**(19), 4425–4433 (2009)
26. Zaumseil, J., Groves, C., Winfield, J.M., Greenham, N.C., Sirringhaus, H.: Electron-hole recombination in uniaxially aligned semiconducting polymers. Adv. Funct. Mater. **18**(22), 3630–3637 (2008)
27. Zaumseil, J., Donley, C.L., Kim, J.S., Friend, R.H., Sirringhaus, H.: Efficient top-gate, ambipolar, light-emitting field-effect transistors based on a green-light-emitting polyfluorene. Adv. Mater. **18**(20), 2708–2712 (2006)
28. Gwinner, M.C., Khodabakhsh, S., Song, M.H., Schweizer, H., Giessen, H., Sirringhaus, H.: Integration of a rib waveguide distributed feedback structure into a light-emitting polymer field-effect transistor. Adv. Funct. Mater. **19**(9), 1360–1370 (2009)
29. O'Neill, M., Kelly, S.M.: Ordered materials for organic electronics and photonics. Adv. Mater. **23**(5), 566–584 (2011)

30. Funahashi, M.: Development of liquid-crystalline semiconductors with high carrier mobilities and their application to thin-film transistors. Polym. J. **41**(6), 459–469 (2009)
31. Pisula, W., Zorn, M., Chang, J.Y., Mullen, K., Zentel, R.: Liquid crystalline ordering and charge transport in semiconducting materials. Macromol. Rapid Commun. **30**(14), 1179–1202 (2009)
32. Van De Craats, A.M., Warman, J.M., Fechtenkötter, A., Brand, J.D., Harbison, M.A., Müllen, K.: Record charge carrier mobility in a room-temperature discotic liquid-crystalline derivative of hexabenzocoronene. Adv. Mater. **11**(17), 1469–1472 (1999)
33. Pisula, W., Menon, A., Stepputat, M., Lieberwirth, I., Kolb, U., Tracz, A., Sirringhaus, H., Pakula, T., Müllen, K.: A zone-casting technique for device fabrication of field-effect transistors based on discotic hexa-perihexabenzocoronene. Adv. Mater. **17**(6), 684–688 (2005)
34. Xiao, S., Myers, M., Miao, Q., Sanaur, S., Pang, K., Steigerwald, M.L., Nuckolls, C.: Molecular wires from contorted aromatic compounds. Angew. Chem. Int. Ed. **44**(45), 7390–7394 (2005)
35. Tracz, A., Jeszka, J.K., Watson, M.D., Pisula, W., Mullen, K., Pakula, T.: Uniaxial alignment of the columnar super-structure of a hexa (alkyl) hexa-peri-hexabenzocoronene on untreated glass by simple solution processing. J. Am. Chem. Soc. **125**(7), 1682–1683 (2003)
36. Tsao, H.N., Pisula, W., Liu, Z., Osikowicz, W., Salaneck, W.R., Müllen, K.: From ambi- To unipolar behavior in discotic dye field-effect transistors. Adv. Mater. **20**(14), 2715–2719 (2008)
37. Iino, H., Hanna, J.: Ambipolar charge carrier transport in liquid crystals. Opto-Electron. Rev. **13**(4), 295–302 (2005)
38. Ponomarenko, S.A., Kirchmeyer, S., Elschner, A., Alpatova, N.M., Halik, M., Klauk, H., Zschieschang, U., Schmid, G.: Decyl-end-capped thiophene-phenylene oligomers as organic semiconducting materials with improved oxidation stability. Chem. Mater. **18**(2), 579–586 (2006)
39. Van Breemen, A.J.J.M., Herwig, P.T., Chlon, C.H.T., Sweelssen, J., Schoo, H.F.M., Setayesh, S., Hardeman, W.M., Martin, C.A., De Leeuw, D.M., Valeton, J.J.P., Bastiaansen, C.W.M., Broer, D.J., Popa-Merticaru, A.R., Meskers, S.C.J.: Large area liquid crystal monodomain field-effect transistors. J. Am. Chem. Soc. **128**(7), 2336–2345 (2006)
40. Vlachos, P., Mansoor, B., Aldred, M.P., O'Neill, M., Kelly, S.M.: Charge-transport in crystalline organic semiconductors with liquid crystalline order. Chem. Commun. **23**, 2921–2923 (2005)
41. Oikawa, K., Monobe, H., Nakayama, K.I., Kimoto, T., Tsuchiya, K., Heinrich, B., Guillon, D., Shimizu, Y., Yokoyama, M.: High carrier mobility of organic field-effect transistors with a thiophene-naphthalene mesomorphic semiconductor. Adv. Mater. **19**(14), 1864–1868 (2007)
42. Funahashi, M., Zhang, F., Tamaoki, N.: High ambipolar mobility in a highly ordered smectic phase of a dialkylphenylterthiophene derivative that can be applied to solution-processed organic field-effect transistors. Adv. Mater. **19**(3), 353–358 (2007)
43. Zhang, F., Funahashi, M., Tamaoki, N.: High-performance thin film transistors from semiconducting liquid crystalline phases by solution processes. Appl. Phys. Lett. **91**(6), 063515 (2007)
44. Zhang, F., Funahashi, M., Tamaoki, N.: Thin-film transistors based on liquid-crystalline tetrafluorophenylter thiophene derivatives: thin-film structure and carrier transport. Org. Electron. Phys. Mater. Appl. **10**(1), 73–84 (2009)
45. Liu, J., Zhang, R., Osaka, I., Mishra, S., Javier, A.E., Smilgies, D.M., Kowalewski, T., McCullough, R.D.: Transistor paint: environmentally stable N-alkyldithienopyrrole and bithiazole-based copolymer thin-film transistors show reproducible high mobilities without annealing. Adv. Funct. Mater. **19**(21), 3427–3434 (2009)
46. Ebata, H., Izawa, T., Miyazaki, E., Takimiya, K., Ikeda, M., Kuwabara, H., Yui, T.: Highly soluble [1]benzothieno[3,2-b]benzothiophene (BTBT) derivatives for high-performance, solution-processed organic field-effect transistors. J. Am. Chem. Soc. **129**(51), 15732–15733 (2007)
47. Izawa, T., Miyazaki, E., Takimiya, K.: Molecular ordering of high-performance soluble molecular semiconductors and re-evaluation of their field-effect transistor characteristics. Adv. Mater. **20**(18), 3388–3392 (2008)
48. Iino, H., Hanna, J.I.: Availability of liquid crystallinity in solution processing for polycrystalline thin films. Adv. Mater. **23**(15), 1748–1751 (2011)

49. Fujiwara, T., Locklin, J., Bao, Z.: Solution deposited liquid crystalline semiconductors on a photoalignment layer for organic thin-film transistors. Appl. Phys. Lett. **90**(23), 232108 (2007)
50. O'Neill, M., Kelly, S.M.: Photoinduced surface alignment for liquid crystal displays. J. Phys. D: Appl. Phys. **33**(10), R67–R84 (2000)
51. Smits, E.C.P., Mathijssen, S.G.J., Van Hal, P.A., Setayesh, S., Geuns, T.C.T., Mutsaers, K.A.H.A., Cantatore, E., Wondergem, H.J., Werzer, O., Resel, R., Kemerink, M., Kirchmeyer, S., Muzafarov, A.M., Ponomarenko, S.A., De Boer, B., Blom, P.W.M., De Leeuw, D.M.: Bottom-up organic integrated circuits. Nature **455**(7215), 956–959 (2008)
52. Hikmet, R.A.M., Lub, J.: Anisotropic networks and gels obtained by photopolymerisation in the liquid crystalline state: synthesis and applications. Prog. Polym. Sci. Oxf **21**(6), 1165–1209 (1996)
53. Kelly, S.M.: Anisotropic networks, elastomers and gels. Liq. Cryst. **24**(1), 71–82 (1998)
54. Hikmet, R.A.M., Lub, J., Broer, D.J.: Anisotropic networks formed by photopolymerization of liquid-crystalline molecules. Adv. Mater. **3**(7–8), 392–394 (1991)
55. Broer, D.J., Boven, J., Mol, G.N., Challa, G.: In-situ photopolymerization of oriented liquid-crystalline acrylates, 3. Oriented polymer networks from a mesogenic diacrylate. Makromol. Chem. **190**, 2255–2268 (1989)
56. Kelly, S.M.: Anisotropic networks. J. Mater. Chem. **5**(12), 2047–2061 (1995)
57. McCulloch, I., Coelle, M., Genevicius, K., Hamilton, R., Heckmeier, M., Heeney, M., Kreouzis, T., Shkunov, M., Zhang, W.: Electrical properties of reactive liquid crystal semiconductors. Jpn. J. Appl. Phys. **47**(1 PART 2), 488–491 (2008)
58. McCulloch, I., Zhang, W., Heeney, M., Bailey, C., Giles, M., Graham, D., Shkunov, M., Sparrowe, D., Tierney, S.: Polymerisable liquid crystalline organic semiconductors and their fabrication in organic field effect transistors. J. Mater. Chem. **13**(10), 2436–2444 (2003)
59. Huisman, B.H., Valeton, J.J.P., Nijssen, W., Lub, J., Ten Hoeve, W.: Oligothiophene-based networks applied for field-effect transistors. Adv. Mater. **15**(23), 2002–2005 (2003)
60. Baldwin, R.J., Kreouzis, T., Shkunov, M., Heeney, M., Zhang, W., McCulloch, I.: A comprehensive study of the effect of reactive end groups on the charge carrier transport within polymerized and nonpolymerized liquid crystals. J. Appl. Phys. **101**(2) (2007)
61. Farrar, S.R., Contoret, A.E.A., O'Neill, M., Nicholls, J.E., Richards, G.J., Kelly, S.M.: Nondispersive hole transport of liquid crystalline glasses and a cross-linked network for organic electroluminescence. Phys. Rev. B **66**(12) (2002). doi:doi:12510710.1103/PhysRevB.66.125107
62. Hoang, M.H., Cho, M.J., Kim, D.C., Kim, K.H., Shin, J.W., Cho, M.Y., Js, J., Choi, D.H.: Photoreactive Ï€-conjugated star-shaped molecules for the organic field-effect transistor. Org. Electron. Phys. Mater. Appl. **10**(4), 607–617 (2009)
63. Arias, A.C., MacKenzie, J.D., McCulloch, I., Rivnay, J., Salleo, A.: Materials and applications for large area electronics: solution-based approaches. Chem. Rev. **110**(1), 3–24 (2010)
64. Tsao, H.N., Cho, D., Andreasen, J.W., Rouhanipour, A., Breiby, D.W., Pisula, W., Müllen, K.: The influence of morphology on high-performance polymer field-effect transistors. Adv. Mater. **21**(2), 209–212 (2009)
65. Zeng, L., Yan, F., Wei, S.K.H., Culligan, S.W., Chen, S.: Synthesis and processing of monodisperse oligo(fluorene-co-bithiophene)s into oriented films by thermal and solvent annealing. Adv. Funct. Mater. **19**(12), 1978–1986 (2009)

Index

A
AC frequency (effect of), 70, 77
Acrylate, 78, 157–159, 161, 164, 167, 168, 205, 206, 209–211, 263
Aggregation, 3–5, 11, 177–181
Alq_3, 197, 198, 200
Ambipolar, 46, 47, 145, 161, 163, 165, 168, 232, 255, 258, 259
Anisotropic molecular shape, 45
Anisotropy, 42, 44–45, 70, 147, 149, 168, 184–187, 255
Aromatic amines, 197–200, 203, 205, 206, 211, 212

B
Binding energy, 58, 150, 151, 162, 174, 219, 225
Birefringence, 10, 21–23, 174, 184

C
Calamitic, vi, 2–4, 6–11, 31, 32, 35, 42, 147, 159–164, 168
  liquid crystals, 40, 60, 91, 97–99, 146, 157, 183, 184, 214, 234
Capacitor plates (CV), 153
Carbazole derivatives, 132
CBP, 198, 201
CDM. *See* Correlated Disorder Model (CDM)
Charge carrier
  mobility, 40, 45, 50, 52, 53, 66–91, 99, 109, 111, 128, 131, 134, 145, 168, 189, 203, 247, 249, 251, 264
  transport, 39–62, 147, 231
Charge mobility, 158, 229, 262
Charge recombination, 232, 255
Charge separation, 221, 222, 224, 229–231, 240
Chiral liquid crystals: light emission, 187–189
Chiral nematic (cholesteric) liquid crystals (mesophases), 8
Cholesteric, 8, 9, 39, 40, 187, 209
Col_h (columnar hexagonal), 46, 98
Col_r (columnar rectangular), 46, 98
Columnar liquid crystals (mesophases), 98, 99, 121, 125, 128, 257–259
Columnar liquid crystals for transistors, 257–259
Columnar phases, 3, 11, 15–17, 21, 22, 28, 29, 35, 40, 42, 45, 46, 49, 51, 56, 59–61, 66, 68, 71, 75, 85, 91, 98, 99, 106, 108–110, 112, 113, 117, 225, 258
Columnar triphenylene discotic liquid crystals, 158
Complementary Polytopic Interaction (CPI), 80, 88
Conduction channel, 43, 44, 49, 54, 57
Conductive polymer, 174, 199
Copper phthalocyanine, 46, 199
Correlated Disorder Model (CDM), 148, 150, 159
CPI. *See* Complementary Polytopic Interaction (CPI)
Crosslinking, 146, 147, 158, 159, 161, 164, 166–169, 186, 205–208, 210, 211, 214, 232, 240, 263
CV. *See* Capacitor plates (CV)

D
Developable domain, 21, 30
Diene, 159, 161–163, 167, 186, 211, 235, 263, 264

R.J. Bushby et al. (eds.), *Liquid Crystalline Semiconductors*, Springer Series in Materials Science 169, DOI 10.1007/978-90-481-2873-0,

Disclination, 6, 10, 11, 19–21, 23, 30, 55
Discotic
columnar mesophases, 28, 40, 46, 49, 51, 98
dimers, 108–110
glasses, 108
liquid crystals, 40, 60, 65, 66, 68, 70, 85, 91, 98–100, 113, 121, 157, 158, 225, 240, 241, 257
oligomers, 106, 108–110
photovoltaics, 225–232, 240, 241
reactive mesogens, 146, 157–159
systems (liquid crystal), 35, 88
Dynamic Scattering Mode, 39

E
Electronic conduction, vi, 40–45, 53, 54, 57, 58, 232
Electron injection, 157, 175, 197, 214
Electron transport, 46, 57, 123, 154, 163, 167–169, 181, 197, 200, 238
layer, 145
Emissive layer, 207
Exciton dissociation, 224
Extrinsic conduction, 41, 57

F
Fan texture, 21, 22, 25, 2729
FET. *See* Field Effect Transistor (FET)
Field Effect Transistor (FET), vi, 74, 75, 81, 128, 151, 155–157, 159, 161–163, 168, 181, 229, 241, 254, 264
Flash Photolysis Time Resolved Microwave Conductivity (FP-TRMC), 70
Fluorene oligomers, 204, 205, 210
Fluorescence, 112, 114, 176–177, 203, 210, 231
Focal conic, 20–27
Förster energy transfer, 183, 204
FP-TRMC. *See* Flash Photolysis Time Resolved Microwave Conductivity (FP-TRMC)
Fullerenes, 225, 231–232, 239–241

G
Gaussian Disorder Model (GDM), 148–150, 158, 161–163
Gay-Berne potential, 148, 150
Glass transition temperature (Tg), 200, 204, 263
Grain boundaries, 11, 55, 56, 70, 85, 146, 254, 255

H
Hexaazatriphenylene, 110–112
Hexabenzocoronene derivatives, 75, 77, 83, 109, 229, 257
Hexakis-hexyloxytriphenylene (HAT6), 40, 45
Highest Occupied Molecular Orbital (HOMO) levels, 43, 157, 174, 219
Hole conduction, 40
Hole injection, 157, 174, 256
Hole transport, 46, 157–161, 164, 167, 169, 181, 197, 199, 200, 208, 211, 220, 238, 241, 259
layer, 203, 206
Homeotropic alignment, 73, 229, 230, 262
Homogeneous alignment, 18, 258

I
Impurities, 41, 43–46, 54, 55, 57, 66, 68, 73, 74, 100, 167, 168
Indium tin oxide (ITO), 174, 181, 197–199, 203, 204, 208, 209, 213, 214
Intrinsic conduction, 40, 41
Ionic conduction, 40–44, 54, 56, 57, 232
Iridium organometallic complexes, 197
ITO. *See* Indium tin oxide (ITO)

L
Lamellar polymers, 249–254
Langmuir Blodgett, 203
Lasers
chiral, 187–190
thin film nematic, 190
Light-emitting polymers, 183, 185, 208
Light-emitting transistor, 255–256
Liquid Crystal Displays (LCDs), 1, 7, 8, 18, 24, 39, 73, 173, 174, 202, 203, 214 *See also* Correlated Disorder Model (CDM)
Liquid crystalline
chemical networks, 147, 157, 164, 169
physical gels, 157, 164–165, 169
polyfluorenes, 203, 254
polymer networks, vi, 145–169, 205, 214, 215
polymers for photovoltaics, 157, 236–239
semiconductors, vi, 39–62, 65–92, 164, 168, 174, 176–183, 254–255
Liquid crystal polymers for photovoltaics, 157

Lowest Unoccupied Molecular orbital (LUMO) levels, 43, 54, 157, 174, 175, 219, 220, 224, 256
Low-molar mass (small molecule), 146, 199
Lyotropic, 3–6, 22, 99

M
MeH-PPV, 201
Methacrylate, 211, 235, 263, 264
Miller-Abrahams (MA) hopping, 148
Mobility effect
  charge carriers, 40, 45, 50, 52, 53, 66–91, 99, 109, 111, 128, 131, 134, 145, 168, 176, 189, 203, 247, 249, 251, 264
  field disorder, 88
  lattice reorganisation energy, 89–90
  orbital overlap, 89
  positional disorder, 58, 88
  size of the core, 60, 90–91
Monte Carlo (MC) simulations, 58, 59, 148
Mott-Gurney equation, 73, 151

N
Nano-graphene, 127
n-doping, 66
Nematic
  liquid crystals (mesophases), 11
  photovoltaics, 232, 241
  reactive mesogens, 159, 210, 232–236, 240, 241, 263
  texture, 18–20
Nematic discotic (ND), 14
n-type conductor, 74

O
Oligofluorene, 40, 47, 177, 180, 182, 184, 204, 210, 254
One dimensional transport, 45
Optical texture, 10, 15–29
Organic Field Effect Transistors (OFETs), vi, 247–264
  principles, 248–249
Organic Light Emitting Diodes (OLEDs)
  brightness, 197, 211, 215
  efficiency, 174–176, 180, 181, 186, 197, 198, 200, 206, 214, 215
  principle of operation, 197
Organic photovoltaics, principle of operation, 219
Organic semiconductor, applications, 185–186
Orientation layers, 203, 205, 211, 214
Oriented transistors, 189
Out-coupling, 174, 183
Oxetane, 162, 163, 167, 168, 205, 206, 208, 263

P
p-doping, 66
Perylene derivatives, 114, 115, 119, 225
Perylene materials, 115, 116, 181
2-Phenylbenzothiazole (7O-PBT-S12), 40, 44
2-Phenylnaphthalene derivatives, 49, 50, 52, 55, 56, 59
Phenylquaterthiophene, 40, 53
Phosphorescence, 176, 181–184, 190
Photoalignment, 212–214, 232, 262
Photoconductivity, 40, 152
Photocrosslinkable, 202, 206, 209, 211
Photocurrent, 45, 54–56, 70, 71, 152–154, 219, 222, 224
Photo-embossing, 185–187
Photo-mask, 207
Photopolymerisation, 186, 206, 232, 233, 235, 241, 263
Phthalocyanine derivatives, 60, 74, 75, 82
Planar alignment, 73, 184, 232
Plastic crystal, 2, 3, 13, 50, 91
Plastic electronics, 247, 262
Polarised emission, 197–215
Polarised light, 147, 186
Polarisation ratio, 203–205, 211, 213
Poly(1,4-phenylenevinylene) (PPV), 201, 203, 205, 211
Poly(3,4-ethylenedioxythiophene) (PEDT), 174, 199
Polygonal texture, 21, 26
Polymeric liquid crystals, 40, 43, 44
Polymer Light-Emitting Diodes (PLEDs), 202–205, 207
Polymer liquid crystals for transistors, 249–254
Polymer network, vi, 145–169, 186, 189, 205, 207, 211, 214, 215, 233, 263
Porphyrin derivatives, 86
p-type conductor, 74
Pulse Radiolysis Time Resolved Microwave Conductivity (PRTRMC), 48, 69–70, 111
Pyrene derivatives, 112, 114

Q
Quantum efficiency, 176, 180, 181, 197, 200, 221, 224
Quaterthiophene (QT), 52, 60, 163, 167, 168

R
Reactive mesogens (RMs), vi, 145–169, 186, 197, 203, 205–207

S
Sanditic liquid crystal polymer, 237
Semiconducting polymers, 146, 150, 249, 264
Singlet, 175, 176, 181, 200
  emitters, 198
Small molecules, 40–44, 146, 148, 164, 199, 202
Smectic
  liquid crystals (mesophases), 46, 52, 54
  liquid crystals for transistors, 259–264
  reactive mesogens, 263
  texture, 20–21
Solar cells, 116, 128, 134, 224, 225, 231, 238–240
Solar concentrator, 239–240
Structural defects, 15, 55–56, 99, 223

T
Thermotropic, 3, 4, 6, 8, 22, 24, 97, 99, 145, 164, 236
Time of Flight (TOF), 45, 47, 60, 70–72, 132, 151–155, 157, 159, 161, 163, 168, 238, 259, 263, 264
Trapping, 41, 43, 54, 70, 71, 158, 163–165, 167, 169
Tricycloquinazolene derivatives, 119, 120
2,4,7-Trinitrofluorenone (TNF), 46
Triphenylene derivatives, 51, 56, 75–81
Triplet emitters, 197
Triplets, 175, 176, 181, 189, 198, 200
Trisoxadiazole derivatives, 86
Two dimensional transport, 56

W
White Light OLED (WOLED), 214

X
X-ray diffraction, 9, 15, 27–36

The manufacturer's authorised representative in the EU is Springer Nature Customer Service Centre GmbH, Europaplatz 3, 69115 Heidelberg, Germany. If you have any concerns regarding our products, please contact ProductSafety@springernature.com

Printed and bound by CPI Group (UK) Ltd, Croydon, CR0 4YY
15/07/2026
02167627-0002